Lecture Notes in Mathematics

Edited by A. Dold and B. Eckmann

Subseries: Department of Mathematics, University of Maryland
Adviser: J. Alexander

1041

Lie Group Representations II

Proceedings of the Special Year
held at the University of Maryland, College Park 1982–1983

Edited by R. Herb, S. Kudla, R. Lipsman and J. Rosenberg

Springer-Verlag
Berlin Heidelberg New York Tokyo 1984

Editors

Rebecca Herb
Stephen Kudla
Ronald Lipsman
Jonathan Rosenberg
Department of Mathematics, University of Maryland
College Park, Maryland 20742, USA

AMS Subject Classifications (1980): 22E55, 22E35, 10D40, 10D30

ISBN 3-540-12715-1 Springer-Verlag Berlin Heidelberg New York Tokyo
ISBN 0-387-12715-1 Springer-Verlag New York Heidelberg Berlin Tokyo

Library of Congress Cataloging in Publication Data. (Revised for vol. 2) Main entry under title: Lie group representations. (Lecture notes in mathematics; 1024, 1041) Vol. 2 edited by R. Herb et al. Sponsored by the Dept. of Mathematics, University of Maryland, College Park. 1. Lie groups–Congresses. 2. Representations of groups–Congresses. I. Herb, R. (Rebecca), 1948-. II. University of Maryland, College Park. Dept. of Mathematics. III. Series: Lecture notes in mathematics (Springer-Verlag); 1024, etc.
QA3.L28 no. 1024 [QA387] 510s [512'.55] 83-16871
ISBN 0-387-12725-9 (U.S.: v. 1)

Printing and binding: Beltz Offsetdruck, Hemsbach/Bergstr.
2146/3140-543210

PREFACE

The Department of Mathematics of the University of Maryland conducted a Special Year in Lie Group Representations during the academic year 1982-1983. This volume is the second (of three) containing articles submitted by the main speakers during the Special Year. Most of the invited speakers submitted articles, and virtually all of those appearing here deal with the subject matter on which the authors lectured during their visits to Maryland.

The Special Year program at Maryland represents a thriving departmental tradition—this being the fourteenth consecutive year in which such an event has taken place. As usual, the subject matter was chosen on the basis of active current research and the interests of departmental members. The modern theory of Lie Group Representations is a vast subject. In order to keep the program within bounds, the Special Year was planned around five distinct intensive periods of activity—each one (of three weeks duration) devoted to one of the main branches of current research in the subject. During those periods (approximately) eight distinguished researchers were invited to present lecture series on areas of current interest. Each visitor spent 1-3 weeks in the department and gave 2-5 lectures. In addition, during each period approximately 8-10 other visitors received financial support in order to attend and participate in the Special Year activities. Thus each period had to some extent the flavor of a mini-conference; but the length of the periods, the fact that visitors were provided with office space and the (relatively) low number of lectures per day also left ample time for private discussion and created the atmosphere of "departmental visitor" rather than "conference participant." Furthermore, as part of the Special Year the department was fortunate to have in residence D. Barbasch, J. Bernstein and J.-L. Brylinski for the Fall 1982 semester, and B. Blank for the Spring 1983 semester. These visitors ran semester-long seminars in Group Representations.

All of the activities of the Special Year were enthusiastically supported by the department, its faculty and graduate students.

Although most of the cost of the Special Year was borne by the department, the NSF did provide a generous amount of supplementary support. In particular, the contributions to the additional visitors were entirely funded by NSF. The Mathematics Department is grateful to the Foundation for its support of the Special Year. The Organizing Committee would also like to express its gratitude to the Department for its support. In particular the splendid efforts of Professors W. Kirwan, J. Osborn, G. Lehner, as well as of N. Lindley, D. Kennedy, D. Forbes, M. Keimig, and J. Cooper were vital to the success of the Special Year. The outstanding job of preparation of manuscripts by Berta Casanova, June Slack, Anne Eberly and Pat Pasternack, was of immense help in producing this volume so quickly. Also we are grateful to Springer-Verlag for its cooperation. Finally we are very pleased that so many of our participants provided us with high quality manuscripts, neatly prepared and submitted on time. It is our conviction that the theory of Group Representations has profited greatly from the efforts of all the above people towards the Special Year.

The Editors
August 1983

INTRODUCTION

We have made a serious attempt to group the papers (within the three volumes) according to the Periods in which they were presented and according to subject matter. However we were also influenced by the time at which manuscripts became available, and by a desire to equalize the size of the volumes. This (second) volume contains papers from Period II of the Special Year. The program for that period was as follows:

PERIOD II. The Langlands Program—Arithmetic Groups, Automorphic Representations, Automorphic Forms, L-Groups, Base Change, Orbital Integrals, Adelic and Local Theory

J. Arthur -- The Trace formula and lifting problems
J. Bernstein -- P-Invariant distributions on GL(n)
W. Casselman -- Paley-Wiener theorems, automorphic forms, and cohomology
S. Gelbart -- Automorphic forms on unitary groups
R. Howe -- On a question of Weil: Dual pairs and the poles of L-functions
H. Jacquet -- Residual spectrum, especially for GL(n)
D. Kazhdan -- Harmonic analysis and orbital integrals for p-adic groups
R. Langlands -- Orbital integrals for p-adic groups
I. Piatetski-Shapiro -- L-functions for GSp(4)

Although Paul Sally delivered his lecture series during Period III (see volume I for the program), he was present during Period II, and since his topic fit more closely with those presented here his paper is included in this volume.

The additional participants during Period II of the Special Year were

L. Clozel
Y. Flicker
P. Kutzko
J.-P. Labesse
A. Moy
S. Rallis
F. Rodier
J. Rogawski
P. Sally
F. Shahidi
J. Shalika
B. Speh
A. Terras
J. Tunnell

SPECIAL YEAR DATA

A. The five periods of activity of the Special Year were as follows:

- I. Algebraic Aspects of Semisimple Theory -- Sept. 7, 1982 - Oct. 1, 1982
- II. The Langlands Program -- Nov. 1, 1982 - Nov. 19, 1982
- III. Analytic Aspects of Semisimple Theory -- Jan. 24, 1983 - Feb. 11, 1983.
- IV. The Orbit Method -- Feb. 28, 1983 - March 18, 1983
- V. Applications -- April 18, 1983 - May 6, 1983

B. The speakers and the dates of their visits were:

Period I

Thomas Enright, UCSD (9/7 - 9/22)
Anthony Joseph, Weizmann Institute (9/21 - 9/25)
Bertram Kostant, MIT (9/7 - 9/14)
George Lusztig, MIT (9/7 - 9/11)
Wilfried Schmid, Harvard (9/13 - 9/18)
David Vogan, MIT (9/27 - 10/1)
Nolan Wallach, Rutgers (9/20 - 10/1)

Period II

James Arthur, Toronto (11/1 - 11/19)
William Casselman, British Columbia (11/3 - 11/12)
Stephen Gelbart, Cornell (11/1 - 11/12)
Roger Howe, Yale (11/8 - 11/12)
Hervé Jacquet, Columbia (11/1 - 11/12)
David Kazhdan, Harvard (11/1 - 11/12)
Robert Langlands, IAS (11/1 - 11/12)
Ilya Piatetski-Shapiro, Yale (11/1 - 11/12)

Period III

Mogens Flensted-Jensen, Copenhagen (1/24 - 2/11)
Sigurdur Helgason, MIT (1/24 - 1/28)
Anthony Knapp, Cornell (2/2 - 2/4)
Paul Sally, Chicago (1/24 - 2/11)
V. S. Varadarajan, UCLA (1/24 - 2/11)
Garth Warner, Washington (2/7 - 2/8)
Gregg Zuckerman, Yale (1/24 - 2/4)

Period IV

Lawrence Corwin, Rutgers (3/7 - 3/11)
Michael Cowling, Genova (3/2 - 3/4)
Michel Duflo, Paris, (2/28 - 3/11)
Roger Howe, Yale (3/7 - 3/11)
Henri Moscovici, Ohio State (3/7 - 3/18)
Richard Penney, Purdue (3/7 - 3/11)
Lajos Pukanszky, Penn (3/7, 3/11 - 3/18)
Wulf Rossmann, Ottawa (2/28 - 3/4)
Michèle Vergne, MIT (3/3 - 3/15)

Period V

Lawrence Corwin, Rutgers (4/18 - 4/29)
Bernard Helffer, Nantes (4/18 - 5/6)
Sigurdur Helgason, MIT (4/18 - 4/22)
Roger Howe, Yale (4/18 - 4/22)
Adam Koranyi, Washington Univ. (4/18 - 5/6)
Henri Moscovici, Ohio State (4/25 - 4/30)
Richard Penney, Purdue (4/25 - 5/6)
Linda Rothschild, Wisconsin (4/18 - 4/22)

C. The Organizing Committee for the 1982-1983 Special Year in Lie Group Representations is

Rebecca Herb
Raymond Johnson
Stephen Kudla
Ronald Lipsman (Chairman)
Jonathan Rosenberg

TABLE OF CONTENTS

* For papers with more than one author, an asterisk indicates the author who delivered the lectures.

ON SOME PROBLEMS SUGGESTED BY THE TRACE FORMULA

James Arthur
University of Toronto
Toronto, Ontario, Canada
M55 1A1

In the present theory of automorphic representations, a major goal is to stabilize the trace formula. Its realization will have important consequences, among which will be the proof of functoriality in a significant number of cases. However, it will require much effort, for there are a number of difficult problems to be solved first. Some of the problems, especially those concerning orbital integrals, were studied in [9(e)]. They arise when one tries to interpret one side of the trace formula. The other side of the trace formula leads to a different set of problems. Among these, for example, are questions relating to the nontempered automorphic representations which occur discretely. Our purpose here is to describe some of these problems and to suggest possible solutions.

Some of the problems have in fact been formulated as conjectures. They have perhaps been stated in greater detail than is justified, for I have not had sufficient time to ponder them. However, they seem quite natural to me, and I will be surprised if they turn out to be badly off the mark.

Our discussion will be rather informal. We have tried to keep things as simple as possible, sometimes at the expense of omitting pertinent details. Section 1, which is devoted to real groups, contains a review of known theory, and a description of some problems and related examples. Section 2 has a similar format, but is in the global setting. We would have liked to follow it with a detailed discussion of the trace formula, as it pertains to the conjecture in Section 2. However, for want of time, we will be much briefer. After opening with a few general remarks, we will attempt in Section 3 to motivate the conjecture with the trace formula only in the case of PSp(4). In so doing, we will meet a combinatorial problem which is trivial for PSp(4), but is more interesting for general groups.

I am indebted to R. Kottwitz, D. Shelstad, and D. Vogan for enlightening conversations. I would also like to thank the University of Maryland for its hospitality.

§1. A PROBLEM FOR REAL GROUPS

1.1. The trace formula, which we will discuss presently, is an equality of invariant distributions. The study of such distributions leads to questions in local harmonic analysis. We will begin by looking at one such question over the real numbers.

For the time being, we will take G to be a reductive algebraic group defined over $\mathbb{R}$. For simplicity we shall assume that G is quasi-split. Let $\Pi(G(\mathbb{R}))$ (resp. $\Pi_{temp}(G(\mathbb{R}))$) denote the set of equivalence classes of irreducible representations (resp. irreducible tempered representations) of $G(\mathbb{R})$. In the data which one feeds into the trace formula are functions f in $C_c^\infty(G(\mathbb{R}))$. Since the terms of the trace formula are invariant distributions, we need only specify f by its values on all such distributions.

<u>Theorem 1.1.1</u>: The space of invariant distributions on $G(\mathbb{R})$ is the closed linear span of

$$\{tr(\pi): \Pi_{temp}(G(\mathbb{R})),$$

where $tr(\pi)$ stands for the distribution $f \to tr\pi(f)$.

One can establish this theorem from the characterization [1(a)] of the image of the Schwartz space of $G(\mathbb{R})$ under the (operator valued) Fourier transform. We hope to publish the details elsewhere.

Thus, for the trace formula, we need only specify the function

$$F(\pi) = \operatorname{tr} \pi(f), \qquad \pi \in \Pi_{temp}(G(\mathbb{R})) . \tag{1.1.2}$$

It is clearly important to know what functions on $\Pi_{temp}(G(\mathbb{R}))$ are of this form. The elements in $\Pi_{temp}(G(\mathbb{R}))$ can be given by a finite number of parameters, some continuous and some discrete. Via these parameters, one can define a Paley-Wiener space on $\Pi_{temp}(G(\mathbb{R}))$. It consists of functions which, among other things, are in the classical Paley-Wiener space in each continuous parameter. We would expect this Paley-Wiener space on $\Pi_{temp}(G(\mathbb{R}))$ to be the image of $C_c^\infty(G(\mathbb{R}))$ under the map above. This fact may well be a consequence of recent work of Clozel and Delorme. We shall assume it implicitly in what follows.

There is one point we should mention before going on. The function F can be evaluated on any invariant distribution on $G(\mathbb{R})$. In particular,

$$F(\pi) = \langle \operatorname{tr} \pi, F\rangle = \operatorname{tr} \pi(f)$$

is defined for any irreducible representation π, and not just a tempered one. If $\rho = \oplus \pi_i$ is a finite sum of irreducible representations, we set

$$F(\rho) = \sum F(\pi_i).$$

Now, consider an induced representation

$$\rho_\sigma = \operatorname{Ind}_{P(\mathbb{R})}^{G(\mathbb{R})} (\sigma \otimes id_N),$$

where $P = NM$ is a parabolic subgroup of G (defined over $\mathbb{R}$), σ is a representation in $\Pi_{temp}(M(\mathbb{R}))$, and id_N is the trivial representation of the unipotent radical $N(\mathbb{R})$. Let λ be a complex valued linear function on $\mathfrak{a}_M$, the Lie algebra of the split component of the center of $M(\mathbb{R})$, and let σ_λ be the representation obtained by translating σ by λ. Then ρ_{σ_λ} is in general a nonunitary, reducible representation of $G(\mathbb{R})$. Representations of this form are sometimes called standard representations. The function $F(\rho_{\sigma_\lambda})$, defined by the prescription above, can be obtained by analytic continuation from the purely imaginary values of λ, where the induced representation is tempered. Suppose that π is an arbitrary irreducible, but not necessarily tempered, representation of $G(\mathbb{R})$. It is known (see [15] that $tr(\pi)$ can be written

$$(1.1.3) \qquad tr(\pi) = \sum_\rho M(\pi,\rho)tr(\rho),$$

where ρ ranges over a finite set of standard representations of $G(\mathbb{R})$ and $\{M(\pi,\rho)\}$ is a uniquely determined set of integers. Then $F(\pi)$ is given by

$$F(\pi) = \sum_\rho M(\pi,\rho)F(\rho).$$

Thus, the problem of determining $F(\pi)$ is equivalent to determining the decomposition (1.1.3).

1.2. Among the invariant distributions are the stable distributions, which are of particular interest for global applications. Shelstad has shown [11(c)] that these may be defined either by orbital integrals or, as we shall do, by tempered characters.

We recall the Langlands classification [9(a)] of $\Pi(G(\mathbb{R}))$. Let $\Phi(G/\mathbb{R})$ be the set of admissible maps

$$\phi: W_{\mathbb{R}} \to {}^LG,$$

where $W_{\mathbb{R}}$ is the Weil group of $\mathbb{R}$, and

$${}^LG = {}^LG^0 \times W_{\mathbb{R}}$$

is the L-group of G. The elements in $\Phi(G/\mathbb{R})$ are to be given only up to conjugacy by ${}^LG^0$. To each $\phi \in \Phi(G/\mathbb{R})$ Langlands associates an L-packet $\Pi_\phi = \Pi_\phi^G$ consisting of finitely many representations in $\Pi(G(\mathbb{R}))$. He shows that the representations in Π_ϕ are tempered if and only if the projection of the image of ϕ onto ${}^LG^0$ is bounded. Let $\Phi_{temp}(G/\mathbb{R})$ denote the set of all such ϕ.

<u>Definition 1.2.1</u>: A <u>stable</u> distribution is any distribution, necessarily invariant, which lies in the closed linear span of

$$\{\sum_{\pi\in\Pi_\phi} tr(\pi) : \phi \in \Phi_{temp}(G/\mathbb{R})\} .$$

If F is a function of the form (1.1.2), we can set

$$F(\phi) = \sum_{\pi\in\Pi_\phi} F(\pi)$$

for any $\phi \in \Phi_{temp}(G/\mathbb{R})$. In [11(c)] Shelstad shows that any tempered character on $G(\mathbb{R})$ can be expressed in terms of sums of this form, but associated to some other groups of lower dimension. Given our discussion above, this means that any invariant distribution on $G(\mathbb{R})$ may be expressed in terms of stable distributions associated to other groups. We shall review some of this theory.

The notion of endoscopic group was introduced in [9(c)] and studied further in [11(c)]. Let s be a semisimple element in ${}^LG^0$, defined modulo

$$Z_G = \mathrm{Cent}({}^LG, {}^LG^0),$$

the centralizer of ${}^LG$ in ${}^LG^0$. An endoscopic group $H = H_s$ for G (over $\mathbb{R}$) is a quasi-split group in which ${}^LH^0 = {}^LH_s^0$ equals

$$\mathrm{Cent}(s, {}^LG^0)^0,$$

the connected component of the centralizer of s in ${}^LG^0$. If G is a split group with trivial center, this specifies H uniquely. For then ${}^LG^0$ is a simply connected complex group, in which the centralizer of any semisimple element is connected ([14], Theorem 2.15). The group H is then the unique split group whose L group is the direct product of ${}^L H^0$ with $W_{\mathbb{R}}$. In general, it is required only that each element $w \in W_{\mathbb{R}}$ act on ${}^L H^0$ by conjugation with some element

$$g \times w, \qquad g \in {}^L G^0,$$

in $\mathrm{Cent}(s, {}^L G)$. Since the group $\mathrm{Cent}(s, {}^L G^0)$ is not in general connected, there might be more than one endoscopic group for a given s and ${}^L H_s^0$. Two endoscopic groups H_s and $H_{s'}$ will be said to be equivalent if there is a $g \in {}^L G^0$ such that s equals $gs'g^{-1}$ modulo the product of Z_G with the connected component of Z_{H_s}, and the map

$$\mathrm{ad}(g^{-1}) : {}^L H^0 \to {}^L (H')^0$$

commutes with the action of $W_{\mathbb{R}}$. (Thus, for us an endoscopic group really consists of the element s as well as the group H, and should strictly be called an endoscopic datum. See [9(e)].)

An admissible embedding ${}^L H \subset {}^L G$ of an endoscopic group is one which extends the given embedding of ${}^L H^0$, which commutes with the projections onto $W_{\mathbb{R}}$, and for which the image of ${}^L H$ lies in $\mathrm{Cent}(s, {}^L G)$. We shall suppose from now on that for each endoscopic group we have fixed an admissible embedding ${}^L H \subset {}^L G$, such that the embeddings for equivalent groups are compatible. (The additional restriction this puts on G is not serious. See [9(c)].) We shall say that H is <u>cuspidal</u> if the image of ${}^L H$ in ${}^L G$ lies in no proper parabolic subgroup of ${}^L G$.

<u>Example 1.2.2:</u> Let $G = PSp(4)$. Then

$${}^L G^0 = Sp(4,\mathbb{C}) = \{ g \in GL(4,\mathbb{C}) : \begin{pmatrix} & & & 1 \\ & & 1 & \\ & -1 & & \\ -1 & & & \end{pmatrix} {}^t g^{-1} \begin{pmatrix} & & & -1 \\ & & -1 & \\ & 1 & & \\ 1 & & & \end{pmatrix} = g \}.$$

The only cuspidal endoscopic groups are G and H_s, with

$$s = \begin{pmatrix} 1 & & & \\ & -1 & & \\ & & -1 & \\ & & & 1 \end{pmatrix}. \quad \text{Then}$$

$$ {}^L H_s^0 = \left\{ \begin{pmatrix} * & 0 & 0 & * \\ 0 & * & * & 0 \\ 0 & * & * & 0 \\ * & 0 & 0 & * \end{pmatrix} \right\} \cong SL(2,\mathbb{C}) \times SL(2,\mathbb{C}) , $$

and

$$ H_s \cong PGL(2) \times PGL(2) . $$

For each of these groups we take the obvious embedding of ${}^L H$ into ${}^L G$.

If ϕ is any parameter in $\Phi(G/\mathbb{R})$, define

$$ C_\phi = C_\phi^G = \mathrm{Cent}(\phi(W_{\mathbb{R}}), {}^L G^0) , $$

the centralizer in ${}^L G^0$ of the image of ϕ. Since the homomorphism ϕ is determined only up to ${}^L G^0$ conjugacy, C_ϕ is really only a conjugacy class of subgroups of ${}^L G^0$. However, we can identify each of these subgroups with a fixed abstract group, the identification being canonical up to an inner automorphism of the given group. Set

$$ \mathcal{C}_\phi = C_\phi / C_\phi^0 Z_G , $$

where C_ϕ^0 is the identity component of C_ϕ. Then $\mathcal{C}_\phi$ is a finite group which is known to be abelian. ([11(c)]. See also [5].) It can therefore be canonically identified with an abstract group which depends only on the class of ϕ.

For each $\phi \in \Phi_{temp}(G(\mathbb{R}))$, Shelstad defines a pairing $<,>$ on $\Pi_\phi \times \mathcal{C}_\phi$, such that the map

$$ \pi \to <\pi, \cdot>, \qquad \pi \in \Pi_\phi , $$

is an injection from Π_ϕ into the group $\hat{\mathcal{C}}_\phi$ of characters of $\mathcal{C}_\phi$. Unfortunately, the pairing cannot be defined canonically. However.

Shelstad shows that there is a function c from C_ϕ/Z_G to $\{\pm 1\}$, which is invariant on conjugacy classes, such that

$$c(s)\langle \bar{s},\pi\rangle, \qquad s \in C_\phi/Z_G,\ \pi \in \Pi_\phi,$$

is independent of the pairing. Here, $\bar{s}$ is the projection of s onto C_ϕ. This latter function can be used to map functions on $G(\mathbb{R})$ to functions on endoscopic groups.

Given a parameter $\phi \in \Phi_{temp}(G/\mathbb{R})$ and a semisimple element $s \in C_\phi/Z_G$, one can check that there is a unique endoscopic group $H = H_s$ such that

$$\phi(W_{\mathbb{R}}) \subset {}^LH \subset {}^LG .$$

ϕ then defines a parameter $\phi_1 \in \Phi_{temp}(H/\mathbb{R})$. For a given H, every parameter in $\Phi_{temp}(H/\mathbb{R})$ arises in this way. For any function $f \in C_c^\infty(G(\mathbb{R}))$, Shelstad defines a function $f_H \in C_c^\infty(H(\mathbb{R}))$, unique up to stable distributions on $H(\mathbb{R})$. To do so, it is enough to specify the value

$$f_H(\phi) = \sum_{\pi\in\Pi^H_{\phi_1}} f_H(\pi_1) = \sum_{\pi_1\in\Pi^H_{\phi_1}} \operatorname{tr} \pi_1(f_H) ,$$

for every such ϕ_1. This is done by setting

$$f_H(\phi_1) = c(s) \sum_{\pi\in\Pi_\phi} \langle\bar{s},\pi\rangle \operatorname{tr} \pi(f). \tag{1.2.3}$$

Actually, Shelstad defines f_H by transferring orbital integrals, and then proves the formula (1.2.3) as a theorem. However, we shall take the formula as a definition. Shelstad shows that the mapping $f \to f_H$ is canonically defined up to a sign. (It also depends on the embedding ${}^LH \subset {}^LG$ which we have fixed.) We shall fix the signs in any way, asking only that in the case $H = G$, f_G be consistent with the notation above. That is, $c(1) = 1$.

1.3. It is important for the trace formula to understand how the notions above relate to <u>nontempered</u> parameters ϕ. Shelstad defined the pairings $\langle\bar{s},\pi\rangle$ only for tempered ϕ, but it is easy enough to extend the definition to arbitrary parameters. For one can show that there is a natural way to decompose any parameter ϕ by

$$\phi(w) = \phi_0(w)\phi_+(w), \qquad \phi_0 \in \Phi_{temp}(G/\mathbb{R}), \ \phi_+ \in \Phi(G/\mathbb{R})),$$

so that the images of ϕ_0 and ϕ_+ commute, and so that ϕ itself is tempered whenever $\phi_+(W_{\mathbb{R}}) = \{1\}$. The centralizer in ${}^LG$ of the image of ϕ_+ will be the Levi component ${}^LM$ of a parabolic subgroup of ${}^LG$, and $\Pi^M_{\phi_+}$ will consist of a positive quasi-character ν_+ of $M(\mathbb{R})$. The image of ϕ_0 must lie in ${}^LM$, so that ϕ_0 defines an element in $\Phi_{temp}(M/\mathbb{R})$. There will be a bijection between $\Pi^M_{\phi_0}$ and Π^G_{ϕ}, the elements in Π^G_{φ} being the Langlands quotients obtained from the tempered representations in $\Pi^M_{\phi_0}$ and the positive quasi-character ν_+ of $M(\mathbb{R})$. On the other hand $C^M_{\phi_0}$ equals C^G_{ϕ}, so we can define the pairing on $C^G_{\phi} \times \Pi^G_{\phi}$ to be the one obtained from the pairing on $C^M_{\phi_0} \times \Pi^M_{\phi_0}$.

However, simply defining the pairing for nontempered ϕ is not satisfactory. For it could well happen that the distribution

$$\sum_{\pi\in\Pi_\phi} \mathrm{tr}(\pi)$$

is not stable if the parameter ϕ is not tempered. A related difficulty is that (1.2.3) no longer makes sense if ϕ_1 is not a tempered parameter for H. We shall define a subset of $\Phi(G/\mathbb{R})$ for which these difficulties are likely to have nice solutions. The subset will contain $\Phi_{temp}(G/\mathbb{R})$, and ought also to account for the representations of $G(\mathbb{R})$ which are of interest in global applications.

Let $\Psi(G/\mathbb{R})$ be the set of ${}^LG^0$-conjugacy classes of maps

$$\psi: W_{\mathbb{R}} \times SL(2,\mathbb{C}) \to {}^LG$$

such that the restriction of ψ to $W_{\mathbb{R}}$ belongs to $\Phi_{temp}(G/\mathbb{R})$. For any $\psi \in \Psi(G/\mathbb{R})$ define a parameter ϕ_ψ in $\Phi(G/\mathbb{R})$ by

$$\phi_\psi(w) = \psi\left(w, \begin{pmatrix} |w|^{1/2} & 0 \\ 0 & |w|^{-1/2} \end{pmatrix}\right), \qquad w \in W_{\mathbb{R}}.$$

Here it is helpful to recall that

$$w \to \begin{pmatrix} |w|^{1/2} & \\ & |w|^{-1/2} \end{pmatrix}$$

is the map from $W_{\mathbb{R}}$ to

$$SL(2,\mathbb{C}) = {}^L(PGL(2))^0$$

which assigns the trivial representation to $PGL(2,\mathbb{R})$. Recall also that the unipotent conjugacy classes in any complex group are bijective with the conjugacy classes of maps of $SL(2,\mathbb{C})$ into the group. The unipotent conjugacy classes for complex groups have been classified by weighted Dynkin diagrams. (See [13].) Now any $\psi \in \Psi(G/\mathbb{R})$ can be identified with a pair (ϕ,ρ), in which $\phi \in \Phi_{temp}(G/\mathbb{R})$ and ρ is a map from $SL(2,\mathbb{C})$ into C_ϕ, given up to conjugacy by C_ϕ. From the classification of nilpotents it follows that ρ is determined by its restriction to the diagonal subgroup of $SL(2,\mathbb{C})$. We obtain

<u>Proposition 1.3.1</u>: The map

$$\psi \to \phi_\psi, \qquad \psi \in \Psi(G/\mathbb{R}),$$

is an injection from $\Psi(G/\mathbb{R})$ into $\Phi(G/\mathbb{R})$.

Thus, $\Psi(G/\mathbb{R})$ can be regarded as a subset of $\Phi(G/\mathbb{R})$. It contains $\Phi_{temp}(G/\mathbb{R})$ as the set of $\psi = (\phi,\rho)$ with ρ trivial.

Conjecture 1.3.2: For any $\psi \in \Psi(G/\mathbb{R})$, the representations in Π_{ϕ_ψ} are all unitary.

Suppose that $\psi = (\phi,\rho)$ is an arbitrary parameter in $\Phi(G/\mathbb{R})$. Copying a previous definition we set

$$C_\psi = C_\psi^G = \mathrm{Cent}(\psi(W_{\mathbb{R}} \times SL(2,\mathbb{C})),{}^LG^0)$$

and

$$\mathcal{C}_\psi = \mathcal{C}_\psi^G = C_\psi/C_\psi^0 Z_G .$$

The group C_ψ always equals $\mathrm{Cent}(\rho(SL(2,\mathbb{C}),C_\phi))$, and in particular is contained in C_ϕ. Therefore, there are natural maps $C_\psi \to C_\phi$ and $\mathcal{C}_\psi \to \mathcal{C}_\phi$. It is easy to check that this second map is surjective. In other words, there is an injective map

$$\hat{\mathcal{C}}_{\phi_\psi} \to \hat{\mathcal{C}}_\psi$$

from the (irreducible) characters on $\mathcal{C}_{\phi_\psi}$ to the irreducible characters on $\mathcal{C}_\psi$.

Fix $\psi \in \Psi(G/\mathbb{R})$. Take one of the pairings $\langle , \rangle$ on $\mathcal{C}_{\phi_\psi} \times \Pi_{\phi_\psi}$ discussed above, as well as the associated function c on the conjugacy classes of C_{ϕ_ψ}/Z_G. We pull back c to a function on the conjugacy classes of C_ψ/ Z_G. We conjecture that the set Π_{ϕ_ψ} can be enlarged and the pairing extended so that all the theory for tempered parameters holds in this more general setting.

Conjecture 1.3.3: There is a finite set Π_ψ of irreducible representations of $G(\mathbb{R})$ which contains Π_{ϕ_ψ}, a function

$$\varepsilon_\psi : \Pi_\psi \to \{\pm 1\}$$

which equals 1 on Π_{ϕ_ψ}, and an injective map

$$\pi \to \langle \cdot, \pi \rangle, \qquad \pi \in \Pi_\psi,$$

from Π_ψ into $\hat{C}_\psi$, all uniquely determined, with the following properties.

(i) π belongs to the subset Π_{ϕ_ψ} of Π_ψ if and only if the function $\langle \cdot, \pi \rangle$ lies in the image of $\hat{C}_{\phi_\psi}$ in C_ψ.

(ii) The invariant distribution

$$\sum_{\pi \in \Pi_\psi} \varepsilon_\psi(\pi) \langle 1, \pi \rangle \operatorname{tr}(\pi) \tag{1.3.4}$$

is stable. (If C_ψ is abelian, which is certainly the case most of the time, the distribution is

$$\sum_{\pi \in \Pi_\psi} \varepsilon_\psi(\pi) \operatorname{tr}(\pi),$$

which except for the signs $\varepsilon_\psi(\pi)$ is just the sum of the characters in the packet Π_ψ.) We shall denote the value of this distribution on the function (1.1.2) by $F(\psi)$.

(iii) Let s be a semisimple element in C_ψ/Z_G. Let $H = H_s$ be the unique endoscopic group such that

$$\psi(W_{\mathbb{R}} \times SL(2,\mathbb{C})) \subset {}^L H \subset {}^L G,$$

so that, in particular, ψ defines a parameter in $\Psi(H/\mathbb{R})$. Then if $f \in C_c^\infty(G(\mathbb{R}))$, and $\bar{s}$ is the image of s in C_ψ,

$$f_H(\psi) = c(s) \sum_{\pi \in \Pi_\psi} \varepsilon_\psi(\pi) \langle \bar{s}, \pi \rangle \operatorname{tr} \pi(f).$$

It is not hard to check the uniqueness assertion of this conjecture. The third condition states that

$$\hat{\chi}_{(\psi,x)}(f) = c(s)^{-1} f_{H_s}(\psi)$$

depends only on the projection x of s onto C_ψ, and that for any irreducible character Θ in $\hat{C}_\psi$,

$$(1.3.5)\qquad \frac{1}{|C_\psi|}\sum_{x\in C_\psi}\hat{\chi}_{(\psi,x)}(f)\overline{\Theta(x)} = \begin{cases}\varepsilon_\psi(\pi)\,\mathrm{tr}\,\pi(f), & \text{if } \Theta = \langle\cdot,\pi\rangle \text{ for some } \pi\in\Pi_\psi,\\ 0, & \text{otherwise}.\end{cases}$$

Assume inductively that the distribution (1.3.4) has been defined and shown to be stable whenever G is replaced by a proper endoscopic group $H = H_s$. Since the function f_{H_s} has already been defined on any stable distribution, the numbers $f_{H_s}(\psi)$ and $\hat{\chi}_{(\psi,x)}(f)$, with $\bar{s} = x \neq 1$, then make sense. To define $f_G(\psi)$, take $\Theta = 1$. If π_1 is the representation in Π_{ϕ_ψ} such that $\langle\cdot,\pi_1\rangle$ equals 1, we obtain

$$(1.3.6)\qquad |C_\psi|\,\mathrm{tr}\,\pi_1(f) = \sum_{x\in C_\psi}\hat{\chi}_{(\psi,x)}(f)\ .$$

The distribution

$$f_G(\psi) = \hat{\chi}_{(\psi,1)}(f)$$

is then equal to

$$(1.3.7)\qquad |C_\psi|\,\mathrm{tr}\,\pi_1(f) - \sum_{\substack{x\in C_\psi\\ x\neq 1}}\hat{\chi}_{(\psi,x)}(f)\ .$$

To complete the inductive definition, it is necessary to show it is stable. The formula (3.1.5) would then give the elements in Π_ψ uniquely, but only as <u>virtual characters</u>. The remaining problem is to show that the nonzero elements among them are linearly independent, and that up to a sign (which would serve as the definition of ε_ψ) they are irreducible characters.

The packets Π_ψ should have some other nice properties. For example, one can associate an R-group to any $\psi \in \Psi(G/\mathbb{R})$. Define

R_ψ to be the quotient of C_ψ by the group of components in C_ψ/Z_G which act on the identity component by inner automorphisms. If R_ψ is not trivial, the identity component will also not be trivial. The image of ψ will be contained in a Levi component of a proper parabolic subgroup of ${}^LG$. Let ${}^LM$ be a minimal Levi subgroup of ${}^LG$ which contains the image of ψ. Then ψ also represents a parameter in $\Psi(M/\mathbb{R})$. There is a short exact sequence

$$1 \to C_\psi^M \to C_\psi^G \to R_\psi \to 1.$$

The group R_ψ should govern the reducibility of the induced representations

$$\rho_\sigma = \mathrm{Ind}_{P(\mathbb{R})}^{G(\mathbb{R})} (\sigma \otimes \mathrm{id}_N), \qquad \sigma \in \prod_\psi^M ,$$

where $P = MN$ is a parabolic subgroup of G. Note that ρ_σ is obtained by unitary induction from a representation which is in general not tempered.

Finally, the conjecture should admit extensions in two directions - to real groups which are not necessarily quasi-split, and to pairs (G,α), where α is an automorphism of G (modulo the group of inner automorphisms). Both will eventually be needed to exploit the trace formula in full generality.

1.4. Conjecture 1.3.3 is suggested by the global situation, which we will come to later. I do not have much local evidence. The largest group for which I have been able to verify the conjecture completely is $PSp(4)$. However, even this group is instructive. We shall look at three examples which illustrate why it is the parameters ψ, and not ϕ_ψ, which govern questions of stability of characters. In each case, $\prod_{\phi_\psi}$ will consist of one representation π such that $\mathrm{tr}(\pi)$ is not stable. However, each group C_ψ will be of order two, and the sets Π_ψ will consist of π and another representation. It is only with these larger sets that we obtain a nice theory of stability.

In each example we will consider parameters ψ for $G = PSp(4)$ such that the projection of ψ onto ${}^LG^0$ factors through the endoscopic group

$$ {}^LH^0 = {}^LH_s^0 \cong SL(2,\mathbb{C}) \times SL(2,\mathbb{C}), $$

with

$$ s = \begin{pmatrix} 1 & & & \\ & -1 & & \\ & & -1 & \\ & & & 1 \end{pmatrix} . $$

As we have said, Π_{ϕ_ψ} will consist of one representation π. It will be the Langlands quotient of a nonunitarily induced representation ρ of $G(\mathbb{R})$. We shall let π^H denote the unique representation in the packet $\Pi^H_{\phi_\psi} = \Pi^H_\psi$, and we shall let ρ^H be the nonunitarily induced representation of $H(\mathbb{R})$ of which π^H is the Langlands quotient.

In order to deal with Ψ-parameters on $G(\mathbb{R})$, we must first know something about the Φ-parameters. The L-packets

$$ \Pi_\phi, \qquad \phi \in \Phi(G/\mathbb{R}), $$

contain one or two elements. Those with two elements contain discrete series or limits of discrete series. They are of the form

$$ \Pi_\phi = \{\pi_{Wh}, \pi_{hol}\}, $$

where π_{Wh} has a Whitaker model, and π_{hol} is the irreducible representation of $PSp(4,\mathbb{R})$ which combines the holomorphic and anti-holomorphic (limits of) discrete series for $Sp(4,\mathbb{R})$. We take the pairing $\langle , \rangle$ on $C_\phi \times \Pi_\phi$ so that $\langle \cdot, \pi_{Wh} \rangle$ is the trivial character on $C_\phi \cong \mathbb{Z}/2\mathbb{Z}$, and $\langle \cdot, \pi_{hol} \rangle$ is the nontrivial character. It is not hard to verify that with this choice of pairing, all Shelstad's functions $c(s)$ may be taken to be 1. In our examples, we shall consider only representations with singular infinitesimal character, since these are the most difficult to handle. For this reason, $\{\pi_{Wh}, \pi_{hol}\}$ will now denote the L-packet in $G(\mathbb{R})$ which contains the lowest limits of discrete series. If π^H_{disc} is the lowest discrete series for $H(\mathbb{R})$,

$$\mathrm{tr}\ \pi^H_{disc}(f_H) = \mathrm{tr}\ \pi_{Wh}(f) - \mathrm{tr}\ \pi_{hol}(f) ,$$

for any $f \in C_c^\infty(G(\mathbb{R}))$. On the other hand, it will be clear in each example that

$$\mathrm{tr}\ \rho^H(f_H) = \mathrm{tr}\ \rho(f) ,$$

with ρ and ρ^H as above. As a distribution on $G(\mathbb{R})$, this last expression is stable.

We will prove the conjecture in each example by looking at the expression (1.3.7) for $f_G(\psi)$. If π is the unique representation in Π_{ϕ_ψ}, it will equal

$$2\ \mathrm{tr}\ \pi(f) - f_H(\psi) .$$

To check the stability of this distribution, we will need to express it as a linear combination of <u>standard</u> characters on $G(\mathbb{R})$. To then construct the packet Π_ψ, we will have to rewrite the expression as a linear combination of <u>irreducible</u> characters. The term $2\ \mathrm{tr}\ \pi(f)$ is handled by computing the character formula (1.1.3) for the representation π of $G(\mathbb{R})$. This can be accomplished by reducing to the case of regular infinitesimal character through the procedure in [12] and then using Vogan's algorithm obtained from the Kazdan-Lusztig conjectures [15]. We will only quote the answer. To deal with $f_H(\psi)$, we shall first write the character formula (1.1.3) for the representation π^H of $H(\mathbb{R})$. Since $H(\mathbb{R})$ is isomorphic to $PGL(2,\mathbb{R}) \times PGL(2,\mathbb{R})$, such formulas are well known. We will then lift the resulting standard characters on $H(\mathbb{R})$ to characters on $G(\mathbb{R})$ using the remarks above.

Example 1.4.1: Let ψ be given by the diagram

$$
\begin{array}{ccc}
W_{\mathbb{R}} & \times & SL(2,\mathbb{C}) \\
\downarrow & \overset{\mu}{\searrow} & \downarrow \text{ id} \\
SL(2,\mathbb{C}) & \times & SL(2,\mathbb{C}) \cong {}^{L}H^{0}
\end{array} \quad ,
$$

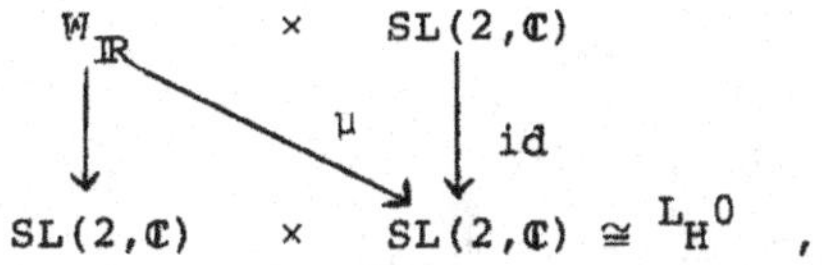

in which the vertical arrow on the left is the parameter for $PGL(2,\mathbb{R})$ which corresponds to the lowest discrete series, and the image of μ in $SL(2,\mathbb{C})$ is contained in $\{\pm 1\}$. The centralizers are given as follows.

$C_{\phi_{\psi}}$	$C_{\phi_{\psi}}$	C_{ψ}	C_{ψ}
$\mathbb{Z}/2\mathbb{Z} \times \mathbb{C}^{\times}$	$\{1\}$	$\mathbb{Z}/2\mathbb{Z} \times \mathbb{Z}/2\mathbb{Z}$	$\mathbb{Z}/2\mathbb{Z}$

$\qquad \cdot$

We write π_{μ} for the representation in $\prod_{\phi_{\psi}}$. As we have agreed, ρ_{μ} then denotes the standard representation of which π_{μ} is the quotient, and π_{μ}^{H} and ρ_{μ}^{H} denote the corresponding representations of $H(\mathbb{R})$. The character formula (1.1.3) is easily shown to be

$$
\operatorname{tr} \pi_{\mu}(f) = \operatorname{tr} \rho_{\mu}(f) - \operatorname{tr} \pi_{Wh}(f) .
$$

On the other hand, from the well known character formula for π_{μ}^{H} we obtain

$$
\begin{aligned}
f_{H}(\psi) &= \operatorname{tr} \pi_{\mu}^{H}(f_{H}) \\
&= \operatorname{tr} \rho_{\mu}^{H}(f_{H}) - \operatorname{tr} \pi_{disc}^{H}(f_{H}) \\
&= \operatorname{tr} \rho_{\mu}(f) - \operatorname{tr} \pi_{Wh}(f) + \operatorname{tr} \pi_{hol}(f) .
\end{aligned}
$$

From our formula for $\operatorname{tr} \pi_{\mu}(f)$ we see that this equals

$$\mathrm{tr}\ \pi_\mu(f) + \mathrm{tr}\ \pi_{hol}(f).$$

Thus, the distribution

$$f_G(\psi) = 2\ \mathrm{tr}\ \pi_\mu(f) - f_H(\psi)$$

on one hand equals

$$\mathrm{tr}\ \pi_\mu(f) - \mathrm{tr}\ \pi_{hol}(f)\ ,$$

but can also be written as

$$\mathrm{tr}\ \rho_\mu(f) - (\mathrm{tr}\ \pi_{Wh}(f) + \mathrm{tr}\ \pi_{hol}(f))\ .$$

From the second expression we see that it is stable. From the first expression we see that the other assertions of the conjecture hold if we define

$$\prod_\psi = \{\pi_\mu, \pi_{hol}\}\ ,$$

$$\varepsilon_\psi(\pi_\mu) = 1, \quad \varepsilon_\psi(\pi_{hol}) = -1\ ,$$

and

$$\langle \cdot, \pi_\mu \rangle = 1, \quad \langle \cdot, \pi_{hol} \rangle = -1\ .$$

We could have defined ψ so that the vertical arrow on the left corresponded to a higher discrete series of $PGL(2,\mathbb{R})$. Everything would have been the same except that $\{\pi_{Wh}, \pi_{hol}\}$ would stand for a pair of discrete series of $G(\mathbb{R})$. These examples are the local analogues of the nontempered cusp forms of $PSp(4)$ discovered by Kurakawa [7]. (See also [9(d),§3].)

<u>Example 1.4.2</u>: Define ψ by the diagram

$$\begin{array}{ccc}
\mathbb{R} & \times & SL(2,\mathbb{C}) \\
\mu_1 \downarrow \quad \mu_2 \searrow \quad \text{id} \nearrow \quad \text{id} \downarrow & & \\
SL(2,\mathbb{C}) & \times & SL(2,\mathbb{C}) \cong {}^{I}H^0 \ ,
\end{array}$$

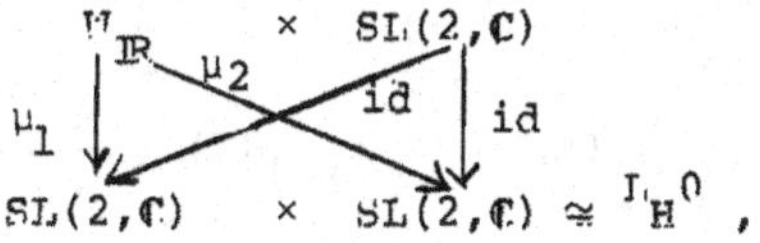

in which the images of μ_1 and μ_2 are contained in $\{\pm 1\}$, and $\mu_1 \neq \mu_2$. The centralizers are

C_{ϕ_ψ}	C_{ϕ_ψ}	C_ψ	C_ψ
$\mathbb{C}^X \times \mathbb{C}^X$	$\{1\}$	$\mathbb{Z}/2\mathbb{Z} \times \mathbb{Z}/2\mathbb{Z}$	$\mathbb{Z}/2\mathbb{Z}$

We write π_{μ_1,μ_2} for the representation in $\prod_{\phi_\psi}$, and follow the notation above. The character formula (1.1.3) can be calculated to be

$$\operatorname{tr} \pi_{\mu_1,\mu_2}(f) = \operatorname{tr} \rho_{\mu_1,\mu_2}(f) - \operatorname{tr} \rho_{\mu_1}(f) - \operatorname{tr} \rho_{\mu_2}(f) + \operatorname{tr} \pi_{Wh}(f) \ .$$

On the other hand, from the character formula for $\pi^H_{\mu_1,\mu_2}$ we obtain

$$f_H(\psi) = \operatorname{tr} \pi^H_{\mu_1,\mu_2}(f_H)$$

$$= \operatorname{tr} \rho^H_{\mu_1,\mu_2}(f_H) - \operatorname{tr} \rho^H_{\mu_1}(f_H) - \operatorname{tr} \rho^H_{\mu_2}(f_H) + \operatorname{tr} \pi^H_{disc}(f_H)$$

$$= \operatorname{tr} \rho_{\mu_1,\mu_2}(f) - \operatorname{tr} \rho_{\mu_1}(f) - \operatorname{tr} \rho_{\mu_2}(f) + \operatorname{tr} \pi_{Wh}(f) - \operatorname{tr} \pi_{hol}(f)$$

$$= \operatorname{tr} \pi_{\mu_1,\mu_2}(f) - \operatorname{tr} \pi_{hol}(f) \ .$$

Thus, the distribution

$$f_G(\psi) = 2 \operatorname{tr} \pi_{\mu_1,\mu_2}(f) - f_H(\psi)$$

on one hand equals

$$\operatorname{tr} \pi_{\mu_1,\mu_2}(f) + \operatorname{tr} \pi_{hol}(f) ,$$

but can also be written as

$$\operatorname{tr} \rho_{\mu_1,\mu_2}(f) - \operatorname{tr} \rho_{\mu_1}(f) - \operatorname{tr} \rho_{\mu_2}(f) + (\operatorname{tr} \pi_{Wh}(f) + \operatorname{tr} \pi_{hol}(f)) .$$

From the second expression we see that it is stable. From the first expression we obtain the other assertions of the conjecture if we define

$$\Pi_\psi = \{\pi_{\mu_1,\mu_2}, \pi_{hol}\} ,$$

$$\varepsilon_\psi(\pi_{\mu_1,\mu_2}) = 1 = \varepsilon_\psi(\pi_{hol}) ,$$

and

$$\langle \cdot, \pi_{\mu_1,\mu_2} \rangle = 1, \qquad \langle \cdot, \pi_{hol} \rangle = -1 .$$

This example is the local analogue of the nontempered cusp forms discovered by Howe and Piatetski-Shapiro [3].

<u>Example 1.4.3</u>: Define ψ as in the last example, except now take $\mu_1 = \mu_2 = \mu$. This example is perhaps the most striking. It is different from the previous two in that ψ factors through a Levi subgroup ${}^LM$ of a proper parabolic subgroup of ${}^LG$. (It is the maximal parabolic subgroup ${}^LP = {}^LM{}^LN$ whose unipotent radical is abelian.) This shows up in the fact that C_ψ is infinite.

C_{ϕ_ψ}	C_{ϕ_ψ}	C_ψ	C_ψ
$GL(2,\mathbb{C})$	$\{1\}$	$O(2,\mathbb{C})$	$\mathbb{Z}/2\mathbb{Z}$

We write $\pi_{\mu,\mu}$ for the representation in $\prod_{\phi_\psi}$, and follow the notation above. The character formula for $\pi_{\mu,\mu}$ is the most complicated of the three to compute. It is

$$\operatorname{tr} \pi_{\mu,\mu}(f) = \operatorname{tr} \rho_{\mu,\mu}(f) - \operatorname{tr} \rho_\mu(f) - \operatorname{tr} \pi_{hol}(f) .$$

We also have

$$\begin{aligned} f_H(\psi) &= \operatorname{tr} \pi^H_{\mu,\mu}(f_H) \\ &= \operatorname{tr} \rho^H_{\mu,\mu}(f_H) - 2 \operatorname{tr} \rho^H_\mu(f_H) + \operatorname{tr} \rho^H_{disc}(f) \\ &= \operatorname{tr} \rho_{\mu,\mu}(f) - 2 \operatorname{tr} \rho_\mu(f) + \operatorname{tr} \pi_{Wh}(f) - \operatorname{tr} \pi_{hol}(f) . \end{aligned}$$

From our formula for $\operatorname{tr} \pi_{\mu,\mu}(f)$ and the formula for $\operatorname{tr} \pi_\mu(f)$ in Example 1.4.1, we see that this equals

$$\operatorname{tr} \pi_{\mu,\mu}(f) - \operatorname{tr} \pi_\mu(f) .$$

Thus, the distribution

$$f_G(\psi) = 2 \operatorname{tr} \pi_{\mu,\mu}(f) - f_H(\psi)$$

on one hand equals

$$\operatorname{tr} \pi_{\mu,\mu}(f) + \operatorname{tr} \pi_\mu(f) ,$$

but can also be written as

$$\operatorname{tr} \rho_{\mu,\mu}(f) - (\operatorname{tr} \pi_{Wh}(f) + \operatorname{tr} \pi_{hol}(f)) .$$

From the second expression we see that it is stable. From the first expression we obtain the other assertions of the conjecture for the endoscopic groups G and H if we define

$$\prod_\psi = \{\pi_{\mu,\mu}, \pi_\mu\} ,$$

$$\varepsilon_\psi(\pi_{\mu,\mu}) = 1 = \varepsilon_\psi(\pi_\mu) ,$$

and

$$\langle \cdot , \pi_{\mu,\mu} \rangle = 1, \qquad \langle \cdot , \pi_\mu \rangle = -1 .$$

In this example we have a third endoscopic group to consider - the Levi subgroup M, which we can identify with $GL(2)$. Since ψ factors through ${}^L M$, it defines a parameter in $\Psi(M/\mathbb{R})$. To complete the verification of the conjecture we must show that

$$f_M(\psi) = \operatorname{tr} \pi_{\mu,\mu}(f) + \operatorname{tr} \pi_\mu(f) .$$

The packets $\prod^M_{\phi_\psi}$ and $\prod^M_\psi$ both consist of one element, the representation

$$\sigma(m) = \mu(\det(m)), \qquad m \in GL(2,\mathbb{R}).$$

The definitions of Shelstad are set up so that the map

$$f \to f_M$$

is dual to induction. Therefore, we will be done if we can show that the induced representation

$$\rho_\sigma = \operatorname{Ind}^{G(\mathbb{R})}_{P(\mathbb{R})} (\sigma \otimes \mathrm{id}_N)$$

is the direct sum of $\pi_{\mu,\mu}$ and π_μ. Now, σ is a nontempered unitary character of $M(\mathbb{R})$. It is the difference between a nontempered standard character on $GL(2,\mathbb{R})$ and a lowest discrete series on $GL(2,\mathbb{R})$. The induced character $\operatorname{tr}(\rho_\sigma)$ is the difference between the corresponding two induced standard characters. The first is just $\operatorname{tr}(\rho_{\mu,\mu})$. The second is a tempered character on $G(\mathbb{R})$ which is reducible; its constituents are π_{Wh} and π_{hol}. Therefore, our induced character equals

$$\mathrm{tr}(\rho_{\mu,\mu}) - (\mathrm{tr}(\pi_{Wh}) + \mathrm{tr}(\pi_{hol})) ,$$

which, as we have seen above, is just

$$\mathrm{tr}(\pi_{\mu,\mu}) + \mathrm{tr}(\pi_{\mu}) .$$

It follows that

$$\rho_{\sigma} = \pi_{\mu,\mu} \oplus \pi_{\mu} ,$$

as required.

Notice that $C_{\psi} \cong O(2,\mathbb{C})$ acts on $C_{\psi}^{0} \cong SO(2,\mathbb{C})$ by outer automorphism. Consequently,

$$R_{\psi} = C_{\psi} \cong \mathbb{Z}/2\mathbb{Z} .$$

Therefore the order of the R group is equal to the number of irreducible constituents of the induced representation

$$\rho_{\sigma} = \mathrm{Ind}_{P(\mathbb{R})}^{G(\mathbb{R})} (\sigma \otimes \mathrm{id}_{N}) ,$$

as we would hope. Observe that the analogue of the R group for the parameter ϕ_{ψ} is trivial. Thus, we see a further example of behaviour which is tied to the parameter ψ rather than ϕ_{ψ}.

This suggests a concrete problem.

<u>Problem 1.4.4</u>: Let $\mathfrak{a}_M$ be the Lie algebra of the split component of the center of $M(\mathbb{R})$. The Weyl group of $\mathfrak{a}_M$ is in this case isomorphic to R_{ψ}. Let w be a representative in $G(\mathbb{R})$ of its non-trivial element. It is known that the corresponding intertwining operator between $\rho_{\sigma_{\lambda}}$ and $\rho_{\sigma_{-\lambda}}$ can be normalized according to the prescription in [9(b),Appendix II]. Let $N(w)$ be the value of the normalized intertwining operator at $\lambda = 0$. It is a unitary operator whose square is 1. Its definition is canonical up to a choice of the

representative w in $G(\mathbb{R})$. The problem is to show that $N(w)$ is not a scalar, and more precisely, to show that if the determinant of w is positive, then

$$\operatorname{tr}(N(w)\rho_\sigma(f)) = \operatorname{tr} \pi_{\mu,\mu}(f) - \operatorname{tr} \pi_\mu(f)$$

$$= \sum_{\pi\in\Pi_\psi} \langle \bar{s},\pi\rangle \operatorname{tr} \pi(f) \quad .$$

§2. A GLOBAL CONJECTURE

2.1. The conjecture we have just stated can be made for any local field F. If F is non-Archimedean, however, the Weil group must be replaced by the group

$$W_F' = W_F \times SL(2,\mathbb{C})$$

introduced in [9(d)]. If G is a reductive quasi-split group defined over F, $\Phi(G/F)$ must be taken to be the set of equivalence classes of maps

$$W_F \times SL(2,\mathbb{C}) \to {}^LG ,$$

while $\Phi_{temp}(G/F)$ will be the subset of those maps whose restriction to W_F has bounded image, when projected onto ${}^LG^0$. In order to define the parameters ψ we must add on another $SL(2,\mathbb{C})$. We take $\Psi(G/F)$ to be the set of ${}^LG^0$ conjugacy classes of maps

$$\psi: W_F \times SL(2,\mathbb{C}) \times SL(2,\mathbb{C}) \to {}^LG$$

such that the restriction of ψ to the product of W_F with the first $SL(2,\mathbb{C})$ belongs to $\Phi_{temp}(G/F)$. For any such ψ, the parameter

$$\phi_\psi(w,\sigma) = \psi\left(w,\sigma,\begin{pmatrix}|w|^{1/2} & 0\\ 0 & |w|^{-1/2}\end{pmatrix}\right), \qquad w\in W_F,\ \sigma\in SL(2,\mathbb{C}),$$

belongs to $\phi(G/F)$.

The conjecture also has a global analogue. Let F be a global field with adèle ring $\mathbb{A}$, and let G be a reductive group over F. If G is not split, there are minor complications in the definitions related to endoscopic groups. (See [9(e)].) To avoid discussing them we shall simply take G to be split. Then the global definitions connected with endoscopic groups follow exactly the local ones we have given.

The conjecture will describe the automorphic representations which are "tempered" in the global sense; that is, representations which occur in the direct integral decomposition of $G(\mathbb{A})$ on $L^2(G(F)\backslash G(\mathbb{A}))$. However, we cannot use the global Weil group if we want to account for all such representations. For even $GL(2)$ has many cuspidal automorphic representations which will not be attached to two dimensional representations of the Weil group. The simplest way to state the global conjecture is to use the conjectural Tannaka group, discussed in [9(d)]. If certain properties hold for the representations of $GL(n)$, Langlands points out that there will be a complex, reductive pro-algebraic group $G_{\Pi_{temp}(F)}$ whose n-dimensional (complex analytic) representations parametrize the automorphic representations of $GL(n,\mathbb{A})$ which are tempered at each place. For each place v, there will also be a complex, reductive pro-algebraic group $G_{\Pi_{temp}(F_v)}$, equipped with a map

$$G_{\Pi_{temp}(F_v)} \to G_{\Pi_{temp}(F)} \quad ,$$

whose n-dimensional representations parametrize the tempered representations of $GL(n,F_v)$. The composition of this map with an n-dimensional representations of $G_{\Pi_{temp}(F)}$ will give the F_v-constituent of the corresponding automorphic representation.

The sets $\Psi(G/F_v)$ which we have defined could also be described as the set of ${}^LG^0$ conjugacy classes of maps

$$\psi_v: G_{\Pi_{temp}(F_v)} \times SL(2,\mathbb{C}) \to {}^LG$$

The centralizer in ${}^LG^0$ of the image of ψ_v is the same as the centralizer of the image of the corresponding parameter associated to the Weil group. In other words,

$$C_{\psi_v} = \mathrm{Cent}(\psi_v({}^G\Pi_{temp}(F_v) \times SL(2,\mathbb{C})), {}^LG^0)$$

and

$$\mathcal{C}_{\psi_v} = C_{\psi_v}/C^0_{\psi_v} Z_G \ .$$

We make the same definitions globally. Assuming the existence of the groups ${}^G\Pi_{temp}(F)$ and ${}^G\Pi_{temp}(F_v)$, let $\Psi(G/F)$ be the set of ${}^LG^0$ conjugacy classes of maps

$$\psi: {}^G\Pi_{temp}(F) \times SL(2,\mathbb{C}) \to {}^LG \ .$$

If $\psi \in \Psi(G/F)$ is any such global parameter, set

$$C_\psi = \mathrm{Cent}(\psi({}^G\Pi_{temp}(F) \times SL(2,\mathbb{C})) \ , \ {}^LG^0) \ ,$$

and

$$\mathcal{C}_\psi = C_\psi/C^0_\psi Z_G \ .$$

The composition of the map

$${}^G\Pi_{temp}(F_v) \times SL(2,\mathbb{C}) \to {}^G\Pi_{temp}(F) \times SL(2,\mathbb{C})$$

with ψ gives a parameter $\psi_v \in \Psi(G/F_v)$. There are natural maps

$$\mathcal{C}_\psi \to \mathcal{C}_{\psi_v}$$

and

$$C_\psi \to C_{\psi_v}$$

Assume that the analogue of the local Conjecture 1.3.3 holds for each field F_v. Fix $\psi \in \Psi(G/F)$. Then for any place v we have a finite set Π_{ψ_v}, a function ε_{ψ_v} on Π_{ψ_v}, a pairing

$$\langle x_v, \pi_v \rangle, \qquad \pi_v \in \Pi_{\psi_v},\ x_v \in C_{\psi_v},$$

and a function c_v on the conjugacy classes of C_{ψ_v}/Z_G. Define the global packet Π_ψ to be the set of irreducible representations $\pi = \otimes_v \pi_v$ of $G(\mathbb{A})$ such that for each v, π_v belongs to Π_{ψ_v}. Define the global pairing

$$\langle x, \pi \rangle = \prod_v \langle x_v, \pi_v \rangle$$

and the global function

$$\varepsilon_\psi(\pi) = \prod_v \varepsilon_{\psi_v}(\pi_v)$$

for $\pi = \otimes_v \pi_v$ in Π_ψ and x in C_ψ with image x_v in C_{ψ_v}. Almost all the terms in each product should equal 1. It is reasonable to expect that for any element $s \in C_\psi/Z_G$, with image s_v in C_{ψ_v}/Z_G,

$$\prod_v c_v(s_v) = 1.$$

If this is so, the global pairing will be canonical.

<u>Conjecture 2.1.1</u>: (A) The representations of $G(\mathbb{A})$ which occur in the spectral decomposition of $L^2(G(F)\backslash G(\mathbb{A}))$ occur in packets parametrized by $\Psi(G/F)$. The representations in the packet corresponding to ψ will occur in the discrete spectrum if and only if C_ψ is finite.

(B) Suppose that C_ψ is finite. Then there is a positive integer d_ψ and a homomorphism

$$\xi_\psi: C_\psi \to \{\pm 1\}$$

such that the multiplicity with which any $\pi \in \Pi_\psi$ occurs discretely in $L^2(G(F)\backslash G(\mathbb{A}))$ equals

$$\frac{d_\psi}{|C_\psi|} \sum_{x \in C_\psi} \langle x,\pi\rangle\, \xi_\psi(x).$$

In particular, if C_ψ and each C_{ψ_v} are abelian, the multiplicity of π is d_ψ if the character $\langle\cdot,\pi\rangle$ equals ξ_ψ, and is zero otherwise.

2.2. Some comments are in order. First of all, the introduction of the Tannaka groups would seem to put the conjecture on a rather shaky foundation. However, everything may be formulated without them. The set $\Psi(G/F)$ is the same as the collection of pairs (ϕ,ρ), where $\phi \in \Phi_{temp}(G/F)$ and ρ is a map from $SL(2,\mathbb{C})$ into C_ϕ, given up to conjugacy by C_ϕ. Included in the conjecture (and also implicit in [9(d)]) is the assertion that $\Phi_{temp}(G/F)$ is the set of L equivalence classes of automorphic representations of $G(\mathbb{A})$ which are tempered at every place. We could simply take this as the definition of $\Phi_{temp}(G/F)$. To avoid mentioning the Tannaka group at all, we would need to define C_ϕ for each ϕ in $\Phi_{temp}(G/F)$. For then C_ψ would just be the centralizer of the image of ρ in C_ϕ. If one grants the existence of certain liftings, one can show that C_ϕ is equal to the centralizer in ${}^LG^0$ of an embedded L-group in ${}^LG$.

Notice that the conjecture does not specify whether an automorphic representation which occurs in the discrete spectrum is cuspidal or not. Indeed, it is quite possible for a global packet Π_ψ to contain one representation which is cuspidal and another which occurs in the residual discrete spectrum. (See [2] and also Example 2.4.1 below.) I do not know whether there will be a simple explanation for such behaviour.

Multiplicity formulas of the sort we conjecture first appeared in [8]. The integer d_ψ was needed there, even for subgroups of $Res_{E/F}(GL(2))$, to account for distinct global parameters which were everywhere locally equivalent. The sign characters ξ_ψ are more mysterious. Suppose that ${}^LG^0$ is the set of fixed points of an outer automorphism of $GL(n,\mathbb{C})$. Then one can observe the existence of such characters from the anticipated properties of the twisted trace formula for $GL(n)$. The character will be 1 if ψ corresponds to a pair (ϕ,ρ) with ρ trivial; that is, if the representations in Π_ψ are tempered at each local place. In general, however, ξ_ψ will not be trivial, and will be built out of the orders at 1/2 of certain L-functions of ϕ. Incidentally, in the examples I have looked at, both local and global, the groups C_ψ have all been abelian. The extrapolation to nonabelian C_ψ is no more than a guess. In fact if C_ψ is nonabelian, the functions $\langle\ ,\pi\rangle$ may turn out to be only class functions on C_ψ, and not irreducible characters."

2.3. Let us look at a few examples. Consider first the group $G=GL(n)$. The centralizer of any reductive subgroup of ${}^LG^0=GL(n,\mathbb{C})$ is connected. This means that the packet Π_ψ (both local and global) should each contain only one representation. The groups C_ϕ will be of the form

$$GL(n_1,\mathbb{C}) \times \ldots \times GL(n_r,\mathbb{C}) ,$$

so that a parameter ψ will consist of the tempered parameter ϕ and a map of $SL(2,\mathbb{C})$ into this group. The representations in Π_ψ should belong to the discrete spectrum (modulo the center of $G(\mathbb{A})$) if and only if C_ψ equals $\mathbb{C}^\times$. This will be the case precisely when C_ϕ equals $GL(n_1,\mathbb{C})$ and ρ is the irreducible n_1 dimensional representation of $SL(2,\mathbb{C})$. Then n_1 will necessarily divide n, $n = n_1m$, and ϕ will be identified with a cuspidal automorphic representation of $GL(m,\mathbb{A})$, embedded diagonally in $GL(n)$. This prescription for the discrete spectrum of $GL(n,\mathbb{A})$ (modulo the center) is exactly what is expected. (See [4].) It is only for $GL(n)$

(and closely related groups such as $SL(n)$) that the distinction between the cuspidal spectrum and the residual discrete spectrum will be so clear.

The multiplicity formula of the conjecture is compatible with the results of Labesse and Langlands [8] for $SL(2)$. More recently, Flicker [2] has studied the quasi-split unitary group in three variables. The conjecture, or rather its analogue for non-split groups, is compatible with his results.

Langlands has shows [9(b), Appendix 3] that for the split group G of type G_2 there is an interesting automorphic representation which occurs in the discrete noncuspidal spectrum. Its Archimedean component is infinite dimensional, of class one and is not tempered. The existence of such a representation is predicted by our conjecture. ${}^LG^0$ is just the complex group of type G_2. It has three unipotent conjugacy classes which meet no proper Levi subgroup. These correspond to the principal unipotent classes of the embedded subgroups

$$ {}^LH_i^0 \to {}^LG^0 \qquad\qquad i = 1,2,3, $$

where

$$ {}^LH_1^0 = {}^LG^0 $$

$$ {}^LH_2^0 \cong SL(2,\mathbb{C}) \times SL(2,\mathbb{C})/\{\pm 1\} , $$

and

$$ {}^LH_3^0 \cong SL(3,\mathbb{C}) . $$

Let $\psi_i = (\phi,\rho_i)$ be the parameter in $\Psi(G/F)$ such that ϕ is trivial and ρ_i is the composition

$$ SL(2,\mathbb{C}) \to {}^LH_i^0 \to {}^LG^0 , $$

in which the map on the left is the one which corresponds to the principal unipotent class in ${}^{L}H_i^0$. The packet Π_{ψ_1} contains one element, the trivial representation of $G(\mathbb{A})$. It is the packet Π_{ψ_2} which should contain the representation discovered by Langlands. The remaining representations in Π_{ψ_2} which occur in the discrete spectrum, as well as all such representations in Π_{ψ_3}, are presumably cuspidal.

2.4. Finally, consider the global analogues for $PSp(4)$ of the three examples we discussed in §1. The global conjecture cannot be proved yet for this group, for there remain unsolved local problems. However, Piatetski-Shapiro has proved the multiplicity formulas of the first two examples below by different methods. (See [10(a)], [10(b)], [10(c)].) Using L-functions and the Weil representation, he reduced the proof to a problem which had been solved by Waldspurger [16].

In each example ψ will be given by the diagram for the corresponding local example in §1 except that $W_{\mathbb{R}}$ is to be replaced by the Tannaka group $G_{\Pi_{temp}(F)}$ or, as suffices in these examples, by the global Weil group W_F. Each μ will be a Grössencharacter of order 1 or 2, since the one dimensional representations of $G_{\Pi_{temp}(F)}$, W_F and $F^{x}\backslash \mathbb{A}^{x}$ all co-incide. In each example the integer d_ψ will be 1.

Example 2.4.1: This is the example of Kurakawa. Take the diagram in Example 1.4.1, letting the vertical arrow on the left parametrize a cuspidal automorphic representation $\tau = \otimes_v \tau_v$ of $PGL(2,\mathbb{A})$. As in the local case, we have

$$C_\psi \cong \mathbb{Z}/2\mathbb{Z}, \qquad C_\psi \cong \mathbb{Z}/2\mathbb{Z} .$$

The character ξ_ψ should be 1 or -1 according to whether the order at $s = 1/2$ of the standard L function $L(s,\tau)$ is even or odd. Our conjecture states that a representation π in the packet Π_ψ occurs in the discrete spectrum if and only if the character $\langle\pi,\cdot\rangle$ on C_ψ equals ξ_ψ. The local centralizer group C_{ψ_v} will be of order 2 or 1 depending on whether the representation τ_v of $PGL(2,F_v)$ belongs to the local discrete series or not. Suppose that τ_v belongs to the local discrete series at r different places. Then the global pocket Π_ψ will contain 2^r representations. Exactly half of them will occur in the discrete spectrum of $L^2(G(F)\backslash G(\mathbb{A}))$. (If $r = 0$, the one representation in Π_ψ will occur in the discrete spectrum if and only if $\xi_\psi = 1$.)

For a given complex number s, consider the representation

$$(x,a) \rightarrow \tau(x)\mu(a)|a|^{\frac{s}{2}}, \qquad x \in PGL(2,\mathbb{A}),\ a \in \mathbb{A}^x,$$

of $PGL(2,\mathbb{A}) \times \mathbb{A}^x$. It is an automorphic representation of a Levi subgroup of G which is cuspidal modulo the center. The associated induced representation of $G(\mathbb{A})$ will have a global intertwining operator, for which we can anticipate a global normalizing factor equal to

$$(L(\tfrac{s}{2},\tau)L(s,1_F))(L(-\tfrac{s}{2},\tau)L(-s,1_F))^{-1}.$$

From the theory of Eisenstein series and the expected properties of the local normalized intertwining operators, one can show that Π_ψ will have a representation in the residual discrete spectrum if and only if the function above has a pole at $s = 1$. This will be the case precisely when $L(1/2,\tau)$ does not vanish. Thus, the number of cuspidal automorphic representations in the packet Π_ψ should equal 2^{r-1} or $2^{r-1} - 1$, depending on whether $L(1/2,\tau)$ vanishes or not.

<u>Example 2.4.2</u>: This is the example of Howe and Piatetski-Shapiro. Take the diagram in Example 1.4.2 with $\mu_1 \neq \mu_2$. Then

$$C_\psi \cong \mathbb{Z}/2\mathbb{Z} \times \mathbb{Z}/2\mathbb{Z}, \qquad \mathcal{C}_\psi \cong \mathbb{Z}/2\mathbb{Z}.$$

The character ξ_ψ should always be 1. Our conjecture states that a representation $\pi \in \prod_\psi$ will occur in the discrete spectrum if and only if the character $\langle\cdot,\pi\rangle$ equals 1. Each local centralizer group C_{ψ_v} will be isomorphic to $\mathbb{Z}/2\mathbb{Z}$. It follows that the packet $\prod_\psi$ will contain infinitely many representations, and infinitely many should occur discretely in $L^2(G(F)\backslash G(\mathbb{A}))$.

<u>Example 2.4.3</u>: Take the diagram in Example 1.4.2 with $\mu_1 = \mu_2$. Then

$$C_\psi \cong O(2,\mathbb{C}), \qquad \mathcal{C}_\psi \cong \mathbb{Z}/2\mathbb{Z}.$$

Each local centralizer group C_{ψ_v} will be isomorphic to $\mathbb{Z}/2\mathbb{Z}$, so the packet $\prod_\psi$ will contain infinitely many representations. However, since C_ψ is infinite, the conjecture states that none of them will occur discretely in $L^2(G(F)\backslash G(\mathbb{A}))$.

§3. THE TRACE FORMULA

3.1. The conjecture of §2 can be motivated by the trace formula, if one is willing to grant the solutions of several local problems. We hope to do this properly on some future occasion, but at the moment even this is too large a task. We shall be content here to discuss a few problems connected with the trace formula, and to relate them to the conjecture in the example we have been looking at - the group $PSp(4)$. For a more detailed description of the trace formula, see the paper [1(b)] and the references listed there.

Let G be as in §2, but for simplicity, take F to be the field of rational numbers $\mathbb{Q}$. The trace formula can be regarded as an equality

$$(3.1.1) \qquad \sum_{o \in O} I_o(f) = \sum_{\chi \in X} I_\chi(f), \qquad f \in C_c^\infty(G(\mathbb{A})),$$

of invariant distributions on $G(\mathbb{A})$. The distributions on the left are parametrized by the semisimple conjugacy classes in $G(\mathbb{Q})$, while those on the right are parametrized by cuspidal automorphic representations associated to Levi components of parabolic subgroups of G. Included in the terms on the left are orbital integrals on $G(\mathbb{A})$ (the distributions in which the semisimple conjugacy class in $G(\mathbb{Q})$ is regular elliptic) and on the right are the characters of cuspidal automorphic representations of $G(\mathbb{Q})$ (the distributions in which the Levi subgroup is G itself). In general the terms on the left are invariant distributions which are obtained naturally from <u>weighted</u> orbital integrals on $G(\mathbb{A})$. The terms on the right are simpler, and can be given by a reasonably simple explicit formula. (See [1(b)]).

The goal of [9(c)] was to begin an attack on a fundamental problem - to stabilize the trace formula. The endoscopic groups for G are quasi-split groups defined over $\mathbb{Q}$: they can be regarded as endoscopic groups over the completions $\mathbb{Q}_v$ of $\mathbb{Q}$. As in §1, we suppose that for each endoscopic group H we have fixed an admissible embedding ${}^LH \subset {}^LG$ which is compatible with equivalence. We also assume that the theory of Shelstad for real groups has been extended to an arbitrary local field. Then for any function $f \in C_c^\infty(G(\mathbb{A}))$ and any endoscopic group H we will be able to define a function f_H in $C_c^\infty(H(\mathbb{A}))$. For example, if f is of the form $\otimes_v f_v$, we simply set

$$f_H = \otimes_v f_{v,H}$$

However, f_H will be determined only up to evaluation on stable distributions on $H(\mathbb{A})$. To exploit the trace formula, it will be necessary to express the invariant distributions which occur in terms of stable distributions on the various groups $H(\mathbb{A})$.

Kottwitz [6] has introduced a natural equivalence relation, called <u>stable conjugacy</u>, on the set of conjugacy classes in $G(\overline{\mathbb{Q}})$ on the regular semisimple classes. If $\mathcal{O}$ is the set of all semisimple conjugacy classes in $G(\mathbb{Q})$, let $\overline{\mathcal{O}}$ be the set of stable conjugacy classes in $\mathcal{O}$. For any $\bar{\sigma} \in \overline{\mathcal{O}}$, set

$$I_{\bar{\sigma}}(f) = \sum_{\sigma \in \bar{\sigma}} I_{\sigma}(f), \qquad f \in C_c^\infty(G(\mathbb{A})).$$

If H is an endoscopic group for G, it can be shown that there is a natural map

$$\overline{\mathcal{O}}_H \to \overline{\mathcal{O}}$$

from the semisimple stable conjugacy classes of $H(\mathbb{Q})$ to those of $G(\mathbb{Q})$. One of the main results of [9(e)] was a formula

$$(3.1.2) \qquad I_{\bar{\sigma}}(f) = \sum_H \iota(G,H) \sum_{\{\bar{\sigma}_H \in \overline{\mathcal{O}}_H : \bar{\sigma}_H \to \bar{\sigma}\}} S^H_{\bar{\sigma}_H}(f_H),$$

for any $f \in C_c^\infty(G(\mathbb{A}))$ and any class $\bar{\sigma} \in \overline{\mathcal{O}}$ consisting of regular elliptic elements. For each endoscopic group H, $\iota(G,H)$ is a constant and $S^H_{\bar{\sigma}_H}$ is a stable distribution on $H(\mathbb{A})$. The sum over H (as well as all such sums below) is taken over the equivalence classes of cuspidal endoscopic groups for G.

<u>Problem 3.1.3</u>: Show that the formula (3.1.2) holds for an arbitrary stable conjugacy class $\bar{\sigma}$ in $\overline{\mathcal{O}}$.

This problem is similar in spirit to that posed by Conjecture 1.3.3. It is not necessary to construct the stable distributions $S^H_{\bar{\sigma}_H}$. One would assume inductively that they had been defined for any $H \neq G$. (Of course we could not continue to work within the limited category we have adopted for this exposition - namely, G is a split group with embeddings ${}^LH \subset {}^LG$.) The problem would then amount to showing that the invariant distribution

$$f \to I_{\bar{o}}(f) - \sum_{H \neq G} \iota(G,H) \sum_{\{\bar{o}_H \to \bar{o}\}} S^H_{\bar{o}_H}(f_H)$$

was stable. However, this assertion is still likely to be quite difficult. The problem does not seem tractable, in general, without a good knowledge of the Fourier transforms of the distributions $I_{\bar{o}}$.

In any case, assume Problem 3.1.3 has been solved. Define

$$I(f) = I^G(f) = \sum_{\bar{o} \in \bar{\mathcal{O}}} I_{\bar{o}}(f) ,$$

and

$$S(f) = S^G(f) = \sum_{\bar{o} \in \bar{\mathcal{O}}} S^G_{\bar{o}}(f) ,$$

for any $f \in C_c^\infty(G(\mathbb{A}))$. The expression for $I(f)$ is just equal to each side of the trace formula (3.1.1). It is clear that it converges absolutely. The same cannot be said of the expression for $S(f)$. The problem is discusses in [9(e),VIII.5]. We must make the assumption that there are only finitely many H such that $f_H \neq 0$. (See Lemma 8.12 of [9(e)].) This is certainly true if G is adjoint for then there are only finitely many endoscopic groups (up to equivalence, of course). Since the constant $\iota(G,G)$ equals 1, we obtain

$$\sum_{\bar{o} \in \bar{\mathcal{O}}} S^G_{\bar{o}}(f)$$

$$= \sum_{\bar{o} \in \bar{\mathcal{O}}} (I_{\bar{o}}(f) - \sum_{H \neq G} \iota(G,H) \sum_{\{\bar{o}_H \in \bar{\mathcal{O}}_H : \bar{o}_H \to \bar{o}\}} S^H_{\bar{o}_H}(f_H))$$

$$= \sum_{\bar{o}} I_{\bar{o}}(f) - \sum_{H \neq G} \iota(G,H) \sum_{\bar{o}_H \in \bar{\mathcal{O}}_H} S^H_{\bar{o}_H}(f_H)$$

$$= I(f) - \sum_{H \neq G} \iota(G,H) S^H(f_H) ,$$

if we assume inductively that the expression used to define S^H converges absolutely whenever $H \neq G$. It follows that the expression for $S^G(f)$ converges absolutely, and S^G is a stable distribution on $G(\mathbb{A})$. Moreover,

$$I(f) = \sum_H \iota(G,H) S^H(f_H) , \tag{3.1.4}$$

for any $f \in C_c^\infty(G(\mathbb{A}))$.

3.2. An identity (3.1.4) could be used to yield interesting information about the discrete spectrum of G, since there is an explicit formula for

$$I(f) = \sum_{\chi \in X} I_\chi(f) . \tag{3.2.1}$$

The formula is given as a sum of integrals over vector spaces $i\mathfrak{a}_M^* / i\mathfrak{a}_G^*$, where $P = MN$ is a parabolic subgroup of G (defined over $\mathbb{Q}$), A_M is the split component of the center of the Levi component M of P, and $\mathfrak{a}_M$ is the Lie algebra of $A_M(\mathbb{R})$. The most interesting part of the formula is the term for which the integral is actually discrete; in other words, for which $P = G$. It is only this term that we shall describe.

Suppose that $P = MN$ is a parabolic subgroup and that σ is an irreducible unitary representation of $M(\mathbb{A})$. Let ρ_σ be the induced representation

$$\operatorname{Ind}_{P(\mathbb{R})}^{G(\mathbb{R})} (L^2_{\mathrm{disc}}(A_M(\mathbb{R})^0 M(\mathbb{Q})\backslash M(\mathbb{A}))_\sigma \otimes \mathrm{id}_N) ,$$

where id_N is the trivial representation of the unipotent radical $N(\mathbb{A})$, and $L^2_{\mathrm{disc}}(A_M(\mathbb{R})^0 M(\mathbb{Q})\backslash M(\mathbb{A}))_\sigma$ is the σ-primary component of the subrepresentation of $M(\mathbb{A})$ on $L^2(A_M(\mathbb{R})^0 M(\mathbb{Q})\backslash M(\mathbb{A}))$ which decomposes discretely. Let $W(\mathfrak{a}_M)$ be the Weyl group of $\mathfrak{a}_M$, and let

$W(\mathfrak{a}_M)_{reg}$ be the subset of elements in $W(\mathfrak{a}_M)$ whose space of fixed vectors is $\mathfrak{a}_G$. For any w in $W(\mathfrak{a}_M)$ let $T(w)$ be the (unnormalized) global intertwining operator from ρ_σ to $\rho_{w\sigma}$. For any function $f \in C_c^\infty(G(\mathbb{A}))$, define

$$(3.2.2) \qquad I_+(f) = I_+^G(f)$$

$$= \sum_{\{(M,\sigma)\}} |W(\mathfrak{a}_M)|^{-1} \sum_{w \in W(\mathfrak{a}_M)_{reg}} |\det(1-w)_{\mathfrak{a}_M/\mathfrak{a}_G}|^{-1} \operatorname{tr}(T(w)\rho_\sigma(f)) ,$$

where the first sum is over pairs (M,σ) as above, with M given up to $G(\mathbb{Q})$ conjugacy. Then I_+ is the "discrete part" of the explicit formula for (3.2.1). Here we have obscured a technical complication for the sake of simplicity. It is not known that the sum over σ in (3.2.2) converges absolutely (although one expects it to do so). In order to insure absolute convergence, one should really group the summands in (3.2.2) with other components of $I(f)$ in a way that takes account of the decomposition on the right hand side of (3.2.1).

We expect to be able to isolate the various contributions of (3.1.4) to the distribution I_+. This would mean that we could find (for every G) a stable distribution S_+^G on $G(\mathbb{A})$ such that

$$(3.2.3) \qquad I_+(f) = \sum_H \iota(G,H) S_+^H(f_H) ,$$

for any $f \in C_c^\infty(G(\mathbb{A}))$. Said another way, the distribution

$$f \to I_+(f) - \sum_{H \neq G} \iota(G,H) S_+^H(f_H) ,$$

would be stable. Now this is actually a rather concrete assertion. The distribution I_+ is certainly given by a concrete formula, and the distributions S_+^H are defined inductively in terms of the formulas for I_+^H. Moreover, Kottwitz has recently evaluated the constants

$\iota(G,H)$. We will not give the general formula, but if G and $H = H_s$ are both split groups, $\iota(G,H)$ equals

$$|Z_H/Z_G|^{-1} \ |\mathrm{Norm}(sZ_G, {}^LG^0)/{}^LH^0|^{-1} ,$$

where $\mathrm{Norm}(sZ_G, {}^LG^0)$ denotes the group of elements σ in ${}^LG^0$ which normalize the coset sZ_G.

A formula like (3.2.3) will have interesting implications for the discrete spectrum of G. Consider the one dimensional automorphic representations of the various endoscopic groups H. Our examples for $PSp(4,\mathbb{R})$ suggest that for $H \neq G$, the contributions of such one dimensional representations to the right hand side of (3.2.3) will not be stable distributions of f. They will have to correspond to something in the formula (3.2.2) for $I_+(f)$. Suppose that some one dimensional representations cannot be accounted for by any terms in (3.2.2) indexed by (M,σ), with $M \neq G$. Then they will have to correspond to terms with $M = G$. In other words, they ought to give rise to interesting nontempered automorphic representations of $G(\mathbb{A})$ which occur in the discrete spectrum.

It is implicit in our conjecture that we should index the one dimensional automorphic representations of $H(\mathbb{A})$ by maps

$$W_{\mathbb{Q}} \times SL(2,\mathbb{C}) \to {}^LH,$$

in which the image of $W_{\mathbb{Q}}$ in ${}^LH^0$ commutes with ${}^LH^0$ and the image of $SL(2,\mathbb{C})$ corresponds to the principal unipotent in ${}^LH^0$. (For the correspondence between unipotent conjugacy classes and representations of $SL(2,\mathbb{C})$, see [13].) It is of course easy to do this. What is not clear is why we should do it. Why introduce an $SL(2,\mathbb{C})$ when the one dimensional representations of $H(\mathbb{A})$ can be described perfectly well without it? According to the conjecture, the $SL(2,\mathbb{C})$ factor will be essential in describing the corresponding automorphic representations of $G(\mathbb{A})$. In particular, a one dimensional automorphic representation of $H(\mathbb{A})$ should give rise to automoprhic representations of $G(\mathbb{A})$ which occur discretely (modulo the center of $G(\mathbb{A})$) if and only if the image of $W_{\mathbb{Q}} \times SL(2,\mathbb{C})$ under composition

$$W_{\mathbb{Q}} \times SL(2,\mathbb{C}) \to {}^LH \to {}^LG$$

lies in no proper Levi subgroup of ${}^LG$. We shall examine this question for $PSp(4)$.

3.3. Consider the example of $G = PSp(4)$. As a reductive group over $\mathbb{Q}$, G has only two cuspidal endoscopic groups (up to equivalence) - G itself, and

$$H = H_s \cong PGL(2) \times PGL(2) ,$$

with

$$s = \begin{pmatrix} 1 & & & \\ & -1 & & \\ & & -1 & \\ & & & 1 \end{pmatrix}.$$

Let us look at the formula (3.2.3) in this case. The constant $\iota(G,G)$ equals 1. The group

$$\mathrm{Norm}(sZ_G, {}^LG^0)/{}^LH^0$$

has order 2, the nontrivial element being the coset of the matrix

$$\begin{pmatrix} 0 & 1 & 0 & 0 \\ 1 & 0 & 0 & 0 \\ 0 & 0 & 0 & 1 \\ 0 & 0 & 1 & 0 \end{pmatrix}.$$

Since

$$Z_H / Z_G \cong \mathbb{Z}/2\mathbb{Z} ,$$

we have

$$\iota(G,H) = \frac{1}{4} .$$

The group H has no proper cuspidal endoscopic group. This means that S_+^H equals I_+^H, and so is given by the formula (3.2.2). Formula (3.2.3) is then equivalent to the assertion that the distribution

$$f \to I_+^G(f) - \frac{1}{4} I_+^H(f_H) \qquad f \in C_c^\infty(G(\mathbb{A})),$$

is stable. Since the distribution

$$f \to I_+^H(f_H)$$

is neither stable nor tempered, the assertion would give interesting information about the discrete spectrum of G.

The one dimensional automorphic representations of H are just

(3.3.1) $$(h_1,h_2) \to \mu_1(\det h_1)\mu_2(\det h_2) , \qquad h_1,h_2 \in PGL(2,\mathbb{A})$$

where μ_1 and μ_2 are Grössencharacters whose images are contained in $\{\pm 1\}$. For any such representation define

$$\psi: W_{\mathbb{Q}} \times SL(2,\mathbb{C}) \to SL(2,\mathbb{C}) \times SL(2,\mathbb{C}) \times W_{\mathbb{Q}} \cong {}^L H$$

by

$$\psi(w,\sigma) = (\mu_1(w')\sigma, \mu_2(w')\sigma, w) ,$$

where w' is the projection of w onto the commutator quotient of $W_{\mathbb{Q}}$, and each $\mu_i(w')$ is identified with a central element in $SL(2,\mathbb{C})$. As we did for real groups, we define a map

$$\phi_\psi : W_{\mathbb{Q}} \to {}^L H$$

as the composition of the map

$$w \to \left(w, \begin{pmatrix} |w|^{1/2} & 0 \\ 0 & |w|^{-1/2} \end{pmatrix}\right), \qquad w \in W_{\mathbb{Q}},$$

with ψ. Then the global L-packet $\Pi^H_{\phi_\psi}$ equals Π^H_ψ, and contains exactly one element, the representation (3.3.1). By composing with the natural embedding ${}^LH \subset {}^LG$, we identify each ψ with a mapping of $W_{\mathbb{Q}} \times SL(2,\mathbb{C})$ into ${}^LG$. In this way we obtain parameters in $\Psi(G/\mathbb{Q})$. They are just the ones considered in Examples 2.4.2 and 2.4.3.

The contribution of ψ and H to the right hand side of (3.2.3) equals the product of $\frac{1}{4}$ with the character of the representation (3.3.1) evaluated at f_H. Assume that the Examples 1.4.2 and 1.4.3 for $G(\mathbb{R})$ carry over to each local group $G(\mathbb{Q}_v)$. Then to the local parameters $\psi_v \in \Psi(G/\mathbb{Q}_v)$, obtained from ψ, we have the local packets Π_{ψ_v}. On these packets, the signs ε_{ψ_v} are all 1. If

$$f = \otimes_v f_v, \qquad f_v \in C_c^\infty(G(\mathbb{Q}_v)),$$

the contribution of ψ and H to (3.2.3) is just

$$\begin{aligned} \tfrac{1}{4} f_H(\psi) &= \tfrac{1}{4} \prod_v f_{v,H}(\psi_v) \\ &= \tfrac{1}{4} \prod_v \Big(c_v(s_v) \sum_{\pi_v \in \Pi_{\psi_v}} \langle \bar{s}_v, \pi_v\rangle \operatorname{tr} \pi_v(f_v)\Big), \end{aligned}$$

where s_v is the image of s in C_{ψ_v}/Z_G and $\bar{s}_v$ is its projection onto C_{ψ_v}. This becomes

$$\text{(3.3.2)} \qquad \tfrac{1}{4} \prod_v \Big(\sum_{\pi_v \in \Pi_{\psi_v}} \langle \bar{s}_v, \pi_v\rangle \operatorname{tr} \pi_v(f_v)\Big)$$

if we assume the product formula

$$\prod_v c_v(s_v) = 1 .$$

Suppose that $\mu_1 = \mu_2 = \mu$. The conjecture requires that (3.3.2) should be cancelled by a term in (3.2.2) indexed by (M,σ) with $M \neq G$. The projection of the image of ψ onto ${}^LG^0$ is conjugate to

(3.3.3) $$\left\{ \begin{pmatrix} h & 0 \\ 0 & h' \end{pmatrix} : h \in SL(2,\mathbb{C}) \right\} ,$$

a subgroup of

$${}^LM^0 = \left\{ \begin{pmatrix} g & 0 \\ 0 & \alpha(g) \end{pmatrix} \cdot g \in GL(2,\mathbb{C}) \right\} ,$$

where

$$h' = \begin{pmatrix} 1 & 0 \\ 0 & -1 \end{pmatrix} h \begin{pmatrix} 1 & 0 \\ 0 & -1 \end{pmatrix} ,$$

and

$$\alpha(g) = \begin{pmatrix} 0 & 1 \\ 1 & 0 \end{pmatrix} {}^tg^{-1} \begin{pmatrix} 0 & 1 \\ 1 & 0 \end{pmatrix} .$$

But ${}^LM^0$ is the identity component of the L-group of a Levi subgroup M of G which is isomorphic to $GL(2)$. Set

$$\sigma(m) = \mu(\det(m)) , \qquad m \in GL(2,\mathbb{A}).$$

Then σ can be regarded as an automorphic representation of M which occurs discretely (modulo the center of $M(\mathbb{A})$). It is the pair (M,σ) whose contribution to (3.2.2) we will compare with (3.3.2).

Let w be a representative in $G(\mathbb{Q})$ of the nontrivial element of the Weyl group $W(\mathfrak{a}_M)$. The representation σ is a lift to $GL(2)$ of an automorphic representation of $PGL(2)$. It is fixed by $ad(w)$. The contribution of (M,σ) to the formula (3.2.2) for $I_+(f)$ is

(3.3.4) $$\frac{1}{4}\,\mathrm{tr}(T(w)\rho_\sigma(f)) ,$$

since

$$|W(\mathfrak{a}_M)|^{-1}\ |\det(1-w)_{\mathfrak{a}_M}|^{-1} = \frac{1}{2}\cdot\frac{1}{2} = \frac{1}{4} .$$

We can expect a decomposition

$$T(w) = m(w)\prod_v N_v(w)$$

of $T(w)$ into local normalized intertwining operators. (See [9(b), p. 282].) If ϕ_1 is the three dimensional representation of $W_{\mathbb{Q}}$ obtained by composing ϕ_ψ with the adjoint representation of the group (3.3.3), and $\tilde{\phi}_1$ is its contragradient, the global normalizing factor $m(w)$ equals

$$\lim_{s\to 0}\frac{L(s,\tilde{\phi}_1)}{L(-s,\phi_1)} .$$

One checks that it equals 1. Therefore, (3.3.4) equals

$$\frac{1}{4}\prod_v \mathrm{tr}\ (N_v(w)\rho_{\sigma_v}(f_v)) ,$$

where σ_v is the character $\mu_v(\det(\cdot))$ on $GL(2,\mathbb{Q}_v)$, with μ_v the local component of the Grössencharacter μ. With a resolution to Problem 1.4.4, or rather its analogue for each place v, the expression would become

$$\frac{1}{4}\prod_v \Big(\sum_{\pi_v\in\Pi_{\psi_v}}\langle\bar{s}_v,\pi_v\rangle \mathrm{tr}\ \pi_v(f_v)\Big) .$$

This is just (3.3.2).

Thus, when $\mu_1 = \mu_2 = \mu$, so that ψ factors through a Levi subgroup, the contribution of ψ and H to (3.2.3) would be completely cancelled by a term in (3.2.2) with $M \neq G$. This suggests that such ψ contribute nothing to the discrete spectrum of $G(\mathbb{A})$, as predicted by the conjecture.

3.4. In order for the two terms above to cancel, it was essential that

$$\iota(G,H) = |W(\mathfrak{a}_M)|^{-1} |\det(1-w)_{\mathfrak{a}_M}|^{-1} ,$$

the common value, we recall, being $\frac{1}{4}$. This fact may be interpreted as a combinatorial property of the complex group

$$C_\psi = O(2,\mathbb{C}) .$$

The generalization of this property will be a key to affecting similar cancellations for arbitrary groups. We shall describe it.

Let C be the set of complex points of a complex reductive algebraic group. We do not assume that C is connected. Let C^0 be the identity component of C. Let T^0 be a Cartan subgroup of C^0, and let W be the normalizer of T^0 in C, modulo T^0. Then W is an extension of

$$W^0 = W \cap C^0 ,$$

the Weyl group of (C^0,T^0). It acts on T^0 and on its Lie algebra. Let W_{reg} be the set of elements in W for which 1 is not an eigenvalue. If w is any element in W, set

$$\varepsilon(w) = (-1)^{n(w)} ,$$

where $n(w)$ equals the number of positive roots of (C^0,T^0) which are mapped by w to negative roots. ($\varepsilon(w)$ is independent of how the positive roots are chosen.) For each connected component x of C we define

$$i(x) = |W^0|^{-1} \sum_{w \in W_{reg}(x)} \varepsilon(w) |\det(1-w)|^{-1} ,$$

where $W_{reg}(x)$ is the set of elements in W_{reg} induced from points in x. The number $i(x)$ is a sort of scalar analogue of the invariant distribution (3.2.2).

For each component x of C, let $Orb(C^0,x)$ be the set of C^0-orbits of elements in x for which the adjoint map (as a linear operator on the Lie algebra of C^0) is semisimple. If s belongs to any of the orbits, the group

$$C_s = Cent(s,C^0)$$

satisfies the same hypothesis as C. Its conjugacy class in C^0 depends only on the orbit of s. The number

$$|C_s/C_s^0|^{-1}$$

of connected components in C_s also depends only on the orbit of s. It is possible to define uniquely a number $\sigma(C)$, for every group C, which depends only on C^0, and vanishes unless the center of C^0 is finite, such that

$$i(C^0) = \sum_{s\in Orb(C^0,C^0)} |C_s/C_s^0|^{-1}\sigma(C_s) \tag{3.4.1}$$

for every group C. Indeed, there are only finitely many orbits s in $Orb(C^0,C^0)$ such that the center of C_s is finite, so we can define $\sigma(C)$ inductively by this last equation. We see inductively that it depends only on C^C. The numbers $\sigma(C)$ are scalar analogues of the stable distribution defined by (3.2.3).

<u>Theorem 3.4.2</u>: With the possible exclusion of the case that C^0 has exceptional simple factors, we have

$$i(x) = \sum_{s\in Orb(C^0,x)} |C_s/C_s^0|^{-1}\sigma(C_s), \tag{3.4.3}$$

for every component x of C.

The details will appear in [1(c)]. (I have not yet had a chance to look at the exceptional groups.)

Equations (1.3.6), (3.1.2) and (3.4.1) are all in the same spirit. They each provide an inductive definition for a set of objects (stable distributions, for example) in terms of given objects (such as invariant distributions). The inductive definition in each case is by a sum over indices which are closely related to endoscopic groups. Equations (1.3.6) and (3.1.2) should have twisted analogues. These should be true identities, involving the objects defined by the original equations. The twisted analogue of (3.4.1) we have just encountered. It is the formula (3.4.3).

REFERENCES

1. Arthur, J.,
   (a) A theorem on the Schwartz space of a reductive Lie group, Proc. Nat. Acad. Sci. 72 (1975), 4718-4719.
   (b) The trace formula for reductive groups, to appear in Publ. Math. de l'Université Paris VII.
   (c) In preparation.

2. Flicker, Y., L-packets and liftings for U(3), preprint.

3. Howe, R., and Piatetski-Shapiro, I., A counter-example to the "generalized Ramanujan conjecture" for (quasi-) split groups, Proc. Sympos. Pure Math., Vol. 33, Part I, Amer. Math. Soc., Providence R.I. (1979), 315-322.

4. Jacquet, H., article in these proceedings.

5. Knapp, A., Commutativity of intertwining operators for semisimple groups, Comp. Math. 46 (1982), 33-84.

6. Kottwitz, R., Rational conjugacy classes in reductive groups, preprint.

7. Kurakawa, N., Examples of eigenvalues of Hecke operators on Siegel cusp forms of degree two, Inventiones Math., 49 (1978), 149-165.

8. Labesse, J. P., and Langlands R.P., L-indistinguishability for SL(2), Can. J. Math. 31 (1979), 726-785.

9. Langlands, R.P.,
   (a) On the classification of irreducible representations of real algebraic groups, mimeographed notes, Institute for Advanced Study, 1973.
   (b) On the Functional Equations Satisfied by Eisenstein Series, Lecture Notes in Math., 544 (1976).
   (c) Stable conjugacy: definitions and lemmas, Canad. J. Math. 31 (1979), 700-725.
   (d) Automorphic representations, Shimura varieties, and motives. Ein Marchen. Proc. Sympos. Pure Math., Vol. 33, Part 2, Amer. Math. Soc., Providence R.I., (1979), 205-246.
   (e) Les débuts d'une formule des traces stables, Publ. Math. de l'Université Paris VII, No. 13 (1983).

10. Piatestski-Shapiro, I.,
    (a) On the Saito-Kurakawa lifting, preprint.
    (b) Special Automorphic forms on PGSp(4), preprint.
    (c) Article in these proceedings.

11. Shelstad, D.,
    (a) Notes on L-indistinguishability (based on a lecture of R.P. Langlands), Proc. Sympos. Pure Math., Vol. 33, Part 2, Amer. Math. Soc., Providence R.I., (1979), 193-203.
    (b) Embeddings of L-groups, Canad. J. Math. 33 (1981), 513-558.
    (c) L-indistinguishability for real groups, Math. Ann. 259 (1982), 385-430.
    (d) Orbital integrals, endoscopic groups and L-indistinguishability for real groups, to appear in Publ. Math. de l'Université Paris VII.

12. Speh. B., and Vogan, D., Reducibility of generalized principal series representations, Acta. Math. 145 (1980), 227-299.

13. Springer, T., and Steinberg, R., Conjugacy classes, Lecture Notes in Math., 131, 1970, 167-266.

14. Steinberg, R., Torsion in reductive groups, Advances in Math. 15 (1975), 63-92.

15. Vogan. D., Complex geometry and representations of reductive groups, preprint.

16. Waldspurger, J.L., Correspondence de Shimura et quaternions. Preprint.

Note added in proof: The sign function ε_ψ in the local Conjecture 1.3.3 and the sign character ξ_ψ in the global Conjecture 2.1.1 should both have simple formulas.

Suppose that

$$\psi : W_{\mathbb{R}} \times SL(2,\mathbb{C}) \to {}^LG$$

is given as in Conjecture 1.3.3. Then

$$s_\psi = \psi(1, \begin{pmatrix} -1 & 0 \\ 0 & -1 \end{pmatrix})$$

belongs to the centralizer C_ψ. Let $\bar{s}_\psi$ be the image of s_ψ in C_ψ. Then ε_ψ should be given in terms of the pairing on $C_\psi \times \Pi_\psi$ by

$$\varepsilon_\psi(\pi) = \langle \bar{s}_\psi, \pi \rangle, \qquad \pi \in \Pi_\psi.$$

In particular, if the unipotent element

$$\psi(1, \begin{pmatrix} 1 & 1 \\ 0 & 1 \end{pmatrix})$$

in ${}^L G^0$ is even, the function ε_ψ will be identically 1.

Suppose that F is global and

$$\psi : G_{\Pi_{\mathrm{temp}}(F)} \times SL(2,\mathbb{C}) \to {}^L G$$

is given as in Conjecture 2.1.1. Assume that C_ψ is finite. Let $\mathfrak{g}$ be the Lie algebra of ${}^L G^0$, and define a finite dimensional representation

$$r_\psi : C_\psi \times G_{\Pi_{\mathrm{temp}}(F)} \times SL(2,\mathbb{C}) \to GL(\mathfrak{g})$$

by

$$r_\psi(c,w,g) = \mathrm{Ad}(c \cdot \psi(w,g)),$$

for $c \in C_\psi$, $w \in G_{\Pi_{\mathrm{temp}}(F)}$ and $g \in SL(2,\mathbb{C})$. Then there is a decomposition

$$r_\psi = \oplus_{i \in I_\psi} (\xi_i \otimes \phi_i \otimes \rho_i)$$

where ξ_i, ϕ_i and ρ_i are irreducible (finite-dimensional) representations of C_ψ, $G_{\Pi_{\mathrm{temp}}(F)}$ and $SL(2,\mathbb{C})$ respectively. Suppose that for a given i, the representation ϕ_i is equivalent to its contragredient. Then from the anticipated functional equation of the L-function $L(s,\phi_i)$, we see that

$$\varepsilon(\tfrac{1}{2},\phi_i) = \pm 1.$$

Let I_ψ^- be the set of such indices i such that $\varepsilon(\frac{1}{2},\phi_i)$ actually equals -1, and such that in addition, the dimension of ρ_i is even. Then the sign character should be given by

$$\xi_\psi(c) = \prod_{i \in I_\psi^-} \det(\xi_i(c)), \qquad c \in C_\psi.$$

Such a formula (assuming it is true) is rather intruiging. It ties the values of ε-factors at $\frac{1}{2}$ in an essential way to multiplicities of cusp forms, and it also suggests that the adjoint representation of the L-group might play some role in questions of L-indistinguishability.

P-INVARIANT DISTRIBUTIONS ON GL(N) AND THE CLASSIFICATION OF UNITARY REPRESENTATIONS OF GL(N) (NON-ARCHIMEDEAN CASE)

Joseph N. Bernstein*
Department of Mathematics
University of Maryland
College Park, Maryland
20742

§0. INTRODUCTION

P-invariant Pairings

0.1. Let F be a non-archimedean local field, $G = GL(n,F)$, and $P \subset G$ the subgroup of all matrices with the last row equal to $(0,0,\ldots,0,1)$. Many results about representations of G were obtained by studying their restrictions to P (see [GK], [BZ1], [BZ2], [Z1]). In this paper we prove the following important technical result which clarifies the relations between representations of G and their restrictions to P.

<u>Theorem</u> A (see 5.1). Let (π,E) be a smooth irreducible representation of G in a (complex) vector space E, $\tilde{\pi} = (\tilde{\pi},\tilde{E})$ the contragredient representation. Then each P-invariant pairing $B: E \times \tilde{E} \to \mathbb{C}$ is proportional to the standard pairing.

0.2. Theorem A implies the following

<u>Theorem</u> (see 5.4). Each irreducible unitary representation of G remains irreducible when restricted to P.

H. Jacquet noticed that this theorem implies the following result about representations of G.

* Supported in part by NSF grant MCS-8203622.

Corollary. Any representation of G parabolically induced from an irreducible unitary representation of a Levi subgroup of G is always irreducible. In other words, in the case of GL all R-groups are trivial.

Jacquet's proof uses the explicit description, in terms of Mackey's construction, of the restriction of an induced representation to P. We give another proof in 8.2.

0.3. Using theorem 0.1 we prove that any nondegenerate unitarizable irreducible representation (π, E) of G is generic, i.e. the scalar product in E can be written as a standard integral in the Kirillov model of π (see 6.2).

In §6 we generalize this result to nonunitarizable representations. Namely, we prove that the scalar product between an irreducible nondegenerate G-module E and its contragredient $\tilde{E}$ can be written via an integral in their Kirillov models. (This integral does not converge, but there exists a natural regularization procedure for its evaluation, see 6.3-6.4.) This result gives an alternative proof of the uniqueness and the injectivity of the Kirillov model (see 6.5).

In the case of a degenerate irreducible representation (π, E) A. Zelevinsky described in [Z1, §8] a degenerate Kirillov model. If π is unitarizable, we also can write the scalar product in E via an integral (see 7.4, remark). If we had a regularization procedure for a degenerate Kirillov model we would prove an analogous result for any π.

An Algorithm for the Classification of Unitary Representations of GL(n)

0.4. Using theorem A we establish a unitarizability criterion for irreducible G-modules (see 7.4). It claims that an irreducible representation (π, E) of $GL(n)$ is unitarizable iff it is Hermitian and its derivatives $\pi^{(k)}$ satisfy some inequalities (these derivatives $\pi^{(k)}$ are representations of the groups $GL(m)$ with $m < n$, which describe the restriction of π to P, see 7.2).

This criterion gives an algorithm for the classification of irreducible unitary representations of $G = GL(n)$. More precisely, let us start from some classification of irreducible smooth representations of G (we use Zelevinsky's classification, which is based on the detailed study of derivatives of representations of G, see 7.5-7.8). Moreover, suppose we know the multiplicity matrix $m = (m_{ab})$, which describes the decomposition in the Grothendieck group of induced representations into irreducible ones. In terms of this matrix we can calculate all the derivatives for all irreducible representations of G. Now, using the unitarizability criterion, we can identify those irreducible representations of G which are unitarizable (see 7.9).

0.5. In [Z2] A. Zelevinsky described some polynomials $P_{ab}(q)$, analogous to the Kazhdan-Lusztig polynomials, and conjectured that $m_{ab} = P_{ab}(1)$. Later he proved that these polynomials can be expressed in terms of usual Kazhdan-Lusztig polynomials for symmetric groups (not published). Hence, if we believe Zelevinsky's conjecture, we have explicit formulae for m_{ab} in terms of Kazhdan-Lusztig polynomials, i.e. our algorithm becomes quite precise. This leads to a very interesting question about complexity of the set of unitarizable representations. The problem is that Kazhdan-Lusztig polynomials are given by some recursive formulae and apparently there are no explicit formulae for them. Thus it might happen that the description of unitarizable representations can not be given by explicit formulae and only by some inductive procedure. But maybe for the description of unitary representations we do not need the whole complexity of the Kazhdan-Lusztig polynomials (I even do not rule out the possibility that they can be described by simple-minded methods like those in section 8 without using Zelevinsky's conjecture). Then we can suppose that the classification of irreducible unitary representations for any reductive group ($\mathfrak{p}$-adic or real) can be given by reasonably explicit formulae, since the groups GL(n) are more simple but not much more simple than other groups.

In any case, algorithm 0.4 together with Zelevinsky's conjecture reduces this question to a pure combinatorial problem.

0.6. Our proof of theorem A is based on the following geometrical statement.

Theorem B. Any distribution E on G invariant under the adjoint action of the subgroup P is automatically invariant under the adjoint action of the whole group G.

We prove the implication Theorem B $\Rightarrow$ Theorem A using the technique of Gelfand-Kazhdan (see [GK]). Also the proof of theorem B is reminiscent of the proof in [GK]. But there is one essential difference - unlike [GK] we can not consider each G-orbit separately, since there exist G-orbits which have Ad(P)-invariant but not Ad(G)-invariant distributions*). Theorem B means that these distributions can not be extended from these orbits to the whole group G as Ad(P)-invariant distributions. In order to prove this we use the Fourier transform.

0.7. Let me illustrate the method of the proof of theorem B in the case of the group GL(2).

First of all, applying the localization principle 1.4, which is a formalization of Gelfand-Kazhdan's method, we can assume that E is concentrated on the closure of one G-orbit $O_x = \mathrm{Ad}(G)x$. It is easy to check that a P-invariant distribution E on O_x corresponds to a distribution E' on the space $P\backslash G \simeq F^2 \setminus 0$, which is quasiinvariant under the action of the centralizer G_x of x in G, i.e. $\delta(g)E' = \nu(g)E'$, where $g \in G_x$, $\nu(g) = |\det g|$.

If x is semisimple, the distribution E' is proportional to

*) This is the reason why theorems A and B are false for a finite field F.

the Haar measure on F^2, i.e. the corresponding distribution $\mathcal{E}$ is G-invariant. But if x is unipotent, there exists a quasiinvariant distribution $\mathcal{E}'$, concentrated on a line in F^2. The corresponding distribution $\mathcal{E}$ is concentrated on the unipotent subgroup of P_2, which we identify with the affine line F. $\mathcal{E}$ is defined only on the subset $F \setminus 0$ of nontrival unipotent elements and is invariant under the action of the multiplicative group F^*. We claim that it can not be extended to an F^*-invariant distribution on F.

Indeed, for any F^*-invariant distribution $\mathcal{Q}$ on F, its Fourier transform $\hat{\mathcal{Q}}$ is quasiinvariant. Using this it is easy to check that $\hat{\mathcal{Q}}$ is proportional to a Haar measure on F, and hence $\mathcal{Q}$ is concentrated at 0.

In the general case, for $GL(n)$ with $n > 2$, the proof is analogous up to the last statement. This statement - that $\hat{\mathcal{Q}}$ is proportional to a Haar measure - we deduce from theorem B for a group $GL(m)$ with $m < n$. This finishes the proof.

0.8. Let me give a brief description of the contents of the paper.

In chapter I (sections 1-4) we study invariant distributions. Section 1 contains a brief review of general properties of distributions. In section 2 we formulate theorem B and give several equivalent reformations which we use in the inductive proof. Section 3 contains the proof of the theorem. Some technical details, including the existence of orbital integrals in positive characteristic, are proved in section 4.

Chapter II (sections 5-9) gives applications to representation theory. Section 5 contains proofs of theorem A and related theorems A', A", and the proof of theorem 0.2. In section 6 we discuss corollaries of theorem A for Kirillov models. Section 7 describes an algorithm for the classification of irreducible unitary representations. In 7.1-7.4 we prove a unitarizability criterion for G-modules. In 7.5-7.9 we recall Zelevinsky's classification and formulate the algorithm.

In section 8 we discuss some miscellaneous results about irreducibility and unitarizability of G-modules. In 8.1-8.2 we prove some irreducibility criteria, based on unitarizability. In 8.3-8.7 we show that the algorithm describing unitary representations works essentially with discrete data. In other words, we show how to handle complementary series. In particular in 8.7 we establish some nice inequalities for unitarizable representations which are stronger than the inequalities in the unitarizability criterion 7.4.

In 8.8-8.9 we consider two examples of applications of the algorithm 7.9. Example 8.9 gives the classification of nondegenerate unitary representations of G. In 8.10 we formulate a conjecture that duality preserves unitarizability.

In section 9 we prove the unitarizability criterion 7.3 for P-modules which we use in section 7.

0.9. This paper arose from an attempt to answer the question by H. Jacquet and T. Shalika, whether each nondegenerate unitary irreducible representation is generic (i.e. is topologically irreducible when restricted to P). Relatively soon I understood that this can be proved using methods of [BZ2]. But these methods, even combined with [Z1], do not allow us to prove an analogous statement for the degenerate case (see 0.2). The only way to prove it which I see is to use theorems B and A.

Only much later I realized that the most interesting application of theorem A is an algorithm for the classification of unitary representations. I think that theorems A and B, criterion 7.4 and, with some modifications, algorithm 7.9 remain true for an archimedean field F. I even almost have a proof and I hope to overcome some technical problems which appear in the proof.

<u>Remark</u>. In [K] A.A. Kirillov tried to prove theorem 0.2 for the

Archimedean case, using essentially the same ideas. But his proof was incorrect and his means were absolutely insufficient for the proof.

0.10. I thank I.I. Piatetski-Shapiro, who turned my attention to the problem. I am very grateful to A. Zelevinsky for numerous fruitful discussions of representations of $\mathfrak{p}$-adic groups.

I thank J. Rosenberg and G. Zettler who read some preliminary versions of the paper and made useful remarks. I thank R. Herb, R. Lipsman and J. Rosenberg for organizing this wonderful Special Year.

I would like to thank the faculty of the mathematical department of the University of Maryland, and especially its former chairman W. Kirwan, who helped me a lot in my first year in this country.

CHAPTER I

THEOREMS ON INVARIANT DISTRIBUTIONS

1. PRELIMINARIES: GENERAL PROPERTIES OF DISTRIBUTIONS (SEE [BZ1, §1, 6])

Distributions On ℓ-Spaces

1.1. Let X be an ℓ-space, i.e. a Hausdorff topological space which has a basis consisting of open compact subsets. Denote by $S(X)$ the Schwartz space of X, i.e. the space of locally constant functions $f: X \to \mathbb{C}$ of compact support. Any linear functional E on $S(X)$ is called a distribution on X. We consider the weak topology on the space $S^*(X)$ of distributions on X.

Recall that the weak topology on the algebraic dual E^* of a vector space E is defined as the weakest topology, compatible with the linear structure, such that the set $e^{\perp} = \{ e^* \in E^* | \langle e^*, e\rangle = 0\}$ is open for each $e \in E$. For any linear subspace $L \subset E$ its orthogonal complement $L^{\perp}$ is closed and $(L^{\perp})^{\perp} = L$. For any linear subspace $W \subset E^*$ the space $(W^{\perp})^{\perp}$ coincides with the closure of W.

1.2. Let Z be a closed subset of X, $U = X\setminus Z$. We have natural exact sequences (see [BZ1; §1]):

$$(*)\qquad 0 \to S(U) \to S(X) \to S(Z) \to 0$$

$$(**)\qquad 0 \to S^*(Z) \xrightarrow{i} S^*(X) \xrightarrow{res} S^*(U) \to 0$$

(i=extension by zero; res = restriction of distributions $E \to E|_U$).

For any distribution $E \in S^*(X)$ there exists a minimal closed subset $\mathrm{supp}\, E \subset X$, called the support of E, such that $E|_{X\setminus \mathrm{supp}\, E} = 0$.

Using (**) we will identify $S^*(Z)$ with the subspace $S^*_Z(X) \subset S^*(X)$ consisting of distributions supported on Z. In particular, if Y is a locally closed subset of X (i.e. Y is open in its closure $\bar{Y}$) and E is a distribution supported on $\bar{Y}$ we will define the restriction $E|_Y$ by

$$E|_Y = (E|_{\bar{Y}})|_Y.$$

1.3. Let G be an ℓ-group and $\gamma: G \times X \to X$ a (left) continuous action of G on X. We denote by the same symbol γ the (left) actions of G on $S(X)$ and $S^*(X)$ given by

$$(\gamma(g)f)(x) = f(\gamma(g^{-1})x), \langle\gamma(g)E,f\rangle = \langle E,\gamma(g^{-1})f\rangle,$$
$$g \in G,\ x \in X,\ f \in S(X),\ E \in S^*(X).$$

Let χ be a character of G, i.e. a locally constant homomorphism $\chi: G \to \mathbb{C}^*$. We call a distribution $E \in S^*(X)$ χ-invariant under the action of G (or (G,χ)-invariant) if $\gamma(g)E = \chi(g)E$ for all $g \in G$. The space of (G,χ)-invariant distributions we denote by $S^*(X)^{G,\chi}$ (or simply $S^*(X)^G$ if $\chi = 1$).

Localization Principle

1.4. Let $q: X \to T$ be a continuous map of ℓ-spaces. Then $S(X)$ and hence $S^*(X)$ become $S(T)$-modules. For any $t \in T$ consider the fiber $X_t = q^{-1}(t)$ and identify the space $S^*(X_t)$ with the subspace $S^*_{X_t}(X) \subset S^*(X)$ of distributions concentrated on this fiber.

Localization principle. Let W be a closed subspace of $S^*(X)$ which is an $S(T)$-submodule. Then W is generated by distributions concentrated on fibers, i.e. the sum of subspaces $W^t = W \cap S^*(X_t)$, $t \in T$, is dense in W.

The following corollary is crucial for our proof.

Corollary. Let an ℓ-group G act on the space X preserving each fiber X_t, and let P be a subgroup of G. Suppose that for each $t \in T$ all P-invariant distributions on X_t are G-invariant, i.e. $S^*(X_t)^P = S^*(X_t)^G$. Then any P-invariant distribution on X

is G-invariant, i.e. $S^*(X)^P = S^*(X)^G$.

Indeed, $S^*(X)^P$ is a closed $S(T)$- submodule of $S^*(X)$ and hence it is generated by subspaces $S^*(X_t)^P$. Since $S^*(X_t)^P = S^*(X_t)^G \subset S^*(X)^G$ we have $S^*(X)^P \subset S^*(X)^G$, i.e. $S^*(X)^P = S^*(X)^G$.

Proof of localization principle. Let M be an $S(T)$-module. We say that M is unital if $M = S(T)\cdot M$. For any point $t \in T$ put

$$J_t = \{f \in S(T) \mid f(t) = 0\}, \quad M_t = M/J_t\cdot M .$$

The space M_t is called the fiber of M at the point t. For any $m \in M$ we denote by m_t its image in M_t.

Lemma (see the proof in [BZ1; 1.13, 1.14, 2.36])

(i) Subquotients of a unital $S(T)$-module are unital. The functor $M \to M_t$ is exact. If $m \in M$ and $m \neq 0$ then for some point $t \in T$ $m_t \neq 0$.

(ii) $S(X)$ is a unital $S(T)$-module and the natural morphism $S(X)_t \to S(X_t)$ is an isomorphism.

We will prove the following result:

(*) Let M be a unital $S(T)$-module and $W \subset M^*$ be a closed $S(T)$-submodule. Then W is generated by subspaces $W^t = W \cap (M_t)^*$ for $t \in T$. The localization principle is a particular case of (*) for $M = S(X)$, since $S(X)_t = S(X_t)$.

Put $L = W^\perp \subset M$, $N = M/L$. It is clear that L and N are $S(T)$-modules. Since W is closed it is isomorphic to N^*. Moreover, for each t, $W^t = (N_t)^* \subset N^*$. Consider the space $W' = \oplus W^t = \oplus (N_t)^* \subset$ $\subset N^*$ and its orthogonal complement in N. If $n \in W'^\perp$ then for any t, $n \in (N_t)^{*\perp}$, i.e. $n_t = 0$. Statement (i) of the lemma implies that $n = 0$, i.e. $W'^\perp = 0$. Therefore the closure of W' coincides with $W'^{\perp\perp} = 0^\perp = N^* = W$. This proves (*).

Frobenius Reciprocity

1.5. Let an ℓ-group G act on an ℓ-space X and let χ be a character of G. Sometimes we can reduce the study of χ-invariant distributions on X to the study of distributions on a smaller space. Namely, suppose we could find a continuous G-equivariant map $p: X \to Z$, where Z is a homogeneous G-space. For simplicity assume that we have a quasiinvariant measure μ on Z, i.e. $\mu \in S^*(Z)^{G,\nu}$ for some character ν. Fix such a measure $\mu \neq 0$ and fix a point $z_0 \in Z$.

Put $X_0 = p^{-1}(z_0) \subset X$, $H = \mathrm{Stab}(z_0,G) \subset G$.

Lemma. There exists a canonical isomorphism $\Psi_\mu: S^*(X_0)^{H,\chi\nu^{-1}} \to$ $\to S^*(X)^{G,\chi}$. If $E_0 \in S^*(X_0)^{H,\chi\nu^{-1}}$, then $\mathrm{supp}\,\Psi_\mu(E_0) = G\,\mathrm{supp}\,(E_0)$. The morphism Ψ_μ can be written explicitly:

$$\langle\Psi_\mu(E_0),f\rangle = \int_Z (\chi\nu^{-1})(g_z)\cdot\langle E_0,\gamma(g_z)f\rangle\, d\mu(z),$$

where $f \in S(X)$ and $g_z \in G$ is an element such that $g_z(z) = z_0$. This lemma is an easy consequence of Frobenius reciprocity (see [BZ1; 2.21-2.36]).

Remark 1. If Z does not have a quasiinvariant measure one can nevertheless prove an analogue of the lemma. Namely, consider the character $\nu = \Delta_G|_H \cdot \Delta_H^{-1}$ of the group H (here Δ is the module of a group). Then there exists an isomorphism

$$\Psi: \quad S^*(X_0)^{H,\chi\nu^{-1}} \to S^*(X)^{G,\chi}$$

(see [BZ1, 2.21-2.36]).

Remark 2. Frobenius reciprocity in particular implies that all G-invariant distributions on Z are proportional.

§2. REFORMULATIONS OF THE MAIN THEOREM

2.1. We fix a nonarchimedean local field F and put $G = G_n = GL(n,F)$,

$$P = P_n = \{ g = (g_{ij}) \in G_n | g_{ni} = \delta_{ni} \text{ for all } i\}.$$

Denote by Ad the adjoint action of G on itself.

Our aim is the following

<u>Theorem</u> B. Let E be a distribution on G invariant under the adjoint action of the subgroup P. Then it is invariant under the adjoint action of the whole group G.

Statements $X(n)$, $Y(n)$ and $Y^*(n)$

2.2. Let $X = X_n = Mat(n,F)$ be the algebra of $n \times n$ matrices. We define the adjoint action of G on X by $Ad(g)x = gxg^{-1}$. In the proof of the theorem we can assume that $\operatorname{supp} E \subset G$ is closed in X (for instance, we can multiply E by some locally constant compactly supported function of $\det(g)$). Hence we can consider E as a distribution on X. Therefore we should prove for each n the following

<u>Statement $X(n)$</u>. $S^*(X_n)^{P_n} = S^*(X_n)^{G_n}$.

We will prove the statement by induction on n. In the proof we will use some reformulations of $X(n)$, which are interesting by themselves.

2.3. Denote by $A = A_n$ the space $Mat(1,n;F)$ of row-vectors of length n and fix a standard basis $e_1,\dots,e_n$ in A. Let δ be the standard action of G on A given by $\delta(g)a = ag^{-1}$.

Denote by ν the character of the group $G = G_n$ given by $\nu(g) = |\det g|$, where $|\ |$ is the standard norm on the field F. Fix a Haar measure μ on A. It is clear that $\mu \in S^*(A)^{G,\nu}$.

Consider the ℓ-space $Y = Y_n = A_n \times X_n$ and the action $\gamma = \delta \times Ad$ of G on Y. The measure μ gives a canonical morphism $\mu\colon S^*(X) \to S^*(Y)$ by $\mu(E) = \mu\otimes E$. It is clear that $\mu(S^*(X)^G) \subset S^*(Y)^{G,\nu}$.

We claim that statement $X(n)$ implies (and in fact is equivalent to) the following.

<u>Statement $Y(n)$</u>. The morphism $\mu: S^*(X_n)^G \to S^*(Y_n)^{G,\nu}$ is an isomorphism.

Put $A' = A\setminus 0$, $Y' = A' \times X \subset Y$. Consider the morphism $\mu': S^*(X)^G \to S^*(Y')^{G,\nu}$, given by $\mu'(E) = \mu'\otimes E$, where $\mu' = \mu|_{A'}$. Since G acts transitively on A' and $\mathrm{Stab}(e_n,G)$ coincides with P_n, we can apply Frobenius reciprocity (see 1.5). It gives an isomorphism $\Psi_\mu: S^*(X)^P \simeq S^*(Y')^{G,\nu}$. The explicit formula for Ψ_μ given in 1.5 shows that $\mu' = \Psi_\mu \circ i$, where $i: S^*(X)^G \to S^*(X)^P$ is the natural imbedding. Hence statement $X(n)$ implies the following statement.

<u>$Y'(n)$</u>. $\mu': S^*(X)^G \to S^*(Y')^{G,\nu}$ is an isomorphism.

In order to prove the implication $X(n) \Rightarrow Y(n)$ it remains to prove that $S^*(Y')^{G,\nu} \approx S^*(Y)^{G,\nu}$. Since $Y' = Y\setminus X$, where $X = 0 \times X \subset Y$, we have an exact sequence $0 \to S^*(X) \to S^*(Y) \overset{res}{\to} S^*(Y') \to 0$ and hence the morphism $res: S^*(Y)^{G,\nu} \to S^*(Y')^{G,\nu}$.

Fix an element z in the center of G such that*) $\nu(z) \neq 1$ and define an endomorphism α of $S^*(Y)$ by $\alpha(E) = \gamma(z)E - E$. Since z acts trivially on X, $\gamma(z)$ is the identity on $S^*(X)$, i.e. $\alpha(S^*(X)) = 0$. Hence we can consider α as a morphism $\alpha: S^*(Y') \to S^*(Y)$. It is clear that on ν-invariant distributions operators $\alpha\circ res$ and $res\circ\alpha$ are multiplications by the nonzero constant $\nu(z) - 1$. Hence res gives an isomorphism $res: S^*(Y)^{G,\nu} \simeq S^*(Y')^{G,\nu}$.

2.4. Let $A^* = A^*_n$ be the dual space of $A = A_n$. It can be described as a space $\mathrm{Mat}(n,1;F)$ of column-vectors of length n. The action δ^* of G on A^* is given by $\delta^*(g)a^* = ga^*$.

Consider the ℓ-space $Y^* = Y^*_n = X_n \times A^*_n$ and the action $\gamma^* = \mathrm{Ad} \times \delta^*$ of G on Y^*. We identify X with a closed subset

*) This trick does not work for a finite field F.

$X \times 0 \subset Y^*$ and denote by $i: S^*(X) \to S^*(Y^*)$ the natural inclusion.

We claim that the statement $Y(n)$ implies (and in fact is equivalent to) the following

<u>Statement $Y^*(n)$</u>. The morphism $i: S^*(X_n)^{G_n} \to S^*(Y_n^*)^{G_n}$ is an isomorphism. In other words, each G-invariant distribution on Y^* is concentrated on X.

In order to prove this implication we fix a nontrivial additive character ψ of F and consider the Fourier transform $\Phi: S(Y) \to S(Y^*)$ given by $\Phi(f)(x,a^*) = \int_A f(a,x)\,\psi(\langle a^*,a\rangle)\,d\mu(a)$.

The usual theory of Fourier transform implies that Φ is an isomorphism. Let $\Phi^*: S^*(Y^*) \to S^*(Y)$ be the dual isomorphism. It is easy to check that Φ^* gives an isomorphism $S^*(Y^*)^G \to S^*(Y)^{G,\nu}$ and that the morphism $\Phi^* \circ i: S^*(X) \to S^*(Y^*) \to S^*(Y)$ coincides with the morphism μ, defined in 2.3. Hence the statement $Y(n)$ implies that $i: S^*(X)^G \to X^*(Y^*)^G$ is an isomorphism, i.e. the statement $Y^*(n)$.

§3. PROOF OF THE STATEMENT X(n)

The Geometric Structure of G- and P-orbits on X.

3.1 Consider the following invariants of the matrix $x \in X_n$:

t_x = characteristic polynomial of x (deg $t_x = n$)

K_x = [span of $e_n, e_n x, \dots, e_n x^n$] $\subset A$

$k_x = \dim K_x$ = minimal k such that $e_n x^k$ is a linear combination of $e_n, e_n x, \dots, e_n x^{k-1}$

τ_x = characteristic polynomial of the operator x on K_x (i.e. $\tau_x = \lambda^k + a_1\lambda^{k-1} + \dots + a_k$, where $k = k_x$, and $e_n x^k + \Sigma a_i e_n x^{k-i} = 0$).

By definition τ_x is the minimal monic polynomial such that $e_n \tau_x(x) = 0$. We call the matrix x <u>P-regular</u> if $\tau_x(x) = 0$. It is clear, that t_x is constant along G-orbits, k_x and τ_x are constant along P-orbits. Besides, the function $x \mapsto k_x$ is upper semicontinuous.

Geometric lemma (see the proof in 4.1-4.2).

a) For any polynomial t the set $X_t = \{x \in X | t_x = t\}$ contains a finite number of G-orbits.

b) Each G-orbit $\mathcal{O}$ contains a finite number of P-orbits.

c) Each G-orbit $\mathcal{O}$ contains a unique P-orbit $\mathcal{O}_P$ open and dense in $\mathcal{O}$. Namely $\mathcal{O}_P = \{x \in \mathcal{O} | x$ is P-regular$\}$.

Proof of the Statement $X(n)$

3.2. We will prove $X(n)$ by induction on n, i.e. we assume $X(m)$ to be true for $m < n$. We fix a P_n-invariant distribution $E \in S^*(X)^{P_n}$ and prove that E is G_n-invariant. Put $S = \operatorname{supp} E \subset X$.

Let T be the space of polynomials of degree n and $q: X \to T$ the characteristic map $q: x \mapsto t_x$. Using the localization principle 1.4 we can (and will) assume that $S \subset X_t$ for some t. Then by 3.1 a,b, S contains a finite number of P-orbits. We will proceed by induction on the number of P-orbits in S.

Key lemma. S contains an open P-orbit $\mathcal{O}_P$ which consists of P-regular elements.

Let us deduce $X(n)$ from the lemma. Consider the G-orbit $\mathcal{O} = \operatorname{Ad}(G)\mathcal{O}_P$. Since $\mathcal{O}_P$ consists of P-regular elements it is open and dense in $\mathcal{O}$, i.e. $\bar{\mathcal{O}} = \bar{\mathcal{O}}_P \subset S$ (see 3.1c). We will use the following statement which we will prove in 4.3.

Statement. For any G-orbit $\mathcal{O} \subset X$ there exists a G-invariant distribution $\mu_{\mathcal{O}}$ such that $\operatorname{supp} \mu_{\mathcal{O}} = \bar{\mathcal{O}}$.

Consider the restrictions of the distributions E and $\mu_{\mathcal{O}}$ on the P-orbit $\mathcal{O}_P$ (this makes sence since $\mathcal{O}_P$ is open in $\operatorname{supp} E = S$ and $\operatorname{supp} \mu_{\mathcal{O}} = \bar{\mathcal{O}} \subset S$). They both are P-invariant and nonzero. Hence for some $c \in \mathbb{C}^*$, $E|_{\mathcal{O}_P} = c \cdot \mu_{\mathcal{O}}|_{\mathcal{O}_P}$ (see 1.5). This means that the distribution $E_0 = E - c\mu_{\mathcal{O}}$ restricts to zero on $\mathcal{O}_P$. The distribution E_0 is P-invariant and $\operatorname{supp} E_0 \subset S \setminus \mathcal{O}_P$ contains strictly fewer

P-orbits than $S = \mathrm{Supp}\, E$. By induction E_o is G-invariant and therefore E is G-invariant.

3.3. Proof of the key lemma.

For each $i = 1, 2, \ldots, n$ put $X_i = \{x \in X \mid k_x = i\}$, $\overline{X}_i = \{x \in X \mid k_x \le i\}$. The sets $\overline{X}_i$ are closed and X_i is open in $\overline{X}_i$.

Let k be the minimal index such that $S \subset \overline{X}_k$. Consider the distribution $E' = E|_{X_k}$ and put $S' = \mathrm{Supp}\, E' = S \cap X_k$. Then S' is a nonempty open subset of S. Since S' contains a finite number of P-orbits it contains an open P-orbit $\mathcal{O}_P$. Hence it is sufficient to prove the following statement:

(*) Any P-invariant distribution E' on X_k has support $S' = \mathrm{supp}\, E'$ consisting of P-regular elements, provided it contains a finite number of P-orbits.

We can study P-invariant distributions on X_k using Frobenius reciprocity 1.5. Consider the natural map $\pi\colon X_k \to A^{k-1}$ where $A^{k-1} = \{(a_1, \ldots, a_{k-1}) \mid a_i \in A = \mathrm{Mat}(1,n;F)\}$ given by $\pi(x) = (e_n x, e_n x^2, \ldots, e_n x^{k-1})$. This map is P-equivariant and its image Z is an open subset of A^{k-1} given by $Z = \{(a_1, \ldots, a_{k-1}) \mid e_n, a_1, \ldots, a_{k-1}$ are linearly independent$\}$. Z is a homogeneous P-space and it has a quasiinvariant measure $\mu_Z = \mu^{k-1}$, which is ν^{k-1}-invariant with respect to P.

Put $z = (e_{n-1}, \ldots, e_{n-k+1}) \in Z$, $X = \pi^{-1}(z)$, $H = \mathrm{Stab}(z,P)$. Then by Frobenius reciprocity the distribution E' corresponds to an (H, ν^{1-k})-invariant distribution E'' on X such that $S' = \mathrm{Ad}(P)S''$, where $S'' = \mathrm{Supp}\, E''$. Hence we should prove:

(**) Any (H, ν^{1-k})-invariant distribution E'' on X has support S'' consisting of P-regular elements, provided it contains a finite number of H-orbits.

3.4. Let us describe H and X in detail. Put $m = n-k$ and let us write the n×n matrix x in a block form $\begin{pmatrix} A_x & B_x \\ C_x & D_x \end{pmatrix}$, where A_x, B_x, C_x and D_x are matrices of sizes $m \times m$, $m \times k$, $k \times m$ and $k \times k$.

By definition

$$H = \{x \in G_n | e_i x = e_i \text{ for } m < i \le n\} = \{x \in G_n | C_x = 0,\ D_x = 1_k\}$$

$$X = \{x \in X_n | e_i x = e_{i-1} \text{ for } m+1 < i \le n \text{ and } e_{m+1} x \in \operatorname{span}(e_n, \dots, e_{m+1})\}$$

$$= \{x \in X_n | C_x = 0 \text{ and } D_x \in W\},$$

where W is the set of $k \times k$ matrices of the form

$$\left(\begin{array}{c|c} ** \dots & * \\ \hline 1_{k-1} & 0 \end{array}\right).$$

Note that each matrix $w \in W$ is completely defined by its characteristic polynomial τ (the coefficients of the upper row coincide with minus coefficients of τ). We will denote it by w_τ.

The function $x \mapsto \tau_x$ is continuous on X (indeed τ_x is the characteristic polynomial of D_x) and constant on H-orbits. Since S" consists of a finite number of H-orbits, τ_x assumes only a finite number of values on S". Fix one of these values τ and put $X_\tau = \{x \in X | \tau_x = \tau\} = \{x \in X | D_x = w_\tau\}$. Then $X_\tau \cap S"$ is open in S", so we can restrict E" to X_τ. Hence in the proof of (**) we can (and will) assume that E" is an (H, ν^{1-k})-invariant distribution on X_τ.

Put $U = V = \mathrm{Mat}(m,k;F)$. We identify X_τ with $X_m \times V$ by $(x, v) \to \begin{pmatrix} x, & v \\ 0, & \omega_\tau \end{pmatrix}$. For any $u \in U$ we denote by u the matrix $\begin{pmatrix} 1 & u \\ 0 & 1 \end{pmatrix} \in H$. All these matrices form a subgroup $U \subset H$. We identify the group G_m with a subgroup of H by $g \to \begin{pmatrix} g & 0 \\ 0 & 1 \end{pmatrix}$. Then H is a semidirect product of G_m and U. The action of H on X_τ is given by

$$Ad(g)(x,v) = (gxg^{-1}, gv),\ g \in G_m,\ Ad(u)(x,v) = (x, v - xu + u\omega_\tau)\ ,$$

$u \in U$. Let $u_1, \ldots, u_k$ be the columns of the matrix $u \in U$. Put

$$U^+ = \{u \in U | u_1 = 0\}$$

$$X_\tau^+ = \{(x,v) \in X_\tau | v_1 = v_2 = \ldots = v_{k-1} = 0\}$$

$$X_\tau^o = \{(x,v) \in X_\tau | v = 0\}\ .$$

Lemma (see the proof in 3.6). The natural map $\kappa: U^+ \times X_\tau^+ \to X_\tau$, given by $(u,x) \mapsto Ad(u)x$ is a homeomorphism.

Using κ we will identify X_τ and $U^+ \times X_\tau^+$. Since U^+ acts only on the first factor in $U^+ \times X_\tau^+$ and $E"$ is U^+-invariant, it can be written as $E" = \mu^+ \otimes E^+$, where μ^+ is a Haar measure on U^+ and $E^+ \in S^*(X_\tau^+)$. The measure μ^+ is ν^{1-k} invariant with respect to G_m, hence E^+ is G_m-invariant.

Now let us note that as G_m-spaces X_τ^+ is isomorphic to the space Y_m^*, introduced in 2.4. Since we have assumed that the statements $X(m)$ and hence $Y^*(m)$ are true, the support S^+ of the distribution E^+ is concentrated on $X_\tau^o = \{(x,0)\}$. Hence the subset $S"$ of X_τ satisfies the following conditions.

(i) $S"$ is $Ad(U)$-invariant

(ii) $S" \subset Ad(U^+)X_\tau^o$.

3.5. Now let us prove that conditions (i), (ii) imply that $S"$ consists of P-regular elements, i.e. that for any $x \in S"$, $\tau(x) = 0$.

Consider the set $R = \tau(S")$ and prove that $R = \{0\}$. Since the map $x \to \tau(x)$ commutes with the adjoint action we have

(i)' R is $Ad(U)$-invariant

(ii)' $R \subset Ad(U^+)X_m$, where $X_m = \begin{pmatrix} x & 0 \\ 0 & 0 \end{pmatrix} \supset \tau(X_\tau^o)$.

(we use the fact that $\tau(\omega_\tau) = 0$).

The action of U on R is given by $\mathrm{Ad}(u)(x,\nu) = (x,\nu + xu)$, where $(x,\nu) = \begin{pmatrix} x & \nu \\ 0 & 0 \end{pmatrix} \in R$. If $u \in U^+$, i.e. $u_1 = 0$, then $(xu)_1 = 0$.

Hence (ii)' implies that $\nu_1 = 0$ for all $(x,\nu) \in R$. By (ii)' it is sufficient to prove that any element $(x,0) \in R$ is equal to 0.

Let $u \in U$. Then $\mathrm{Ad}(u)(x,0) = (x,xu) \in R$ and, as we have proved, $(x\cdot u)_1 = 0$. Since it is true for all u we have $x = 0$, Q.E.D.

3.6. Proof of the lemma 3.4.

We have $\kappa(u,(x,\nu)) = (x,\nu')$, where $\nu' = \nu - xu + uw_\tau$.

Let us write this for each column:

$$\nu_1' = u_2$$

$$\nu_2' = -xu_2 + u_3$$

$$\cdots$$

$$\nu_{k-1}' = -xu_{k-1} + u_k$$

$$\nu_k' = \nu_k - xu_k + \Sigma a_i u_i .$$

It is clear that for any $\nu_1',\ldots,\nu_k'$ there exist unique $u_2,\ldots,u_k, \nu_k$ which satisfy this system of equations.

§4. PROOFS OF SOME LEMMAS

Description of G-Orbits in X.

4.1. Let $C = F[\lambda]$ be the algebra of polynomials in one variable. Each element $x \in X$ defines on A a C-module structure by $\lambda \mapsto x$. This gives a one-to-one correspondence

$$\{G\text{-orbits on } X\} \leftrightarrow \{n\text{-dimensional } C\text{-modules } M \text{ up to isomorphism}\}.$$

The centralizer G_x of x in G corresponds to the group $\mathrm{Aut}_C M$.

Fix a monic polynomial $t \in C$ and its decomposition

$t = \tau_1\cdot\ldots\cdot\tau_r$, where τ_i are irreducible monic polynomials, not necessarily distinct. Let $x \in X$ be any matrix, annihilated by t, i.e. $t(x) = 0$. Define the t-invariant $\nu = \nu_x$ of x by $\nu = (\nu_1,\ldots,\nu_r)$, where $0<\nu_1\le\nu_2\le\ldots\le\nu_r = n$ are given by

$$\nu_i = \dim \operatorname{Ker}(\tau_1(x)\cdot\ldots\cdot\tau_i(x)).$$

Lemma. Let $x, y \in X$ be annihilated by t. Then they lie on the same G-orbit iff $\nu_x = \nu_y$.

Indeed, let M be the C-module corresponding to x. We can decompose $M \simeq \bigoplus_\alpha M_\alpha$, were $M_\alpha = C/(f_\alpha^{r(\alpha)})$, f_α are irreducible polynomials, $r(\alpha) > 0$. Since $t(x) = 0$, the polynomial f_α appears in the sequence $\tau_1,\ldots,\tau_r$ at least $r(\alpha)$ times. Denote by b_i the multiplicity of τ_i in the sequence $\tau_1,\tau_2,\ldots,\tau_i$. Then it is clear that $\nu_i-\nu_{i-1} = \dim_F(C/(\tau_i))\cdot\#\{\alpha \mid f_\alpha = \tau_i,\ r(\alpha) \ge b_i\}$. It is easy to see that this formula enables us to reconstruct f_α and $r(\alpha)$, and hence the C-module M up to isomorphism, from the invariants $\nu_1,\ldots,\nu_r$. This proves the lemma. If t is a polynomial of degree n and $x \in X_t$, i.e. $t_x = t$, then $t(x) = 0$ and the lemma implies 3.1a).

Description of P-Orbits in X.

4.2. Consider a G-orbit $\mathcal{O} = \mathrm{Ad}(G)x$ in X and denote by M the corresponding G-module. We have $P\backslash\mathcal{O} \approx P\backslash G/G_x \approx (A\backslash 0)/G_x \approx (M\backslash 0)/\mathrm{Aut}_C M$ (as topological spaces). Hence we can reformulate 3.1b),c) as statements about $\mathrm{Aut}_C M$ orbits in $M\backslash 0$.

Decompose $M = \oplus M_\alpha$, $M_\alpha = C/(f_\alpha^{r(\alpha)})$. Assign to each vector $\xi = \Sigma\xi_\alpha \in M$ invariants $\mu_\alpha = \min\{i \mid f_\alpha(\lambda)^i\xi_\alpha = 0\}$. It is clear that these invariants completely determine ξ up to the action of $\mathrm{Aut}_C M$. This proves 3.1b).

Denote by t the minimal polynomial $t_x^{\min}$ of x, i.e. the polynomial of minimal degree such that $t(x) = 0$, and put $\overline{C} = C/(t)$.

Put $M^o = \{\xi \in M | \mathrm{Ann}(\xi,\overline{C}) = 0\} = \{\xi \in M | \mathrm{Ann}(\xi,C) = \mathrm{Ann}(M,C)\}$.
Statement 3.1c) geometrically means that:

(*) M^o is an open dense subset of M and the group $\mathrm{Aut}_C M$ acts transivively on M^o.

By definition $M \setminus M^o = \cup M_J$, where J runs through nonzero ideals of $\overline{C}$, $M_J = \mathrm{Ann}(J,M)$. Since there are finitely many different ideals J and for any $J \neq 0$ M_J is a proper linear subspace of M, M^o is open and dense in M.

Any vector $\xi \in M^o$ defines an inclusion $\overline{C} \xrightarrow{\sim} \overline{C}\xi \subset M$. Since $\overline{C}$ is an injective $\overline{C}$-module*), $\overline{C}\xi$ is a direct summand of M. The Krull-Schmidt theorem implies that any two vectors $\xi, \xi' \in M^o$ are conjugate under $\mathrm{Aut}_C M$. This prove 3.1c).

Existence of Orbital Integrals

4.3. Proof of the Statement 3.2

In case char $F = 0$ a more general result was proved by Ranga Rao and P. Deligne (see [R]). Using specific properties of $GL(n)$ we will adjust the proof for arbitrary characteristic. Let $x \in X'$, $\mathcal{O} = \mathrm{Ad}(G)x$. We would like to prove the existence of G-invariant distribution $\mu_{\mathcal{O}}$ such that supp $\mu_{\mathcal{O}} = \overline{\mathcal{O}}$.

Denote by t the minimal polynomial of x.

(i) Consider at first the case when t is irreducible. Then $\overline{C} = C/_{(t)}$ is a field and $G_x \approx GL(n/\dim \overline{C}, \overline{C})$ is a unimodular group. Hence on $\mathcal{O}$ there exists a Haar measure $\mu_{\mathcal{O}}$. Besides, in this case $\mathcal{O}$ is closed, since any element $y \in \overline{\mathcal{O}}$ satisfies the equation $t(y) = 0$ and by lemma 4.1 is conjugate to x.

(ii) Now consider a general case. Fix a decomposition $t = \tau_1 \cdot \tau_2 \cdot \ldots \cdot \tau_r$ as in 4.1 and consider the t-invariant $\nu = (\nu_1, \ldots, \nu_r)$ of x (see 4.1). Since t is the minimal polynomial of x we have

*) Indeed, consider the $\overline{C}$-module $L = \mathrm{Hom}_F(\overline{C},F)$. Since $\mathrm{Ann}(L,\overline{C}) = 0$ there exists an inclusion $\overline{C} \to L$. Since $\dim L = \dim \overline{C}$, this inclusion is an isomorphism. Hence L is a projective $\overline{C}$-module, i.e. $\overline{C}$ is an injective $\overline{C}$-module.

$0 < \nu_1 < \nu_2 < \dots < \nu_r = n$.

By a ν-flag we mean a sequence $\phi = (L_1 \subset L_2 \subset \dots \subset L_r)$ of subspaces such that $\dim L_i = \nu_i$. The set Φ_ν of all ν-flags is a compact topological space and the natural action of G on Φ_ν is transitive. To any matrix $y \in \mathcal{O}$ we assign a ν-flag $\phi_y = (L_1,\dots,L_r)$, where $L_i = \ker(\tau_1(y)\cdot \dots \cdot\tau_i(y))$. Consider the space $\Sigma_\nu = X \times \Phi_\nu$ with the natural action of G and put $\mathcal{Q} = \{(y,\phi_y) \in \Sigma_\nu | y \in \mathcal{O}\}$. The natural projection $pr: \Sigma_\nu \to X$ is a proper map and hence it defines the morphism of distributions $pr_*: S^*(\Sigma_\nu) \to S^*(X)$ by $\langle pr_*(E),f\rangle = \langle E,pr^*(f)\rangle$. Therefore it is sufficient to construct a G-invariant distribution $\mu_{\mathcal{Q}}$ such that $\mathrm{Supp}(\mu_{\mathcal{Q}}) = \bar{\mathcal{Q}}$.

(iii) Consider the natural projection $pr_2: \Sigma_\nu \to \Phi_\nu$. In order to construct the G-invariant distribution $\mu_{\mathcal{Q}}$ we will use the Frobenius reciprocity (see 1.5).

Fix the ν-flag $\phi = \phi_x = (L_1 \subset \dots \subset L_r)$ and put $P_\phi = \mathrm{Stab}(\phi,G)$, $\chi = \Delta_{P_\phi} \cdot \Delta_G^{-1} = \Delta_{P_\phi}$, $\mathcal{Q}_\phi = \{y \in \mathcal{O} | \phi_y = \phi\} = pr_2^{-1}(\phi) \cap \mathcal{Q}$. According to 1.5, Remark 1, it is sufficient to construct a (P_ϕ, χ)-invariant distribution $\mu_\phi \in S^*(X)$ such that $\mathrm{Supp}(\mu_\phi) = \bar{\mathcal{Q}}_\phi$.

(iv) Put $\mathcal{P} = \{x \in X | xL_i \subset L_i \text{ for all } i\}$. It is clear that $P_\phi = \mathcal{P}^* = \mathcal{P} \cap G$ and $\mathcal{Q}_\phi \subset \mathcal{P}$. Put $\mathcal{U} = \{x \in X | xL_i \subset L_{i-1} \text{ for all } i\}$, $\mathcal{M} = \mathcal{P}/_{\mathcal{U}}$. It is clear that $\mathcal{U}$ is a nilpotent two-sided ideal of $\mathcal{P}$ and the algebra $\mathcal{M}$ is isomorphic to $X_{m_1} \times \dots \times X_{m_r}$, where $m_i = \nu_i - \nu_{i-1}$. For any $z \in \mathcal{P}$ we denote by $\pi(z) = (z_1,\dots,z_r)$ the corresponding element of $\mathcal{U}$.

Put $U = 1 + \mathcal{U} \subset P_\phi$. It is the unipotent radical of P_ϕ and the quotient $M = P_\phi/U$ is isomorphic to the group of invertible elements of $\mathcal{M}$.

Let $y \in \mathcal{Q}_\phi$, i.e. $\mathrm{Ker}(\tau_1(y)\cdot \dots \cdot\tau_i(y)) = L_i$. Then the element $\pi(y) = (y_1,\dots,y_r)$ satisfies $\tau_i(y_i) = 0$ for all i. Lemma 4.1 implies that $\pi(y)$ lies on the orbit $\mathcal{O}_M$ of the element $\pi(x)$ under the action of the group $M = G_{m_1} \times\dots\times G_{m_r}$. Since $\mathcal{O}_M$ is closed, $\pi(\overline{\mathcal{Q}}_\phi) \subset \mathcal{O}_M$. The morphism $\pi: \overline{\mathcal{Q}}_\phi \to \mathcal{O}_M$ is P_ϕ-equivariant and, again using Frobenius reciprocity and the existence of a P_ϕ-invariant measure on $\mathcal{O}_M$, we can reduce the problem to the construction of the (H,χ)-invariant distribution μ such that $\mathrm{Supp}\,\mu = Q$, where $H = \mathrm{Stab}(\pi(x), P_\phi) = M_x \cdot U$. $Q = \pi^{-1}(\pi(x)) \cap \overline{\mathcal{Q}}_\phi$.

(v) The main observation is that:

$$(*) \quad \pi^{-1}(\pi(x)) \cap \overline{\mathcal{Q}}_\phi = \pi^{-1}(\pi(x)) = x + \mathcal{U}.$$

Now if we denote by μ the Haar measure on $\mathcal{U}$ and consider it as a distribution on $x + \mathcal{U}$, we see that it is $\mathcal{U}$-invariant and (M_x,χ)-invariant. (Indeed, the Haar measure on $\mathcal{U}$ is (M,χ)-invariant, since $\chi = \Delta(P_\phi)|_M = \Delta_M \cdot \mathrm{mod}(U) = \mathrm{mod}\ U = \mathrm{mod}\ \mathcal{U}$.)

In order to prove (*) it is sufficient to prove that for almost all $y \in x + \mathcal{U}$ we have $y \in \mathcal{O}$ and $\phi_y = \phi$. Since $y_i = x_i$, we have $\tau_i(y_i) = 0$, i.e. $\tau_i(y)L_i \subset L_{i-1}$. Hence $\mathrm{Ker}(\tau_1(y)\cdot \ldots \cdot\tau_i(y))$ contains L_i. Since for $y = x$ this kernel coincides with L_i, for almost all y (more precisely, for y from some Zariski open subset of $x + \mathcal{U}$) we have equalities $\mathrm{Ker}(\tau_1(y)\cdot \ldots \cdot\tau_i(y)) = L_i$ for all i. This means that $\phi_y = \phi$, and lemma 4.1 implies that y is conjugate to x, i.e. $y \in \mathcal{O}$. Q.E.D.

4.4 <u>Remark</u>. The distribution $\mu_{\mathcal{O}}$ we have constructed is positive, i.e. $\langle\mu_{\mathcal{O}},f\rangle \geq 0$ for positive functions f. Hence it defines a measure on X. In other words, for any continuous function f on X with compact support the integral $\int_{\mathcal{O}} f(x)\,d\mu_{\mathcal{O}}(x)$ converges absolutely.

CHAPTER II

APPLICATIONS TO REPRESENTATION THEORY

In this chapter we study representations of the group $G = GL(n,F)$ and their restrictions to the subgroup P. We use the notations of [BZ1], [BZ2]. In particular the notation (π,H,E) means a representation π of a group H in a complex vector space E. The representation π is called smooth (algebraic in the terminology of [BZ1], [BZ2]) if the stabilizer of each vector $\xi \in E$ is open in H. The category of smooth representations of H we denote by $Alg(H)$.

§5. P-INVARIANT PAIRINGS OF G-MODULES

Proof of Theorem A.

5.1. Theorem A. Let (π,G,E) be a smooth irreducible representation, $(\tilde{\pi},G,\tilde{E})$ the contragredient representation, $B_o: E \times \tilde{E} \to \mathbb{C}$ the canonical pairing $B_o(\xi,\tilde{\xi}) = \langle\xi,\tilde{\xi}\rangle$. Then any P-invariant pairing $B: E \times \tilde{E} \to \mathbb{C}$ is G-invariant and hence is proportional to B_o.

Proof. It is known that π and $\tilde{\pi}$ are admissible (see [BZ1,3.25]), so $(\pi \otimes \tilde{\pi}, G \times G, E \otimes \tilde{E})$ is admissible and irreducible. Consider the regular representation $(Reg, G \times G, S(G))$ given by $Reg(g_1,g_2)f(g) = f(g_1^{-1}gg_2)$, $f \in S(G)$, $g_1,g_2,g \in G$. We will use the following standard lemma (see 5.6).

Lemma. For any admissible representation (π,G,E) there exists a nonzero morphism of $G \times G$-modules.

$$\pi_\mu: S(G) \to E \otimes \tilde{E}$$

If π is irreducible, π_μ is an epimorphism.

Pairings $B: E \times \tilde{E} \to \mathbb{C}$ correspond to morphisms $E \otimes \tilde{E} \to \mathbb{C}$, i.e. to elements of $(E \otimes \tilde{E})^*$. The morphism $\pi_\mu^* : (E \otimes \tilde{E})^* \to S^*(G)$, adjoint to π_μ, is $G \times G$-invariant. In particular, if a pairing B

is g-invariant for some $g \in G$, then the corresponding distribution $E_B = \pi_\mu^*(B)$ is $Ad(g)$-invariant.

Since π is irreducible, π_μ is onto and π_μ^* is a monomorphism. Hence E_B is $Ad(g)$-invariant iff B is g-invariant.

Thus we have: B is P-invariant $\Rightarrow E_B$ is $Ad(P)$-invariant $\Rightarrow E_B$ is $Ad(G)$-invariant (by Theorem B) $\Rightarrow B$ is G-invariant. Since π is irreducible and admissible, B is proportional to the standard pairing, Q.E.D.

5.2. Let us discuss a slightly different version of theorem A.

Let (π, G, E) be a smooth representation. Fix a Haar measure μ on G and for any $f \in S(G)$ define an operator $\pi(f) = \int_G f(g)\pi(g)d\mu(g)$ $: E \to E$ and put $U_f = \operatorname{Ker} \pi(f)$. Consider the weakest topology on E for which all subspaces U_f are open (the weak topology). Denote by $\hat{E}$ the completion of E in this topology and by $(\hat{\pi}, G, \hat{E})$ the natural representation. It is easy to check that

(i) E is a dense G-submodule of $\hat{E}$ and it coincides with the smooth part of $\hat{E}$.

(ii) For admissible π, $(\hat{\pi}, \hat{E})$ is canonically isomorphic to the representation $(\tilde{\pi}^*, \tilde{E}^*)$ dual to the contragredient representation $(\tilde{\pi}, \tilde{E})$.

One can also define $(\hat{\pi}, \hat{E})$ by $\hat{E} = \operatorname{Hom}_G(S(G), E)$, $(\hat{\pi}(g)\alpha)(f) = \alpha(\operatorname{Reg}(1, g^{-1})f)$. Namely, for any $\hat{\xi} \in \hat{E}$ the morphism $\alpha_{\hat{\xi}}: S(G) \to \hat{E}$, given by $\alpha_{\hat{\xi}}(f) = \hat{\pi}(f)\hat{\xi}$, maps $S(G)$ into the subspace $E \subset \hat{E}$ of smooth vectors.

<u>Theorem A'</u>. Let (π, G, E) be a smooth irreducible representation. Then any P-equivariant morphism $\beta: E \to \hat{E}$ is G-equivariant and, hence, is proportional to the standard inclusion.

This theorem is just a reformulation of theorem A. It can also be proved directly. Indeed, for any $\beta: E \to \hat{E}$, $f \in S(G)$ the operator $\hat{\pi}(f) \circ \beta: E \to \hat{E}$ has a finite dimensional image, which lies in E. Hence for any β we can define the distribution $E_\beta \in S^*(G)$ by $E_\beta(f) = \mathrm{tr}(\hat{\pi}(f)\circ\beta)$.

Thus we have constructed a morphism of $G \times G$-modules $\mathrm{Hom}(E,\hat{E}) \to S^*(G)$. Using the formula $\mathrm{tr}(\hat{\pi}(f)\circ\beta\circ\pi(f')) = \mathrm{tr}(\hat{\pi}(f' * f)\circ\beta)$ it is easy to show that, in case of irreducible π, this is a monomorphism. The rest of the proof is the same as in 5.1.

The same arguments enable us to prove the following generalization which may be of some use.

<u>Theorem A"</u>. Let (π,G,E) be an admissible representation. Suppose E has only one irreducible G-submodule V, i.e. any proper G-submodule of E contains V. Then for any P-equivariant morphism $\beta: E \to \hat{E}$ its restriction to V is proportional to the standard inclusion $V \to E \to \hat{E}$.

Corollaries of Theorem A.

5.3. <u>Proposition</u>. Let (π,G,E) be a smooth irreducible representation, and let B_o be a nonzero G-invariant bilinear (or Hermitian) form on E. Then any P-invariant bilinear (respectively, Hermitian) form B on E is proportional to B_o.

<u>Proof</u>. Since π is admissible and irreducible B_o defines an isomorphism $E \simeq \tilde{E}$ (respectively, $\bar{E} \simeq \tilde{E}$, where $\bar{E}$ is the space complex conjugate to E). The form B defines a P-invariant pairing of E with $E \approx \tilde{E}$ (respectively, of E with $\bar{E} \approx \tilde{E}$).

Theorem A then implies that B is proportional to B_o.

5.4. <u>Theorem</u>. Let (σ,G,H) be a unitary topologically irreducible representation. Then its restriction to P is also topologically irreducible.

<u>Proof</u>. It is sufficient to check that any continuous P-equivariant morphism $\alpha: H \to H$ is a scalar operator.

Let (π,G,E) be the smooth part of (σ,H). It is known (see

[BZ1, 4.21]) that π is irreducible and E is dense in H. Consider the Hermitian form B_α given by $B_\alpha(\xi,\eta) = \langle\alpha\xi,\eta\rangle$, where $\langle,\rangle$ is the scalar product in H. The form B_α is P-invariant and by proposition 5.3. on E we have an equality $B_\alpha(\xi,\eta) = c\langle\xi,\eta\rangle$, $c \in \mathbb{C}$. Since E is dense in H, it implies $\alpha = c\cdot 1_H$, Q.E.D.

5.5. Remark. We can prove a more precise result.

Theorem. Let (σ,G,H) be a continuous representation of G in a complete topological vector space H, (π,G,E) its smooth part. Suppose π is irreducible. Then any continuous P-equivariant morphism $\alpha: H \to H$ is a scalar operator.

Indeed, we have a natural inclusion $H \to \hat{E}$ (see 5.2). By theorem A' $\alpha|_E = c\cdot 1_E$. Since α is continuous, it is a scalar.

5.6. Proof of lemma 5.1.

Let G,H be arbitrary ℓ-groups, (π,G,E) and (ρ,H,V) smooth representations. Define the representation $(H(\pi,\rho), G \times H, \mathrm{Hom}(E,V))$ by $\mathrm{Hom}(E,V) = \mathrm{Hom}_{\mathbb{C}}(E,V)$, $H(\pi,\rho)(g,h)(\alpha) = \rho(h)\circ\alpha\circ\pi(g^{-1})$, $g \in G$, $h \in H$, $\alpha \in \mathrm{Hom}(E,V)$.

We have a natural imbedding $i: \tilde{\pi} \otimes \rho \to H(\pi,\rho)$ given by $i(\tilde{\xi} \otimes \eta)\xi = \langle\tilde{\xi},\xi\rangle\cdot\eta$. Denote by $(h(\pi,\rho), G \times H, h(E,V))$ the smooth part of $H(\pi,\rho)$. Then i is an inclusion $i: \tilde{\pi} \otimes \rho \to h(\pi,\rho)$.

Suppose π is admissible. Then for any open compact subgroup $N \subset G$ the space E^N of N-invariant vectors is finite dimensional and $\tilde{E}^N = (E^N)^*$. This implies that $h(E,V)^N = \mathrm{Hom}(E^N,V) \approx (E^N)^* \otimes V \approx \approx (E \otimes V)^N$, i.e. i give an isomorphism $i = \tilde{\pi} \otimes \rho \xrightarrow{\sim} h(\pi,\rho)$.

Now suppose G is a unimodular ℓ-group and fix a Haar measure on G. Define a morphism of representations

π_μ: (Reg, $G \times G$, $S(G)$) $\to$ ($H(\pi,\pi)$, $G \times G$, $H(E,E)$) by $\pi_\mu(f) = \int_G f(g)\pi(g)d\mu(g)$. Since Reg is a smooth representation its image belongs to $h(\pi,\pi)$. If π is admissible we obtain a nontrivial morphism π_μ: $S(G) \to \tilde{E} \times E$ of $G \times G$-modules. If π is irreducible, then $\tilde{\pi} \otimes \pi$ is irreducible and hence π_μ is an epimorphism. Starting from $(\tilde{\pi},\tilde{E})$ instead of (π,E) we obtain lemma 5.1.

§6. SCALAR PRODUCT IN THE KIRILLOV MODEL

Kirillov Model

6.1. Let $U \subset P$ be the subgroup of unipotent upper triangular matrices. Fix a nonzero additive character $\psi: F \to \mathbb{C}^*$ and define the nondegenerate character θ of U by $\theta(u) = \psi(u_{12}+\ldots+u_{n-1,n})$, where $u = (u_{ij})$.

Consider the smooth induced representation $(\tau,P,\hat{S}) = \mathrm{Ind}_U^P(\theta)$. By definition $\hat{S} = \{ f: P \to \mathbb{C} \mid f(up) = \theta(u)f(p)$ for $u \in U$, $p \in P$ and f is smooth under the right action of P $\}$. Denote by (τ°,P,S) the subrepresentation in the subspace of functions with compact support modulo U.

Analogously define representations $(\tau',P,\hat{S}')$ and $(\tau^{\circ\prime},P,S')$ by replacing θ by θ^{-1}.

Note that τ and τ' are isomorphic. The isomorphism ε is given by $\varepsilon f(p) = f(\varepsilon p)$, where $\varepsilon = (\delta_{ij}\cdot(-1)^{n-i}) \in P$.

Let (π,P,E) be a smooth representation. A <u>Kirillov model</u> for π is by definition a nonzero morphism of P-modules $K: E \to \hat{S}$. A P-module (π,E) which has a Kirillov model is called <u>nondegenerate</u>.

The following facts are proved in [BZ1,§5].

(i) There are the natural isomorphisms $\tau \simeq \widetilde{\tau^{\circ\prime}}$ and $\tau' \simeq \widetilde{\tau^\circ}$. Corresponding pairings are given by integration over $U\backslash P$.

(ii) Any proper P-submodule of $\hat{S}$ contains S. In particular the P-module S is irreducible and any morphism $i: S \to \hat{S}$ of P-modules is proportional to the standard inclusion.

(iii) Let (π,P,E) be a smooth representation, $K: F \to \hat{S}$ its

Kirillov model. Then there exists a morphism of P-modules $i_K: S \to E$ such that $K \circ i_K = 1_S$.

6.2. Fix a right invariant Haar measure $\mu_{U\backslash P}$ on $U\backslash P$ and define the P-invariant scalar product $<\ ,\ >$ on S by

$$(*) \qquad <f,h> = \int_{U\backslash P} f(p)\cdot\bar{h}(p)\,d\mu_{U\backslash P}(p), \quad f,\ h \in S.$$

Let $L^2(S)$ be the completion of S with respect to $<,>$ and $L^2_{sm}(S)$ its smooth part (with respect to the natural action of P). It is clear that $L^2_{sm}(S)$ is naturally imbedded in $\hat{S}$ (it consists of functions $f \in \hat{S}$ such that $\int_{U\backslash P}|f|^2 d\mu$ converges).

<u>Theorem</u>. Let (π,G,E) be a smooth irreducible representation and $K: E \to \hat{S}$ its Kirillov model (i.e. K is a nonzero morphism of P-modules). Suppose π is unitarizable, i.e. there exists a G-invariant positive definite Hermitian form B_0 on E. Then $K(E) \subset L^2_{sm}(S)$ and for some constant $c \in \mathbb{R}^{+*}$

$$<K(\xi),\ K(\eta)> \equiv c \cdot B_0(\xi,\eta).$$

<u>Proof</u>. Consider an inclusion $i_K: S \to E$ (see 6.1(iii)) and put $E^+ = i_K(S)$. Statements 6.1 (i) & (ii) imply that $<,>$ is the only P-invariant Hermitian form on S. It means that we can normalize B_0 such that $B_0(i_K(f),i_K(h)) \equiv <f,h>$.

Let H be the completion of E with respect to the norm given by B_0. Then by theorem 5.4 E^+ is dense in H. Hence i_K can be extended to the isomorphism $i_K^{(2)} = L^2(S) \xrightarrow{\sim} H$.

Consider the inverse map $K': H \to L^2(S)$. Then $K'(E) \subset L^2_{sm}(S) \subset \hat{S}$. Using the uniqueness of the Kirillov model for an irreducible G-module (see [G-K], [Sh], [BZ1] or 6.5) we see that $K = K'$, Q.E.D.

Regularity of the Ψ-Function at $s = 0$.

6.3. We want to generalize theorem 6.2 for nonunitary representations. Let $f \in \hat{S}$, $f' \in \hat{S}'$ (see 6.1). Define formally the function $\Psi(s,f,f')$ of the complex variable s by

$$(*) \qquad \Psi(s;f,f') = \int_{U\backslash P} f(p)f'(p)\nu(p)^s d\mu_{U\backslash P} ,$$

where $\nu(p) = |\det p|$.

If $f \in S$ or $f' \in S'$ this integral converges and is regular as a function of s. Moreover $\Psi(0;f,f')$ gives the canonical pairings $\hat{S}$ with S' and S with $\hat{S}'$.

Let (π,G,E), (π',G,E') be two admissible representations of finite length with Kirillov models $K: E \to \hat{S}$, $K': E' \to \hat{S}'$. Define formally for $\xi \in E$, $\xi' \in E'$ the function $\Psi(s;\xi,\xi')$ by $\Psi(s;\xi,\xi') = \Psi(s;K(\xi), K'(\xi'))$.

<u>Statement</u>. For any $\xi \in E$, $\xi' \in E'$ the integral (*) defining the function $\Psi(s;K(\xi),K'(\xi'))$ is absolutely convergent for $\operatorname{Re} s >> 0$. Furthermore, $\Psi(s;\xi,\xi')$ is a rational function of q^s, where q is the cardinality of the residue field of F. There exists a nonzero polynomial $P(q^s)$ which depends only on π and π' such that $P(q^s)\Psi(s,\xi,\xi')$ is a polynomial of $q^{\pm s}$ for any $\xi \in E$, $\xi' \in E'$.

This statement is standard (see e.g. [JPS]).

6.4. <u>Theorem</u>. Let (π,G,E) be a smooth irreducible representation, $(\tilde{\pi},G,\tilde{E})$ the contragredient representation, $K: E \to \hat{S}$, $K': \tilde{E} \to \hat{S}'$ their Kirillov models.

Then

(i) For any $\xi \in E$, $\tilde{\xi} \in \tilde{E}$ the function $\Psi(s,\xi,\tilde{\xi})$ is regular at $s = 0$.

(ii) There exists $c \in \mathbb{C}^*$ such that $\Psi(s,\xi,\tilde{\xi}) = c \cdot \langle\xi,\tilde{\xi}\rangle$.

<u>Proof</u>. Let k be the maximal order of the poles of all functions $\Psi(s,\xi,\tilde{\xi})$ at $s = 0$. Define a pairing $B: E \times \tilde{E} \to \mathbb{C}$ by

$B(\xi,\tilde{\xi}) = (s^k \cdot \Psi(s,\xi,\tilde{\xi}))|_{s=0}$. It is clear that B is a P-invariant nonzero pairing. By theorem A there exists $c \in \mathbb{C}^*$ such that $B(\xi,\tilde{\xi}) \equiv c \cdot \langle\xi,\tilde{\xi}\rangle$.

Choose a vector $\xi \in E$ such that $K(\xi) \in S$ and $K(\xi) \neq 0$ (it is possible because of 6.1(ii)), and then choose a vector $\tilde{\xi} \in \tilde{E}$ such that $\langle\xi,\tilde{\xi}\rangle \neq 0$. Since the function $\Psi(s,\xi,\tilde{\xi})$ is regular everywhere and $(s^k \cdot \Psi(s,\xi,\tilde{\xi}))|_{s=0} = c \cdot \langle\tilde{\xi},\xi\rangle \neq 0$ we see that $k \leq 0$. Further, again using 6.1(ii) we can find $\xi \in E$, $\tilde{\xi} \in \tilde{E}$ such that $\Psi(s;\xi,\tilde{\xi}) \neq 0$, which gives $k \geq 0$. Hence $k = 0$, Q.E.D.

6.5. <u>Corollary</u>. Let (π,G,E) be a smooth irreducible nondegenerate representation. Then its Kirillov model $K\colon E \to \hat{S}$ is uniquely defined up to a scalar and it is an inclusion.

<u>Proof</u>. The contragredient representation $(\tilde{\pi},G,\tilde{E})$ also has a Kirillov model $K'\colon \tilde{E} \to \hat{S}'$. This can be proved either by using the Gelfand-Kazhdan approach as in [GK] or [BZ1], or by using more simple results about pairings of representations of the group P (see [BZ2, §3]).

Consider the formula from theorem 6.4

$$\psi(0;\xi,\tilde{\xi}) = c \cdot \langle\xi,\tilde{\xi}\rangle , \quad c \in \mathbb{C}^*.$$

Let $\xi \in E$, $K(\xi) \in \hat{S}$. The function $K(\xi)$ is completely determined by its scalar products with all functions $f' \in S'$. Since $S' \subset K'(\tilde{E})$, we see that $K(\xi)$ is determined by the constant c, i.e. all morphisms $K\colon E \to \hat{S}$ are proportional.

If $K(\xi) = 0$ we have $\langle\xi,\tilde{\xi}\rangle = 0$ for all $\tilde{\xi}$, i.e. $\xi = 0$.

This corollary gives an alternative proof of the theorem by Gelfand-Kazhdan (see [GK], [Sh], [BZ1]) and of the conjecture by Gelfand-Kazhdan, proved in [BZ2], [JS].

§7. CLASSIFICATION OF UNITARY IRREDUCIBLE REPRESENTATIONS OF $G = GL(n,F)$ VIA MULTIPLICITIES

Criteria for Unitarizability

7.1. Lemma. Let (π,G,E) be a smooth irreducible representation. Suppose π is Hermitian and $\pi|_P$ is unitarizable. Then π is unitarizable.

Proof. Let B_o be a G-invariant Hermitian form on E and B a P-invariant positive definite form on E. By corollary 5.3 B is G-invariant and proportional to B_o, i.e. π is unitarizable.

Remark. It is sufficient to assume that $\pi|_P$ is semiunitarizable, i.e. that the form B is positive semidefinite and nonzero. Since B is proportional to B_o, it is nondegenerate and hence positive definite.

For a given representation (π,E) it is usually easy to determine whether there exists a G-invariant Hermitian form on E, but it is very difficult to determine whether this form is positive definite. The lemma above allows us to restrict the problem to P. In the next subsections we will formulate an inductive unitarizability criterion for P-modules and deduce from it a unitarizability criterion for G-modules. Using this criterion we will describe an algorithm which classifies all unitary representations of G in terms of multiplicities of induced representations.

7.2. We need some constructions and results from [BZ2].

First define exact functors

$$\Phi^-: \mathrm{Alg}(P_n) \to \mathrm{Alg}(P_{n-1}) \text{ and } \Psi^-: \mathrm{Alg}(P_n) \to \mathrm{Alg}(G_{n-1})$$

as in [BZ2, §3] (see also 8.2). For any smooth representation (π,P,E) we define its derivatives $\pi^{(k)} \in \mathrm{Alg}\, G_{n-k}$, $k = 1,2,\dots,n$, by $\pi^{(k)} = \Psi^-(\Phi^-)^{k-1}\pi$.

The highest number h for which $\pi^{(h)} \neq 0$ is called the <u>depth</u> of π and the representation $\pi^{(h)}$ is called <u>the highest derivative</u> of π. We say that π is <u>homogenous</u> (of depth h) if the depth of any nonzero P-submodule of π is equal to h.

For any representation (π, E) of the group G we define the derivatives $\pi^{(k)}$, $k = 0,1,\dots,n$, by $\pi^{(0)} = \pi$, $\pi^{(k)} = (\pi|_P)^{(k)}$.

For the classification of unitary representations it is convenient to introduce the <u>shifted derivatives</u> $\pi^{[k]} = \nu^{1/2} \cdot \pi^{(k)}$, where $\nu^{1/2}$ is the character of G given by $\nu^{1/2}(g) = |\det g|^{1/2}$. Henceforth we consider multiplication by $\nu^{1/2}$ as an autoequivalence of the category $\mathrm{Alg}(G)$.

7.3. Let us identify the group F^* with the center of the group G_m, $m > 0$. For any irreducible representation (ω, G_m, L) we denote by χ_ω its central character, given by $\omega(\lambda) = \chi_\omega(\lambda) \cdot 1_L$, and by $e(\omega)$ the real number given by $|\chi_\omega(\lambda)| \equiv |\lambda|^{e(\omega)}$, where $|\lambda|$ is the standard norm on F^*. We call the number $e(\omega)$ <u>the central exponent</u> of ω.

For any smooth representation (π, G_m, E) we denote by $e(\pi)$ the set of central exponents of all irreducible subquotients of π. The set $e(\pi) \subset \mathbb{R}$ we call <u>central exponents of π</u>. For example, if π is unitarizable $e(\pi) = \{0\}$.

<u>Unitarizability criterion for P-modules.</u>

Let (π, P_n, E) be a smooth representation, homogeneous of depth h. Suppose π is of finite length. Then π is unitarizable if and only if

(i) $\pi^{[h]}$ is a unitarizable representation of G_{n-h}

(ii) For any $k < h$ $e(\pi^{[k]}) > 0$, i.e. all central exponents of $\pi^{[k]}$ are strictly positive.

We will prove this criterion in §9. We say that a P-module π is <u>P-positive</u> if $e(\pi^{[h]}) = 0$ and for any $k < h$ $e(\pi^{[k]}) > 0$. Then the condition (ii) can be written as (ii)' π is a P-positive representation.

We will use a version of this criterion for semiunitarizable representations, i.e. representations which have a nonzero invariant positive semidefinite Hermitian form.

Proposition. Let (π,P,E) be a smooth representation of depth h, such that $\pi^{[h]}$ is semiunitarizable and π is P-nonnegative, i.e. $e(\pi^{[k]}) \geq 0$ for $k < h$. Then π is semiunitarizable. We will prove this in §9.

7.4. Unitarizability criterion for G-modules.

Let (π,G,E) be a smooth irreducible representation. Then π is unitarizable if and only if

(i) π is Hermitian

(ii) The highest shifted derivative $\pi^{[h]}$ is a unitarizable representation of G_{n-h}

(iii) π is P-positive, i.e. $e(\pi^{[k]}) > 0$ for $k < h$.

Indeed, according to [Z1,6.8] the representation $\pi|_P$ is homogeneous of depth h (and of finite length). Then criterion 7.3 and lemma 7.1 establish the criterion.

Remark. Let (π,G,E) be an irreducible unitarizable representation. Then using 5.3 and the results of §8 one can reprove results of A. Zelevinsky: $\pi|_P$ is homogeneous and its highest derivative $\pi^{[h]}$ is irreducible (and unitarizable). Moreover the considerations of §8 essentially prove that the scalar product in E can be written as an integral in its degenerate Kirillov model (see [Z1,5.2]). It would be interesting to apply an analogous approach to nonunitarizable representations. For nondegenerate π it is done in 6.3-6.5. For degenerate π I could not do it since I do not know an analogue of the regularization procedure, described in 6.3.

Zelevinsky's Classification of Irreducible smooth G-Modules

7.5. Let R_n $(n = 0, 1, 2, \ldots)$ be the Grothendieck group of the category of smooth G_n-modules of finite length (here $G_o = \{e\}$, $R_o = \mathbb{Z}$). The induction functor defines a bilinear morphism $R_n \times R_m \to R_{n+m}$, $(\pi, \rho) \mapsto \pi \times \rho$.

We put $R = \oplus_{n=0}^{\infty} R_n$. Then $\times$ defines on R the structure of a commutative algebra.

For any $\alpha \in \mathbb{C}$ we denote by ν^α an automorphism of the ring R, given by $\nu^\alpha(\pi) = \nu^\alpha \cdot \pi$.

Define a morphism $D: R \to R$ by $D(\pi) = \pi^{(0)} + \pi^{(1)} + \ldots + \pi^{(n)}$ for $\pi \in R_n$ (see 7.2). Then D is a ring homomorphism, i.e. $(\pi \times \rho)^{(k)} = \Sigma_{i+j=k} \pi^{(i)} \times \rho^{(j)}$, and D commutes with ν^α ([Z1, §3]).

We will use another homomorphism $D^{[\ ]} = \nu^{1/2} \circ D$ of the ring R, based on shifted derivatives, i.e. $D^{[\ ]}(\pi) = \Sigma \pi^{[k]}$ (see 7.2).

Denote by $\mathrm{Irr} = \cup \mathrm{Irr}_n$ the subset of R, corresponding to irreducible representations. This subset defines on R the structure of an ordered group. By definition the multiplication $\times$ and morphisms ν^α, D, $D^{[\ ]}$ are positive operators.

7.6. We would like to parametrize the set Irr of irreducible representations in terms of cuspidal representations. So denote by $C = \cup C_n$, $n > 0$, the subset of cuspidal representations in Irr.

The subset $\Delta \subset C_d$ of the form $\Delta = (\rho, \nu\rho, \nu^2\rho, \ldots, \nu^{\ell-1}\rho)$, $\ell > 0$, we call a <u>segment</u>; ℓ is called the <u>length</u> of Δ and the representation $\nu^{(\ell-1)/2}\rho$ the center of Δ. The number d is called the depth of Δ.

The set of all segments $\Delta \subset C$ we denote by S.

<u>Statement</u> ([Z1, §3]). Let $\Delta = (\rho, \nu\rho, \ldots, \nu^{\ell-1}\rho) \subset C_d$ be a segment. Then the representation $\rho \times \nu\rho \times \ldots \times \nu^{\ell-1}\rho$ contains a unique irreducible constituent $\langle\Delta\rangle$ of the depth $d = \mathrm{depth}(\Delta)$.

We define the segments Δ^- and Δ' by $\Delta^- = (\rho, \nu\rho, \ldots, \nu^{\ell-2}\rho)$, $\Delta' = \nu^{1/2} \cdot \Delta^-$. If $\ell = 1$ we put $\Delta^- = \Delta' = \emptyset$. From [Z1, §3] we can educe:

$$\nu^{\alpha}<\Delta> \;=\; <\nu^{\alpha}\Delta>$$

$$D<\Delta> = <\Delta> + <\Delta^{-}>, \qquad D^{[\ ]}(<\Delta>) \;=\; <\nu^{1/2}\Delta> + <\Delta'> ,$$

here $<\Delta> \in \mathrm{Irr}_{\ell d}$, $<\Delta'> \in \mathrm{Irr}_{(\ell-1)d}$ and we put $<\Delta'> = 1 \in R_0$ f $\Delta' = \emptyset$.

Note that the operation $\Delta \to \Delta'$ preserves the center of a segment.

.7. Denote by $\mathcal{O}$ the set of finite multisets in S. In other words, n element $a \in \mathcal{O}$ is a sequence of segments $\Delta_1,\Delta_2,\ldots,\Delta_r \in S$ up o permutation.

For any $a = (\Delta_1,\ldots,\Delta_r) \in \mathcal{O}$ we put depth (a) = Σ depth (Δ_i) $a^{-} = (\Delta_1^{-},\ldots,\Delta_r^{-})$, $a' = (\Delta_1',\ldots,\Delta_r')$ (if for some i $\Delta_i' = \emptyset$ we simply hrow it away).

Define an element $\pi(a) \in R$ by

$$\pi(a) \;=\; <\Delta_1> \times\ldots\times <\Delta_r> .$$

Consider multisets $a_1=a,\ a_2=a_1',\ a_3=a_2',\ldots$ and put h_i = depth (a_i). ormulas in 7.5 and 7.6 imply that h_i = depth $\pi(a_i)$ and the highest hifted derivative $\pi(a_i)^{[h]}$ is isomorphic to $\pi(a_{i+1})$. For large i, $a_i=\emptyset$, i.e. $\pi(a_i)=1\in R_0$. Hence there exists a unique irreducible contituent $<a>$ of $\pi(a)$ such that $<a>^{[h_1][h_2]\ldots[h_i]} = 1$.

Theorem ([Z1,§6, 8.1]).

a) depth $<a>$ = depth a. The highest derivative $<a>^{(h)}$ and he highest shifted derivative $<a>^{[h]}$ are isomorphic to $<a^{-}>$ and $<a'>$.

b) The map $a \mapsto <a>$ gives a one-to-one correspondence $\mathcal{O} \to \mathrm{Irr}$.

c) Elements $\pi(a)$ for $a \in \mathcal{O}$ form a basis of the ring R. In other words R is the polynomial ring over $\mathbb{Z}$ in variables $\Delta \in S$.

Define the matrix $m = (m_{ab} | a,b \in \mathcal{O})$ by

$$\langle a \rangle = \sum m_{ab} \pi(b)$$

This matrix is invertible (even unipotent, see [Z1]) and the inverse matrix describes the decomposition of representations $\pi(a)$ into irreducible components. Because of this we call m <u>the multiplicity matrix</u>.

7.8. It remains to describe central exponents and the Hermitian duality in this classification.

If $\pi \in R_n$, $\sigma \in R_m$ are positive elements, then $e(\pi\times\sigma) = e(\pi) + e(\sigma)$, $e(\nu^\alpha\pi) = e(\pi) + n\alpha$.

For $\Delta = (\rho, \nu\rho, \ldots, \nu^{\ell-1}\rho) \subset \mathcal{C}_d$ we put $e(\Delta) = \ell \cdot e(\text{center } \Delta) = \ell \cdot e(\rho) + d\ell(\ell-1)/2$. Then $e(\langle\Delta\rangle) = e(\Delta)$.

For $a = (\Delta_1, \ldots, \Delta_r)$ we put $e(a) = \Sigma e(\Delta_i)$. Then $e(\langle a\rangle) = e(a)$.

Denote by $+$ the Hermitian conjugation, i.e. the ring homomorphism $+ : R \to R$ given by $\pi \mapsto \pi^+ = \bar{\tilde{\pi}}$ the Hermitian contragredient of π. Then we have

$$(\nu^\alpha\pi)^+ = \nu^{-\bar{\alpha}}(\pi^+), \quad e(\pi^+) = -e(\pi).$$

The morphism $+$ preserves Irr and $\mathcal{C}$. Moreover, if $\rho \in \mathcal{C}_d$ we have $\rho^+ = \nu^{-2e(\rho)/d} \cdot \rho$.

The morphism $+$ acts on $\mathcal{S}$ and $\mathcal{O}$ since $(\rho^+, (\nu\rho)^+, \ldots, (\nu^{\ell-1}\rho)^+) = (\rho^o, \nu\rho^o, \ldots, \nu^{\ell-1}\rho^o)$ with $\rho^o = \nu^{1-\ell}\rho^+$.

Note that center $\Delta^+ = (\text{center } \Delta)^+$.

<u>Statement</u> ([Z1, 7.10])

$$\langle\Delta\rangle^+ = \langle\Delta^+\rangle, \quad \langle a\rangle^+ = \langle a^+\rangle$$

$$(a')^+ = (a^+)'$$

Algorithm for Description of Unitary Representations Via Multiplicities

7.9. Assume we know the multiplicity matrix $m = (m_{ab})$, i.e. we know all multiplicites. Let us describe an algorithm which enables us to find out whether a given irreducible representation $\langle a\rangle$ is unitarizable. We can rewrite the criterion 7.4 in the following way.

Criterion. Let $a \in \mathcal{O}$. Then the representation $\langle a\rangle \in \mathrm{Irr}$ is unitarizable if and only if

(i) $a^+ = a$

(ii) Let $h = \mathrm{depth}(a)$. Then the representation $\langle a\rangle^{[h]} = \langle a'\rangle$ is unitarizable.

(iii) For any $k < h$ in the expression $\langle a\rangle^{[k]} = \Sigma_b m^k_{ab}\, \pi(b)$ all coefficients m^k_{ab} with $e(b) \leq 0$ vanish.

Condition (i) can be checked straight-forwardly. In (iii) we can express all coefficients m^k_{ab} via the multiplicity matrix m, using the formula

$$D^{[\ ]}(\langle\Delta_1\rangle \times \langle\Delta_2\rangle \times\ldots\times \langle\Delta_r\rangle) = \prod(\langle\nu^{1/2}\Delta_i\rangle + \langle\Delta_i'\rangle).$$

So it remains to check (ii). But this is the same problem of unitarizability for a smaller group. Hence after several steps we can find out whether the representation $\langle a\rangle$ is unitarizable or not.

§8. UNITARIZABILITY AND IRREDUCIBILITY: MISCELLANEOUS RESULTS

Some Criteria of Irreducibility

8.1. The algorithm 7.9 shows that knowledge of multiplicities enables us to describe all unitary representations. Conversely it turns out that the study of unitary representations enables us to say something about multiplicities. More precisely, it allows us to give some criteria of irreducibility.

Proposition. Let (π, G, E) be a representation of finite length. Suppose that $\pi \approx \pi^+$ in the Grothendieck group R and the highest shifted derivative $\pi^{[h]}$ is irreducible and unitarizable. Then

a) If π is P-positive, i.e. $e(\pi^{[k]}) > 0$ for $1 \leq k < h$, then π is irreducible (and unitarizable).

b) If π is P-nonnegative, but not P-positive, i.e. $e(\pi^{[k]}) \geq 0$ and for some $k < h$ $e(\pi^{[k]}) \ni 0$, then π is reducible.

<u>Remark</u>. If $\pi^{[h]}$ is reducible, then π is also reducible (see [Z1,8.1]).

<u>Proof</u>.

a) Let $\omega \in \mathrm{Irr}$. By 7.8 ω and ω^+ have the same depth, say d, and their highest shifted derivatives are Hermitian dual, i.e. $e(\omega^{+[d]}) = -e(\omega^{[d]})$. Consider an irreducible constituent ω of the representation π. Then ω^+ is a constituent of $\pi^+ \approx \pi$. If $d = \mathrm{depth}(\omega)$ is less than $h = \mathrm{depth}(\pi)$ then $e(\omega^{[d]}), e(\omega^{+[d]}) \in$ $\in e(\pi^{[d]})$ should be both positive, which is impossible. Hence any constituent ω has depth h and since $\pi^{[h]}$ is irreducible, π has only one constituent, i.e. π is irreducible. By criterion 7.4 π is unitarizable.

b) By 7.3 $\pi|_P$ is semiunitarizable. If π were irreducible, it would be unitarizable (see Remark 7.1), and therefore P-positive, contradicting the condition of the proposition.

8.2. We call a G-module π of finite length **G-positive**, if for any $k < h = \mathrm{depth}(\pi)$, including $k = 0$,

$$e(\pi^{[k]}) > e(\pi^{[h]}) .$$

The formula $(\pi \times \sigma)^{[k]} = \Sigma \pi^{[i]} \times \sigma^{[j]}$, $i + j = k$ implies that $\pi \times \sigma$ is G-positive iff both π and σ are G-positive. Criterion 7.4 and proposition 8.1) imply

<u>Criterion</u>. Let π be a G-module of finite length, $h = \mathrm{depth}(\pi)$. Then π is irreducible and unitarizable iff $\pi = \pi^+$ in R, $\pi^{[h]}$ is irreducible and unitarizable and π is G-positive.

Corollary.

a) If π, σ are irreducible and unitarizable, then $\pi \times \sigma$ is irreducible and unitarizable.

b) If π, σ are irreducible and Hermitian and $\pi \times \sigma$ is unitarizable, then π and σ both are unitarizable.

Proof.

a) By induction, the highest shifted derivative $(\pi \times \sigma)^{[h+d]} = \pi^{[h]} \times \sigma^{[d]}$ is irreducible and unitarizable (here $h = \mathrm{depth}(\pi)$, $d = \mathrm{depth}(\sigma)$). Since $\pi \times \sigma$ is G-positive as a product of G-positive modules, $\pi \times \sigma$ is irreducible and unitarizable.

b) By induction the highest shifted derivatives $\pi^{[h]}$ and $\sigma^{[d]}$ are irreducible and unitarizable. The representation $\pi \times \sigma$ is unitarizable and hence G-nonnegative. Therefore π and σ are G-nonnegative. By proposition 8.1b) they are G-positive and hence unitarizable.

Boundary of the Complementary Series

8.3. Let σ be a smooth irreducible representation of G_n. For any $\alpha \in \mathbb{R}$ denote by $\pi(\alpha) = \pi(\sigma,\alpha)$ the representation of G_{2n}

$$\pi(\sigma,\alpha) = \nu^{\alpha}\sigma \times \nu^{-\alpha}\sigma^{+} = (\nu^{\alpha}\sigma) \times (\nu^{\alpha}\sigma)^{+}.$$

Let $h = \mathrm{depth}(\sigma)$. Define inductively an interval $I(\sigma) \subset \mathbb{R}$ by $I(\sigma) = \{\alpha | \alpha \in I(\sigma^{[h]}),\ \nu^{\alpha}\sigma$ and $(\nu^{\alpha}\sigma)^{+}$ are G-positive$\}$. The latter condition can be written as

(*) For any $0 \le k < h$ $\quad e(\sigma^{[k]}) - e(\sigma^{[h]}) > (h-k)\alpha$ and

$$e(\sigma^{+[k]}) - e(\sigma^{+[h]}) > -(h-k)\alpha .$$

Proposition. If $\alpha \in I(\sigma)$ $\pi(\alpha)$ is irreducible and unitarizable. If α lies on the boundary of $I(\sigma)$ $\pi(\alpha)$ is reducible. If α lies outside of the closure of $I(\sigma)$, $\pi(\alpha)$ is not unitarizable.

Indeed, if $\alpha \in I(\sigma)$ then by induction the highest shifted derivative $\pi(\alpha)^{[2h]} = \pi(\sigma^{[h]},\alpha)$ is irreducible and unitarizable and

$\pi(\alpha)$ is G-positive. By criterion 8.2 π is irreducible and unitarizable.

If α lies on the boundary of $I(\sigma)$, then either $\pi(\alpha)^{[2h]}$ is reducible, or it is irreducible and unitarizable and $\pi(\alpha)$ is G-nonnegative but not G-positive. By proposition 8.1b) $\pi(\alpha)$ is reducible.

If α lies outside of the closure of $I(\sigma)$ then either $\pi(\alpha)^{[2h]}$ is not unitarizable or $\pi(\alpha)$ is not G-nonnegative. In both cases 7.4 implies that $\pi(\alpha)$ is not unitarizable.

Representations $\pi(\alpha)$ for $\alpha \in I(\sigma)$ we call a (generalized) <u>complementary series</u>. Note, that if σ is unitarizable, $I(\sigma)$ is not empty, namely $I(\sigma) \ni 0$, and $I(\sigma)$ is symmetric with respect to 0.

Proposition 8.3 together with corollary 8.2b) allows us to describe all complementary series.

<u>Remark</u>. The length of the interval $I(\sigma)$ is always less than or equal to 1 (if $\sigma \neq 1 \in R_o$). Indeed, if the highest shifted derivative $\sigma^{[h]}$ is not $1 \in R_o$, then $I(\sigma) \subset I(\sigma^{[h]})$ and we can use induction. Now, let $\sigma^{[h]} = 1$, i.e. σ is nondegenerate. Applying condition (*) for $k = 0$ we see that for $\alpha \in I(\sigma)$

$$h\alpha < e(\sigma^{[o]}) = e(\nu^{1/2}\sigma) = h/2 + e(\sigma)$$
$$-h\alpha < e(\sigma^{+[o]}) = h/2 + e(\sigma^{+}) = h/2 - e(\sigma), \quad \text{i.e.} \quad |\alpha - e(\sigma)/h| < 1/2.$$

Reduction to the Discrete Data

8.4. We want to show that the algorithm described in 7.9 essentially works with discrete data.

Let $\Delta_1, \Delta_2 \subset C$ be two segments. We say that Δ_1 and Δ_2 are <u>linked</u> if $\Delta_1 \not\subset \Delta_2$, $\Delta_2 \not\subset \Delta_1$ and the set theoretic union $\Delta_1 \cup \Delta_2 \subset C$ is also a segment.

<u>Statement</u> ([Z1, §8]). Let $a_1 = (\Delta_1^1, \ldots, \Delta_r^1)$, $a_2 = (\Delta_1^2, \ldots, \Delta_s^2)$. Suppose that for any i, j the segments Δ_i^1 and Δ_j^2 are not linked.

Then $\langle a_1\rangle \times \langle a_2\rangle = \langle a_1 + a_2\rangle$, where $a_1 + a_2$ is the union of multisets.

Consider the action of the group $\mathbb{R}$ on C given by $\alpha : \rho \to \nu^{\alpha}\rho$. Orbits of this action we call $\mathbb{R}$-lines in C, and orbits of the subgroup $\mathbb{Z} \subset \mathbb{R}$ we call $\mathbb{Z}$-lines. We say that a multiset a is concentrated on a line $\Pi \subset C$ if all segments of a lie in Π. Then we see that any irreducible representation can be written as $\langle a_1\rangle \times \langle a_2\rangle \times \ldots \times \langle a_k\rangle$, where $a_1,\ldots,a_k$ are concentrated on different $\mathbb{Z}$-lines $\Pi_1,\ldots,\Pi_k$. Hence all problems about multiplicities m_{ab} can be reduced to one line.

From now on we fix a <u>unitary</u> cuspidal irreducible representation $\rho \in C_d$ and consider only segments Δ and multisets a, concentrated on the $\mathbb{R}$-line $\Pi_\rho = \{\nu^{\alpha}\rho,\ \alpha \in \mathbb{R}\}$. We will identify in a natural way $\mathbb{R}$ and Π_ρ, $\alpha \leftrightarrow \rho_\alpha = \nu^{\alpha}\rho$. In particular $\rho_\alpha^{+} = \rho_{-\alpha}$.

Let a be a multiset which we suspect to be unitary. Using the statement above we can consider only the case when a is concentrated on a subset

$$\Pi_\alpha = \{\pm\alpha + \mathbb{Z}\} \subset \mathbb{R} = \Pi_\rho, \quad \text{where} \quad 0 \le \alpha \le 1/2 .$$

We consider 2 cases.

Case 1 (rigid case): $\alpha = 0$ or $\alpha = 1/2$, i.e. Π_ρ is a $\mathbb{Z}$-line.

Case 2 (nonrigid case): $0 < \alpha < 1/2$.

8.5. Let us show that in the "nonrigid" case we can exclude the parameter α.

We can decompose a into the union of multisets a_α and $a_{-\alpha}$ concentrated on $\alpha + \mathbb{Z}$ and $-\alpha + \mathbb{Z}$. By statement 8.4 $\langle a\rangle = \langle a_\alpha\rangle \times \langle a_{-\alpha}\rangle$.

Let us denote by b the multiset $\nu^{-\alpha}a_\alpha$ concentrated on the $\mathbb{Z}$-line $\mathbb{Z}$ and put $\sigma = \langle b\rangle$. Then $\langle a\rangle = \pi(\sigma,\alpha)$ (see 8.3). The representation $\pi(\sigma,\alpha)$ is irreducible and unitarizable for all α in the interval $I(\sigma)$, described in 8.3, and it is reducible when α lies on the boundary of $I(\sigma)$. According to the statement 8.4 both boundary points of $I(\sigma)$ are half-integers. Hence for $\alpha \in (0,1/2)$ either all representations $\pi(\sigma,\alpha)$ are unitarizable, or all of them are not unitarizable.

Let us rewrite condition 8.3 (*) for the unitarizability of $\pi(\sigma,\alpha)$.

$$(*) \quad (h-k)\alpha < e(\sigma^{[k]}) - e(\sigma^{[h]}), \quad -(h-k)\alpha < e(\sigma^{+[k]}) - e(\sigma^{+[h]}) .$$

Since it should be true for all $\alpha \in (0,1/2)$ it is equivalent to the condition

$$(**) \quad e(\sigma^{[k]}) - e(\sigma^{[h]}) \geq (h-k)/2, \quad e(\sigma^{+[k]}) - e(\sigma^{+[h]}) \geq 0 .$$

Hence we have proved the following inductive Criterion.

Criterion. Let $b \in O$ be a multiset of depth h concentrated on $\mathbb{Z}$-line $\mathbb{Z} \subset \Pi_\rho$, and let $\alpha \in (0,1/2)$. Then the representation $\nu^{\alpha}\langle b\rangle \times \nu^{-\alpha}\langle b^+\rangle = \langle \nu^{\alpha} b + \nu^{-\alpha} b^+\rangle$ is unitarizable if and only if $\langle \nu^{\alpha} b^{[h]} + \nu^{-\alpha} b^{[h]+}\rangle$ is unitarizable for all $\alpha \in (0,1/2)$ and (**) holds for $\sigma = \langle b\rangle$.

This criterion does not depend on α. It allows us to formulate an algorithm, dealing only with discrete data, for the classification of unitarizable representations in the nonrigid case. In the rigid case we automatically work with discrete data.

Some Conjectures.

8.6. In 8.5 the length of $I(\sigma)$ is less than or equal to 1. Hence if $I(\sigma) \supset (0,1/2)$ we have 3 possibilities: $I(\sigma) = (0,1)$, $I(\sigma) = (-1/2,1/2)$ or $I(\sigma) = (0,1/2)$. For instance, if σ is unitarizable, $I(\sigma) = (-1/2,1/2)$; if $\nu^{1/2}\sigma$ is unitarizable, $I(\sigma) = (0,1)$ (see corollary 8.2 a)).

Conjecture.

a) if $I(\sigma) = (-1/2,1/2)$ then σ is unitarizable. Respectively, if $I(\sigma) = (0,1)$ then $\nu^{1/2}\sigma$ is unitarizable.

b) If $I(\sigma) = (0,1/2)$, then $\sigma = \sigma_1 \times \sigma_2$, where σ_1 and σ_2 are irreducible representations, $I(\sigma_1) = (-1/2,1/2)$, $I(\sigma_2) = (0,1)$.

This conjecture, if true, reduces the studying of the non-rigid case 2 to the rigid case 1.

8.7. Now one remark about the rigid case.

Lemma. Let b be a multiset concentrated on one $\mathbb{Z}$-line $\mathbb{Z}$ or $1/2+\mathbb{Z}$ and $\sigma = \langle b \rangle$. Suppose that σ is unitarizable. Then for all $0 \le k < h = \mathrm{depth}(\sigma)$ we have

$$(*) \qquad e(\sigma^{[k]}) \ge (h-k)/2 .$$

This inequality refines the condition $e(\sigma^{[k]}) > 0$ of 7.4.

Proof. Since $\sigma \times \sigma$ is irreducible and unitarizable by 8.2a), $I(\sigma) = (-1/2, 1/2)$. Thus for any $\alpha \in (-1/2,1/2)$ we have (see 8.3(*))

$$e(\sigma^{[k]}) = e(\sigma^{[k]}) - e(\sigma^{[h]}) > \alpha(h-k).$$

This gives the inequality (*).

I should confess that inequality (*) reminds me of the inequalities for Kazhdan-Lusztig polynomials.

Unitarizability and Duality

8.8. Let us apply the unitarizability criterion to the case when $\langle a \rangle = \pi(a)$. According to [Z1,4.2] this is true if and only if any two segments Δ_1, Δ_2 of a are not linked. For simplicity consider the case when a is concentrated on one $\mathbb{R}$-line Π_ρ (see 8.4).

Lemma. $\pi(a)$ is irreducible and unitarizable if and only if $a^+ = a$ and for any segment $\Delta \in a$ we have $|\mathrm{center}(\Delta)| < 1/2$ (we consider the representation $\mathrm{center}(\Delta)$ as a real number on $\Pi_\rho \simeq \mathbb{R}$).

Proof. Suppose $\pi(a)$ is irreducible and unitarizable and $\Delta \in a$. Then the highest shifted derivative $\pi(a')$ is also irreducible and unitarizable. If $\Delta' \neq \emptyset$, then by induction $\mathrm{center}(\Delta) = \mathrm{center}(\Delta') > -1/2$. If $\Delta' = \emptyset$, i.e. Δ consists of one element, then the condition that $\langle \Delta \rangle$ is G-positive, i.e.

$e(\Delta^{[o]}) = e(\nu^{1/2}\Delta) > 0$, implies that center$(\Delta) > -1/2$. Since Δ, $\Delta^+ \in a$, we have $|\text{center}(\Delta)| < 1/2$. Conversely, suppose that $a^+ = a$ and $|\text{center}(\Delta)| < 1/2$ for all $\Delta \in a$. Then all representations $\langle\Delta\rangle$ are G-positive and applying criterion 8.2 we can prove by induction that $\pi(a)$ is irreducible and unitarizable.

8.9. In [Z1, 9] A. Zelevinsky described an automorphism t of the ring R, which he called duality. On generators $\langle\Delta\rangle$ it is given by

$$\langle\Delta = (\rho, \nu\rho, \ldots, \nu^{\ell-1}\rho)\rangle \mapsto \langle\Delta\rangle^t = \langle\{\rho\}, \{\nu\rho\}, \ldots, \{\nu^{\ell-1}\rho\}\rangle,$$

where on the right side we consider the multiset of ℓ one-point segments.

One can show that t maps Irr into itself. The representations $\langle\Delta\rangle^t$ play a very important role - they are the so-called square integrable representations.

Lemma. Let $a = (\Delta_1, \ldots, \Delta_r) \in \mathcal{O}$. Then the representation $\pi(a)^t = \langle\Delta_1\rangle^t \times \ldots \times \langle\Delta_r\rangle^t$ is irreducible and unitarizable if and only if $a^+ = a$ and for any segment $\Delta \in a$ we have $|\text{center}(\Delta)| < 1/2$.

Proof. According to [Z1, 9.6] for $\Delta = (\rho, \nu\rho, \ldots, \nu^{\ell-1}\rho)$ we have

$$D(\langle\Delta\rangle^t) = \Sigma\langle\Delta_i\rangle^t,$$

where

$$\Delta_i = (\nu^i\rho, \nu^{i+1}\rho, \ldots, \nu^{\ell-1}\rho), \quad i = 0, 1, \ldots, \ell$$

and we assume $\langle\Delta_\ell\rangle = 1$. Hence $\langle\Delta\rangle^t$ is G-positive if and only if $e(\nu^{1/2}\Delta) > 0$, i.e. center(Δ) $-1/2$. Now criterion 8.2 implies the lemma.

Remark. This lemma gives a classification of nondegenerate irreducible unitarizable representations.

8.10. Lemmas 8.8 and 8.9 make reasonable the following.

Conjecture. Duality t: Irr $\to$ Irr maps unitarizable representations into unitarizable representations.

§9. PROOF OF CRITERION AND PROPOSITION 7.3

9.1. Let (π, P, E) be a smooth representation. We call π <u>Φ^--degenerate</u> if $\Phi^-(\pi) = 0$. We call π <u>Φ^--homogeneous</u> if for any nonzero subrepresentation ρ $\Phi^-(\rho) \neq 0$.

Criterion and proposition 7.3 inductively follows from the following.

<u>Proposition</u>. Let (π, P_{m+1}, E) be a smooth representation of finite length.

a) Suppose π is Φ^--degenerate. Then π is (semi) unitarizable iff $\nu^{1/2} \cdot \Psi^-(\pi) \in \mathrm{Alg}\, G_m$ is (semi) unitarizable.

b) Suppose π is Φ^--homogeneous. Then π is unitarizable iff

(i) $\Phi^-(\pi) \in \mathrm{Alg}\, P_m$ is unitarizable.

(ii) $e(\nu^{1/2}\Psi^-(\pi)) > 0$.

c) Suppose π is Φ^--nondegenerate, $\Phi^-(\pi)$ is semiunitarizable and $e(\nu^{1/2}\Psi^-(\pi)) \geq 0$. Then π is semiunitarizable.

9.2. In the proof of proposition 9.1 we will use the geometric realization of the representation (π, E), described in [BZ1, §5].

The group P_{m+1} is the semidirect product of the subgroups*) $G_m = \{(p_{ij}) \mid p_{ij} = \delta_{ij}$ for $i > m$ or $j > m\}$ and $V = V_m = \{(p_{ij}) \mid p_{ij} = \delta_{ij}$ for $j \leq m\}$. Let us identify V_m with the linear space $\mathrm{Mat}(m,1;F)$ of column-vectors of length m and denote by W the dual space $W = \mathrm{Mat}(1,m;F) = V^*$ of row-vectors. For any $v \in V$ we denote by ψ_v the character of W, given by $\psi_v(w) = \psi\langle v,w\rangle$, where ψ is a fixed nontrivial additive character of F. We denote by δ the natural action of G_m on W, given by $\delta(g)w = wg^{-1}$.

*) V is the unipotent radical of P and G is a Levi component of P.

Statement. Let (π, P_{m+1}, E) be a smooth representation. Then there exists a sheaf F on W with an action δ of the group G_m and an isomorphism $i: E \overset{\sim}{\to} S(F)$ of the space E with the space $S(F)$ of compactly supported sections of F such that

$$i(\pi(g)\xi) = \nu(g)^{1/2}\delta(g)(i(\xi))$$

$$i(\pi(v)\xi) = \psi_v \cdot i(\xi) \qquad g \in G_m, \ v \in V, \ \xi \in E.$$

The triple (F, δ, i) is uniquely defined by (π, E) up to a canonical isomorphism.

This statement is a variant of Mackey's construction. It is proved in [BZ1, 5]. More precisely, in [BZ1] the factor $\nu^{1/2}$ is omitted, so we should apply [BZ1] to $\nu^{1/2}\pi$.

We will identify E with $S(F)$ using the isomorphism i. Put

$$E_o = S_o(F) = \{\phi \in S(F) \mid \text{supp } \phi \subset W\backslash 0\}, \ \pi_o = \pi|_{E_o}$$

For any point $w \in W$ we denote by F_w the stalk of the sheaf F at w. Consider two points $0 \in W$ and $e = (0,\ldots,1) \in W$. It is clear, that

$$\text{Stab}(0, G_m) = G_m, \quad \text{Stab}(e, G_m) = P_m.$$

By definition we have

$$\Psi^-(\pi) = (\delta, G_m, F_o), \quad \Phi^-(\pi) = (\delta, P_m, F_e)$$

(this coincides with the definition in [BZ2, 3]).

From these formulae we see that:

(i) π is Φ^--degenerate $\Leftrightarrow$ F is concentrated at $0 \Leftrightarrow V$ acts trivially on E.

(ii) π is Φ^--homogeneous $\Leftrightarrow$ F has no nonzero section concentrated at 0.

9.3. Proof of the proposition 9.1.

(i) Suppose π is Φ^--degenerate. Then V acts trivially on E and $\pi|_{G_m} = \nu^{1/2}\Psi^-(\pi)$. This implies 9.1a).

(ii) Suppose that the representation $\Phi^-(\pi) = (\delta, P_m, F_e)$ is Hermitian, i.e. the space F_e has a P_m-invariant Hermitian form B_e. Since G_m acts transitively on $W\smallsetminus 0$ and $P_m = \mathrm{Stab}(e, G_m)$, we can define a G-invariant system of Hermitian forms B_w for all $w \in W\smallsetminus 0$.

Now fix a Haar measure μ on W and define the Hermitian form B_o on $E_o = S_o(F)$ by

$$B_o(\phi,\phi') = \int_W B_w(\phi_w,\phi'_w)d\mu(w).$$

Since μ is (G_m,ν)-invariant, the formulae of statement 9.2 describing the representation π imply that the form B_o is G-invariant with respect to π_o. It is clear that B_o is positive definite if and only if B_e is positive definite.

Statement. The correspondence $B_e \mapsto B_o$ is a one-to-one correspondence between P_m-invariant Hermitian forms on (δ, F_e) and P_{m+1}-invariant Hermitian forms on (π_o, E_o).

For the (easy) proof see [BZ2,§3].

(iii) Let us fix a positive definite Hermitian form B_e on F_e and the corresponding form B_o on E_o. We want to find out when we can extend B_o to the positive definite P_{m+1}-invariant form B on (π, E). The answer is given by the following.

Analytic criterion. The form B_o can be extended to the P-invariant positive definite Hermitian form B on E iff for any $\phi \in E = S(F)$ the integral $I_\phi = \int_{W\smallsetminus 0} B_w(\phi_w,\phi_w)d\mu(w)$ converges. Indeed, if all these integrals converge we can define B by $B(\phi,\eta) = \int_W B_w(\phi_w,\eta_w)d\mu(w)$. Conversely, suppose we can extend B_o to B. Consider any open compact subgroup $W^o \subset W$ and denote by V^o the dual subgroup of V (i.e. $V^o = \{v \in V | \psi_v(W^o) = 1\}$). Then V^o is an open compact subgroup of V and we can define an operator

$A: E \to E$ by

$$A = \int_{V^o} \pi(v)\,d\mu_V(v) \; / \; \text{measure}(V^o) .$$

This operator is a projector and in the geometric realization it is given by $A(\phi) = \chi \cdot \phi$, where $\phi \in E = S(F)$ and χ is the characteristic function of W^o.

Since the form B is P_{m+1}-invariant, A is an orthogonal projector with respect to this form. Hence for any $\phi \in E$ we have

$$\|\phi\|_B^2 \geq \|(1-A)\phi\|_B^2 =$$

$$= \|(1-\chi)\phi\|_B^2 = \|(1-\chi)\phi\|_{B_o}^2 = \int_{w \in W \setminus W^o} \|\phi_w\|^2 \, d\mu(w)$$

This implies that the integral I_ϕ converges.

9.4. In order to finish the proof of proposition 9.1b) it remains to check that integrals I_ϕ converge for all $\phi \in E$ iff $e(\nu^{1/2}\Psi^-(\pi)) > 0$.

(i) Denote by ϖ a generator of the maximal ideal of the ring of integers of F (i.e. $|\varpi| = q^{-1}$). We will identify ϖ with the central element $\varpi \cdot 1_m \in G_m$.

Consider the quotient representation $(\pi' = \pi/\pi_o, G_m, E' = E/E_o)$. Since π has finite length, π' also has finite length. The operator $\pi'(\varpi)$ is then finite, i.e. it generates a finite dimensional algebra of operators.

Denote by $\mu_1, \ldots, \mu_r$ the eigenvalues of $\pi'(\varpi)$. By definition $|\mu_i| = q^{-e(\omega)}$ for each i, where $e(\omega)$ is the central exponent of an irreducible subquotient of $\pi' = \nu^{1/2}\Psi^-(\pi)$. Hence we can rewrite condition $e(\nu^{1/2}\Psi^-(\pi)) > 0$ as

(*) $$|\mu_i| < 1 \quad \text{for all } i.$$

(ii) Consider the space C of functions $f: W \setminus 0 \to \mathbb{C}$ such that supp f lies in a compact subset of W and f is locally constant on $W \setminus 0$, and define the representation (δ, G_m, C) by

$$\delta(g)f(w) = \nu(g)f(\delta(g^{-1})w).$$

The restriction of δ on the subspace

$$C_o = \{f \in C | f = 0 \text{ in a neighborhood of } 0\}$$

we denote by δ_o.

The Hermitian form B_e defines a pairing $\beta: E \otimes \overline{E} \to C$ by $\beta(\phi,\eta)(w) = B_w(\phi_w,\eta_w)$, which is a morphism of representations $\beta: \pi \otimes \overline{\pi} \to \delta$. It is clear that $\beta(\phi,\eta) \in C_o$ if $\phi \in E_o$ or $\eta \in E_o$.

Denote by $C_\beta \subset C$ the image of β. This space of functions satisfies the following conditions

(**) (α) $C_\beta \supset C_o$

(β) C_β is the linear span of positive functions $f \in C_\beta$.

(γ) Put $C' = C_\beta/C_o$ and denote by A the action of the operator $\delta(\varpi)$ on this quotient space. Then A generates a finite dimensional algebra of operators and all its eigenvalues are of the form $\mu_i\overline{\mu}_i$.

(δ) For any $\mu = \mu_1,\dots,\mu_r$ there exists a positive function $f \in C_\beta$ such that $f \notin C_o$ and $A(f) = \mu\overline{\mu}f \pmod{C_o}$.

Condition (α) follows from the fact that E, and hence C_β, is invariant with respect to multiplication on locally constant functions on W; (β) follows from the polarization formula.

The pairing β defines an epimorphism $\beta': E' \times \overline{E}' \to C'$ of G_m-modules which proves (γ). If $\phi \in E \setminus E_o$ is a vector, such that $\pi(\varpi)\phi = \mu\phi \pmod{E_o}$, then the function $f = \beta(\phi,\phi)$ satisfies condition (δ).

(iii) <u>Lemma</u>. Let $C_\beta \subset C$ be a subspace, satisfying (α) - (δ). Then

a) if $|\mu_i| < 1$ for all i, then for any $f \in C_\beta$ the integral $I_f = \int_W f(w)d\mu(w)$ converges.

b) If $|\mu| \geq 1$ for one of the $\mu = \mu_i$, then I_f does not converge for the corresponding function $f \in C_\beta$.

This lemma implies that the conditions of criterion 9.3 (iii) are satisfied iff $|\mu_i| < 1$ for all i, that proves 9.1b).

Proof of the lemma. Consider some norm $\| \ \|$ on W, put $W_n = \{ w \in W \mid \|w\| = q^{-n} \}$ and define the function $f(n)$, $n \in \mathbb{Z}$, by

$$f(n) = \int_{W_n} f(w)\, d\mu(w).$$

This reduces the lemma to the analogous lemma about functions on $\mathbb{Z}$, where

$$C = \{ f : \mathbb{Z} \to \mathbb{C} \mid f(n) = 0 \quad \text{if} \quad -n \quad \text{is large} \}.$$

$$C_o = \{ f \in C, \quad \text{supp}\ (f) \quad \text{is finite} \}$$

$$I_f = \Sigma_{-\infty}^{\infty} f(n)$$

$$(\delta(\mathfrak{p}) f)(n) = f(n+1)$$

(we use the fact that $\delta(\mathfrak{p})$ preserves integrals).

Condition (**) (γ) implies that $f(n)$ is an exponential polynomial for large n, i.e. $f(n) = \Sigma P_k(n) \cdot \lambda_k^n$, where λ_i are of the form $\mu_i \bar{\mu}_i$. This implies a).

If $|\mu| \geq 1$ and $Af = \mu\bar{\mu} f (\text{mod } C_o)$, then $f = c \cdot (\mu\bar{\mu})^n$ for large n. Since $f \notin C_o$, the constant c is not equal to 0, and hence I_f does not converge; this proves b).

9.5. Proof of 9.1 c). Let B_e be a nonzero positive semidefinite form on $\Phi^-(\pi)$. As in 9.3, 9.4 we define the pairing $\beta: E \otimes \bar{E} \to C$ and put $C_\beta = \text{Im}\ \rho$. This space satisfies conditions (**) (α) (β) (γ) of 9.4.

Lemma. If $|\mu_i| \leq 1$ for all i then there exists a nonzero positive G_m-equivariant functional $I: C_\beta \to \mathbb{C}$ (here positive means $I(f) \geq 0$ for $f \geq 0$).

The formula $B(\phi,\eta) = I(\beta(\phi,\eta))$ defines a nonzero positive semidefinite P-invariant Hermitian form on E, that proves 9.1c).

Proof of the lemma. Choose some norm $\kappa = \| \ \|$ on W.

For any $s > 0$ the integral $I(\kappa,s;f) = \int_W f(w)\kappa(w)^s d\mu(w)$ converges and defines a positive functional $I(\kappa,s)$ on C_β.

The function $I(\kappa,s;f)$ is a rational function in q^s, and the order of the pole of this function at $s = 0$ is bounded by some number k, which does not depend on f (it does not exceed the degree of the minimal polynomial of the operator A). Let us choose minimal possible k and put

$$I(\kappa;f) = \lim_{s \to 0} s^k I(\kappa,s;f)$$

$I(\kappa)$ is a nonzero positive functional on C_β. In order to prove that it is G_m-equivariant it is sufficient to check that $I(\kappa)$ does not depend on κ.

If $\kappa \geq \kappa'$, then $I(\kappa,s) \geq I(\kappa',s)$ and hence $I(\kappa) \geq I(\kappa')$. Besides, $I(\lambda\kappa;s) = \lambda^s I(\kappa,s)$, i.e. $I(\lambda\kappa) = I(\kappa)$. Since for any other norm κ' we have $\lambda\kappa \geq \kappa' \geq \lambda^{-1}\kappa$ for some $\lambda > 0$, we have

$$I(\kappa) = I(\lambda\kappa) \geq I(\kappa') \geq I(\lambda^{-1}\kappa) = I(\kappa),$$

i.e., $I(\kappa) = I(\kappa')$. This proves the lemma.

REFERENCES

[BZ1]. I.N. Bernstein and A.V. Zelevinsky, Representations of the group GL(n,F), where F is a local non-archimedean field, Uspekhi Mat. Nauk 31, no. 3(1976), 5-70 (= Russian Math. Surveys 31, no. 3 (1976), 1-68).

[BZ2]. I.N. Bernstein and A.V. Zelevinsky, Induced representations of reductive $\mathfrak{p}$-adic groups, I, Ann. Scient. Ecole Norm. Sup. 10 (1977), 441-472.

[GK]. I.M. Gelfand and D.A. Kazhdan, Representations of GL(n,K), in Lie groups and their Representations, Summer School of group representations, Budapest, 1971. New York, Halsted Press, 1975, pp. 95-118.

[JS]. H. Jacquet and J. Shalika, Whittaker models of induced representations, to appear in Pacif. J. of Math.

[JPS]. H. Jacquet, I.I. Piatetski-Shapiro and J. Shalika, Rankin-Selberg convolutions, to appear in Amer. J. of Math.

[K]. A.A. Kirillov, Infinite-dimensional representations of the general linear group, Dokl. Akad. Nauk SSSR 144(1962), 37-39 (= Soviet Math. Dokl. 3 (1962), 652-655).

[R]. R.R. Rao, Orbital integrals in reductive groups, Ann. of Math. 96 (1972), 505-510.

[Sh]. J.A. Shalika, The multiplicity one theorem for GL_n, Ann. of Math. 100 (1974), 171-193.

[Z1]. A.V. Zelevinsky, Induced representation of reductive $\mathfrak{p}$-adic groups II, Ann. Scient. Ecole Norm. Sup. 13(1980), 165-210.

[Z2]. A.V. Zelevinsky, $\mathfrak{p}$-adic analogue of the Kazhdan-Lusztig conjecture, Funck. Anal. i Priložen. 15, no. 2 (1981), 9-21 (= Funct. Anal. and Appl. 15 (1981), 83-92).

AUTOMORPHIC FORMS AND A HODGE THEORY FOR CONGRUENCE SUBGROUPS OF $SL_2(\mathbb{Z})$

William Casselman
Vancouver

Suppose for the moment that Γ is any arithmetic subgroup of $SL_2(\mathbb{R})$. It acts discretely on the upper half-plane H by fractional linear transformations

$$z \to az + b/cz + d ,$$

and the quotient will be a non-singular Riemann surface. If the quotient $X = \Gamma\backslash H$ is compact then we have for its cohomology the usual Hodge theory:

$$H^0(X, \mathbb{C}) = \mathbb{C}$$

$$H^1(X, \mathbb{C}) = H^{1,0} \oplus H^{0,1} = \mathbb{C}^g \oplus \mathbb{C}^g$$

$$H^2(X, \mathbb{C}) = \mathbb{C}$$

where g is the genus of X. If X is not compact then it may be embedded into a compact, non-singular Riemann surface $\overline{X}$ by adding a finite number of points at infinity, called <u>cusps</u>:

$$X \cup \{\text{cusps}\} = \overline{X} .$$

The cohomology of X may be determined simply in terms of that of $\overline{X}$ by looking at the long exact sequence of relative cohomology associated to the embedding $X \to \overline{X}$:

$$H^0(X, \mathbb{C}) = \mathbb{C}$$

$$H^1(X, \mathbb{C}) = H^1(\overline{X}, \mathbb{C}) \oplus \mathbb{C}^{n-1}$$

$$H^2(X, \mathbb{C}) = 0 .$$

Here n is the number of cusps.

The cohomology $H^1(\overline{X}, \mathbb{C})$ is represented on X by cusp forms, holomorphic and anti-holomorphic, of weight 2. For the contribution of $\overline{X}$ to the cohomology of X, in other words, there is a perfectly good Hodge theory: yet another way of saying this is that $H^1(X, \mathbb{C})$ contains a subspace called cuspidal cohomology where cohomology is represented uniquely by harmonic forms. What I will be concerned with in this paper is understanding in a similar way the non-cuspidal cohomology - i.e. the term $\mathbb{C}^{n-1}$. It is clear that there must be some relationship with Eisenstein series of weight 2, which in this case also have dimension $(n-1)$. But then how to take account of the anti-holomorphic Eisenstein series, which are of course also harmonic?

Hodge theory on compact Riemannian manifolds involves something which is potentially much deeper, the analysis of the Laplace operator on C^∞ and even $\mathbb{L}^2$ forms. Similarly, what I have in mind is an analysis of the Laplace operator on X, which has a continuous spectrum when X is non-compact. The appropriate result of this analysis is a theorem resembling classical results of Paley-Wiener rather than an $\mathbb{L}^2$ decomposition. My account of this will take advantage of the opportunity to present a brief introduction to the theory of automorphic forms.

Nearly all of what I will say extends in some form to larger arithmetic groups, in particular to describe the cohomology of arithmetic subgroups of semi-simple rational groups of $\mathbb{Q}$-rank one. This completes work of Harder [1975a,b]. This extension has been joint work with

Birgit Speh. Details will appear sooner or later.

The right way to look at these matters, especially when larger arithmetic groups are concerned, is by means of representation theory. However, I have made this account as elementary as possible, and this subject will generally only lurk in the background. Also, I should add that the Hodge theory I suggest here is quite different from - much more analytical than - Deligne's version (as presented, say, in Griffiths-Schmid [1975]). In the context of quotients of general Hermitian symmetric spaces by discrete groups Γ, the relationship between the two will probably be interesting.

1. Tempered currents

The cohomology $H^{\bullet}(X, \mathbb{C})$ may be identified with the cohomology of the de Rham complex of C^{∞} forms on X, but for various technical reasons what I need to know is that it is also that of the complex of currents on X. Recall that a current is a continuous linear functional on the space of C_c^{∞} forms - see de Rham's book [1960]. I need to know also that the space of currents can be replaced by a certain subspace, that of tempered currents. This reduction is, as will be seen, a crucial step.

Let $\Omega \subseteq H$ be a region of the form $|x| \le a$, $y \ge b > 0$. A Schwartz function on Ω is a C^{∞} function $f\colon \Omega \to \mathbb{C}$ all of whose partial derivatives (in x and y) are $O(y^{-n})$ for all $n > 0$. Similarly a Schwartz form on Ω is one whose coefficients with respect to dx, dy or $dx \wedge dy$ are Schwartz functions.

If $\Gamma \subseteq SL_2(\mathbb{Q})$ is an arithmetic subgroup, let Γ_{∞} be the subgroup of all matrices in Γ of the form $\begin{pmatrix} 1 & x \\ 0 & 1 \end{pmatrix}$. This will be generated by a single element $\begin{pmatrix} 1 & x_0 \\ 0 & 1 \end{pmatrix}$. A basis of neighborhoods of the cusp $i\infty$ is formed by the images in $\Gamma\backslash H$ of the sets $y > y_0$. A Schwartz function on this neighborhood is one which lifts back to a Schwartz function on the region $y > y_0$. If P is any other cusp, then some element g of $SL_2(\mathbb{Z})$ will transform it to $i\infty$. Coordinates (x,y) near P one obtains by pull-back will be called <u>proper</u> coordinates. A Schwartz function in the neighborhood of P is one which corresponds in this way to a Schwartz function on $g\Gamma g^{-1}\backslash H$ in the neighborhood of the cusp $i\infty$. A <u>Schwartz function on all of</u> $\Gamma\backslash H$ is one which is C^∞ on all of $\Gamma\backslash H$ and a Schwartz function in the neighborhood of each cusp. Similarly for a Schwartz form. I will write the space of all such functions on $X = \Gamma\backslash H$ as $S(X)$, that of all Schwartz forms as $\Omega_S^\cdot(X)$.

<u>1.1. Theorem</u>. <u>The inclusion of</u> $\Omega_c^\cdot(X)$ in $\Omega_S^\cdot(X)$ <u>induces an isomorphism of cohomology</u>.

Here Ω_c is the de Rham complex of C_c^∞ forms on X. In other words, the compactly supported cohomology of X may be identified with the cohomology of the Schwartz forms on X. Results like this were first noticed, I believe, by Borel.

One proof goes like this: compactify X by adding whole circles at each cusp, not just points: one point at $i\infty$, for example, for every vertical line $x = \text{const.}$ (modulo x_0). This gives the Borel-Serre compactification of X, which is a manifold with boundary. As a coordinate system on this boundary circle, use (x, y^{-1}): this gives neighborhoods of the component isomorphic to $(\mathbb{R}/\mathbb{Z}) \times [0,c)$ for various c. The Schwartz functions in this neighborhood are precisely those which are C^∞ up to and including the boundary circle, but vanish of infinite order along the boundary. Similarly for Schwartz forms, and other cusps as well, using proper coordinates. Thus the notion of Schwartz form here is a special case of the following definition: let M be any manifold with boundary ∂M. A Schwartz form on M is one which is C^∞ everywhere on M, up to and including the boundary, and vanishes of infinite order on the boundary. If M is locally $\{(x_1,\ldots,x_n) \mid x_n \geq 0\}$ then a C^∞ form on M is one obtained by restriction from $\mathbb{R}^n$. In this situation, both forms with support in the interior of M, which I will call $\Omega_c^{\bullet}(\text{Int } M)$, and Schwartz forms are global sections of the fine sheaves on M defined by their germs. (The stalks of $\Omega_c^{\bullet}(\text{Int } M)$ are null on the boundary.) Since $\Omega_c^{\bullet} \subseteq \Omega_S^{\bullet}$, the natural obvious generalization of 1.1 to arbitrary M is a consequence of a standard sheaf-theoretic result (see Godement [1958], Theorem 4.6.2 on p. 178) and thus:

1.2. Sublemma. <u>At the boundary of</u> M, <u>the cohomology of the complex of germs of Schwartz forms is null.</u>

This is an easy consequence of almost any proof of the Poincaré Lemma.

Dual to the complex $\Omega_c^{\bullet}$ on X is that of currents on X. Define the <u>tempered</u> currents $C^{\bullet}$ to be those currents which extend continuously to $\Omega_S^{\bullet}$. Since $\Omega_c^{\bullet}$ is dense in $\Omega_S^{\bullet}$, the extension must be unique. It is known that the cohomology of the complex of currents is the same as the ordinary cohomology of X, and dual to the cohomology of compact support. As a consequence of Theorem 1.1:

1.3. Corollary. <u>The inclusion of the complex of tempered currents in that of all currents induces an isomorphism of the cohomology of this complex with</u> $H^{\bullet}(X, \mathbb{C})$.

It may seem that replacing, say, the ordinary C^{∞} forms on X by such a large space is a regression, but it will turn out that this is not the case.

2. A decomposition of the Schwartz spaces

One way to obtain forms on $X = \Gamma\backslash H$ is by the construction of Eisenstein series. In the beginning we will need only the simplest version.

First I will formulate a technical observation about cusps. Consider the region $y \geq Y$ in H and its transform by an element $g = \begin{pmatrix} a & b \\ c & d \end{pmatrix} \in SL_2(\mathbb{Z})$, with $c \neq 0$. This image is the interior of a circle which touches the real axis at $g(\infty) = a/c$, and using the formula

$$y(g(z)) = y(z)|cz + d|^{-2}$$

it is not hard to see that <u>the height of this circle is</u> $(c^2Y)^{-1}$.

Now let Γ be an arithmetic subgroup of $SL_2(\mathbb{Q})$. Its intersection with $SL_2(\mathbb{Z})$ has finite index in Γ, so that Γ preserves some lattice in $\mathbb{Z}^2$. Therefore some conjugate of Γ is contained in $SL_2(\mathbb{Z})$, and one may as well assume this to be the case for Γ itself. Hence for any $g \in \Gamma$ with $c \neq 0$, <u>the image of the region</u> $y \geq Y$ <u>under</u> g <u>lies inside the region</u> $y \leq Y^{-1}$.

Also, suppose P is some cusp of Γ not equivalent to $i\infty$ — i.e. in particular a rational number. Let C be a circle of height $Y > 0$ touching the real axis at P. Say $P = g(\infty)$ for some

$$g = \begin{pmatrix} a & b \\ c & d \end{pmatrix} \in SL_2(\mathbb{Z}) .$$

Then for any $\gamma \in \Gamma$ the γ-transform of C is another circle in H touching the real axis. Since $\gamma = \gamma g \cdot g^{-1}$, the previous remarks apply to show that <u>the height of</u> γC <u>is at most</u> $c^2 Y$.

Now let ω be a C^∞ differential form on H satisfying these conditions: (1) it is invariant under translations and (2) its support lies in a band $0 < y_1 \leq y \leq y_2 < \infty$. If $\omega = \phi$ is a function, for example, these conditions amount to requiring that ϕ be the lift to H of a C_c^∞ function in $y \in (0,\infty)$. The <u>Eisenstein series</u> associated to ω is the form

$$E_\omega = \sum_{\Gamma_\infty \backslash \Gamma} \gamma^*(\omega) .$$

Here Γ_∞ is the subgroup $\begin{pmatrix} \pm 1 & * \\ 0 & \pm 1 \end{pmatrix}$ of Γ stabilizing the cusp $i\infty$. (It is slightly bigger than the Γ_∞ I defined earlier.) This sum makes at least formal sense because $\gamma^*(\omega) = \omega$ for $\gamma \in \Gamma_\infty$. It makes real sense because, as the geometrical observations above imply, only a finite number of terms are non-zero. The form E_ω is clearly Γ-invariant, and even of compact support modulo Γ.

Let P be some other cusp of Γ. Let N_P be the subgroup of unipotent elements of $SL_2(\mathbb{R})$ fixing P. If ω is a C^∞ form on $\mathcal{H}$ which is N_P-invariant and of compact support modulo N_P, we can define similarly an Eisenstein series associated to ω.

Assign to $\mathcal{H}$ the $SL_2(\mathbb{R})$-invariant measure coming from the form $y^{-2}dxdy$, and X the quotient measure. Define

$$\Omega_2^{\cdot} = \mathbb{L}^2\text{-forms on } X$$

$$\Omega_{2,Eis}^{\cdot} = \text{closure in } \Omega_2^{\cdot}(X) \text{ of the subspace spanned by the compactly supported Eisenstein series associated to all cusps}$$

$$\Omega_{S,Eis}^{\cdot} = \Omega_S^{\cdot} \cap \Omega_{2,Eis}^{\cdot}$$

$$\Omega_{2,cusp}^{\cdot} = \text{orthogonal complement in } \Omega_2^{\cdot} \text{ of } \Omega_{2,Eis}^{\cdot}$$

$$\Omega_{S,cusp}^{\cdot} = \Omega_S^{\cdot} \cap \Omega_{2,cusp}^{\cdot} .$$

Then of course

$$(2.1) \quad \Omega_2^{\cdot} = \Omega_{2,Eis}^{\cdot} \oplus \Omega_{2,cusp}^{\cdot}$$

by definition. What is not at all obvious, however, is that there is a corresponding decomposition of the Schwartz forms as well. This is what will be proven in the rest of this section, except that for convenience of notation I will look only at functions, rather than forms, and will also assume Γ has one cusp (which I take to be $i\infty$). I will write $\mathbb{L}^2(X)$ instead of $\Omega_2^0(X)$, etc.

The elements of $\mathbb{L}^2_{cusp}$ have a useful characterization. Choose $\phi \in C_c^\infty(0,\infty)$. Then for $F \in \mathbb{L}^2(X)$,

$$
\begin{aligned}
(2.2) \qquad \langle F, E_\phi\rangle &= \iint_{\Gamma\backslash H} F(z)E_\phi(z)y^{-2}dxdy \\
&= \iint_{\Gamma\backslash H} F(z) \sum_{\Gamma_\infty\backslash\Gamma} \phi(y(\gamma(z)))y^{-2}dxdy \\
&= \iint_{\Gamma_\infty\backslash H} F(z)\phi(y(z))y^{-2}dxdy \\
&= \int_0^M dx\int_0^\infty F(x+iy)\phi(y)y^{-2}dy \\
&= M\int_0^\infty F_0(y)\phi(y)y^{-2}dy
\end{aligned}
$$

where

$$(2.3) \qquad F_0(y) = \frac{1}{M}\int_0^M F(x+iy)dx .$$

Here I take Γ_∞ to be generated by $\begin{pmatrix} \pm1 & M \\ 0 & \pm1 \end{pmatrix}$. Of course this makes sense only for almost all y. At any rate, $F \in \mathbb{L}^2_{cusp}$ if and only if this inner product (2.2) is null for all $\phi \in C_c^\infty(0,\infty)$: if and only if $F_0(y) = 0$.

The term F_0 is just one of several in the Fourier expansion of F:

$$(2.4) \qquad F(x+iy) = \sum F_n(y)e^{2\pi inx/M}$$

where

$$(2.5) \qquad F_n(y) = \frac{1}{M}\int_0^M e^{-2\pi inx/M}F(x+iy)dx.$$

The whole group $SL_2(\mathbb{R})$ acts on H, and the isotropy group of i is $K = SO(2)$. Hence functions on $\Gamma\backslash H$ may be identified with functions on $\Gamma\backslash SL_2(\mathbb{R})/K$, or with K-invariant functions on $\Gamma\backslash SL_2(\mathbb{R})$. Let f be a C_c^∞ function on $SL_2(\mathbb{R})$ which is bi-invariant with respect to K. Then for $F \in \mathbb{L}^2(\Gamma\backslash SL_2(\mathbb{R})/K)$ define

$$F*f(g) = \int_{SL_2(\mathbb{R})} F(gx)f(x)dx .$$

This lies again in $\mathbb{L}^2(\Gamma\backslash SL_2(\mathbb{R})/K)$, and is also C^∞. This technical remark allows us to prove

2.1. Lemma. <u>The intersection of</u> $C_c^\infty(X)$ <u>with</u> $\mathbb{L}^2_{cusp}(X)$ <u>is dense</u> <u>in</u> $\mathbb{L}^2_{cusp}(X)$.

Proof. Given $F \in \mathbb{L}^2(X)$ and Y large, let F_Y be the truncation of F in the neighborhood $y \geq Y$ of the cusp - that is to say $F_Y(x+iy) = 0$ for $y \geq Y$ while F and F_Y agree for $y < Y$ in X. As $Y \to \infty$, $F_Y \to F$ in $\mathbb{L}^2(X)$, clearly. Now given F_Y, choose functions $\phi \in C_c^\infty(K\backslash SL_2(\mathbb{R})/K)$ which are non-negative, of total integral one, and of support tending to K. Then $F_Y*\phi$ will be compactly supported in X and tend towards F_Y in $\mathbb{L}^2(X)$. If F lies in $\mathbb{L}^2_{cusp}(X)$ so does F_Y and $F_Y * \phi$, which proves the lemma.

This result yields a useful characterization of the decomposition of $F \in \mathbb{L}^2(X)$ according to (2.1). Say

(2.6) $$F = F_{cusp} + F_{Eis} .$$

Then (a) for every $\phi \in C^\infty_{c,cusp}(X)$

(2.7) $$\langle F,\phi\rangle = \langle F_{cusp},\phi\rangle$$

and (b) for every $\phi \in C^\infty_{c,Eis}(X)$

(2.8) $$\langle F,\phi\rangle = \langle F_{Eis},\phi\rangle .$$

It follows from Lemma 2.1 that these properties determine the decomposition (2.6). In other words, if $F_c \in \mathbb{L}^2_{cusp}(X)$ and $F_E \in \mathbb{L}^2_{Eis}$ such that (2.7)-(2.8) hold for F_c and F_E, respectively, then $F_c = F_{cusp}$ and $F_E = F_{Eis}$.

2.2 Theorem. <u>The space</u> $S(X)$ <u>is the direct sum of</u> $S_{cusp}(X)$ <u>and</u> $S_{Eis}(X)$. <u>If</u> $f \in S(X)$ <u>and</u> $f = f_{cusp} + f_{Eis}$ <u>according to (2.1), then both</u> f_{cusp} <u>and</u> f_{Eis} <u>lie in</u> $S(X)$.

Proof. It must be shown that $f \in S(X)$ implies $f_{cusp} \in S(X)$. Since $f \in S(X)$, so is every $\Delta^n f$, where Δ is the Laplacian of H:

$$\Delta = y^2(\partial^2/\partial x^2 + \partial^2/\partial y^2) .$$

The differential operator Δ commutes with $SL_2(\mathbb{R})$ on H, and in particular with N_∞, so that if $\phi \in C^\infty_{c,cusp}$ so is every $\Delta^n\phi$. Similarly for $\phi \in C^\infty_{c,Eis}$. From this and the remarks about (2.6)-(2.8) it follows that the components of $\Delta^n f$ according to (2.1) are the same as the distributions $\Delta^n(f_{cusp})$, $\Delta^n(f_{Eis})$. Thus, let $F = f_{cusp}$. Then $F \in \mathbb{L}^2(X)$ and so are all the $\Delta^n F$. By Sobolev's lemma applied locally, since Δ is elliptic, F is at least a C^∞ function on X. Also, the constant term of F is null.

Let now Y be a large positive number. Let $f(y)$ be a C^∞ function on $(0,\infty)$ with the property that it is null for $y \le Y$, 1 for $y \ge Y+1$. Then the function $\Phi = f\cdot F$ satisfies these three conditions: (1) it is a C^∞ function on $\Gamma_\infty\backslash H$ with support in the region $y \ge Y$; (2) all $\Delta^n\Phi$ lie in $\mathbb{L}^2(\Gamma_\infty\backslash H)$; (3) $\Phi_0 \equiv 0$.

2.3. Proposition. <u>Any function</u> Φ <u>satisfying these three conditions is, along with all its derivatives</u> $(\partial^{n+m}/\partial x^m \partial y^n)\Phi$, <u>rapidly decreasing at</u> $i\infty$.

The proof will be long, but elementary. A similar result is true when Γ is allowed to be any arithmetic subgroup of a reductive algebraic group, but the only proof I know in this case uses a well known, relatively recent, result of Dixmier-Malliavin [1978] in representation theory. The proof I give here is close to arguments about Sobolev spaces, and this result of Dixmier-Malliavin can often be used to replace such arguments.

First I will show that Φ itself vanishes rapidly.

Express

$$\Phi = \sum \Phi_n(y)e^{2\pi inx/M} .$$

Then

$$\|\Phi\|^2 = \iint |\Phi(z)|^2 y^{-2} dxdy \tag{2.7}$$

$$= M\cdot \sum \int_0^\infty |\Phi_n(y)|^2 y^{-2} dy$$

and

$$\Delta\Phi = \sum y^2(\Phi_n'' - 4\pi^2 n^2 M^{-2}\Phi_n)e^{2\pi inx/M} . \tag{2.8}$$

This motivates the next step. Let $D = D_\lambda$ be the operator $\phi \mapsto y^2(\phi'' - \lambda^2)$ (for $\lambda \geq 0$) and consider a function $\phi(y)$ satisfying (a) $\phi \in C^\infty(0,\infty)$ with support on $(1,\infty)$; (b) all the $y^{-1}D^\ell\phi$ lie in $\mathbb{L}^2(0,\infty)$. For example, each Φ_n satisfies these, with $\lambda = 2\pi|n|/M$. Let $\psi = D\phi$. Then

$$\phi'' - \lambda^2\phi = y^{-2}\psi$$

or

$$\phi = c_1e^{\lambda y} + c_2e^{-\lambda y} - \frac{1}{2\lambda}\int_1^\infty e^{-\lambda|y-x|}\psi(x)x^{-2}dx. \tag{2.9}$$

for constants c_1, c_2. The integral, ϕ, and $c_2e^{-\lambda y}$ are all $\mathbb{L}^2$, so c_1 must be 0 and

$$(2.10) \qquad c_2 = C = \frac{1}{2\lambda}\int_1^\infty e^{-\lambda x}\psi(x)x^{-2}dx .$$

The integral in (2.9) must now be estimated.

2.4 Lemma. _Let_ $\lambda_0 > 0$ _be given. There exist constants_ C_0, y_0 _depending only on_ λ_0 _such that_

$$\int_1^y e^{-\lambda(y-x)}x^{-n}dx \le \frac{C_0}{\lambda y^n}$$

for all $\lambda \ge \lambda_0$, $y \ge y_0$.

This follows from integration by parts

$$\int_1^y e^{-\lambda(y-x)}x^{-n}dx = \frac{1}{\lambda}\left[\frac{1}{y^n} - e^{-\lambda(y-1)} + n\int_1^y e^{-\lambda(y-x)}x^{-(n+1)}dx\right]$$

together with

$$\int_1^y e^{-\lambda(y-x)}x^{-(n+1)}dx \le \frac{1}{y^n}$$

2.5 Lemma. _For every integer_ $\ell > 0$ _there exists a constant_ C_ℓ _such that_

$$|\phi(y)| \le \frac{C_\ell\|D^\ell\phi\|}{\lambda^{1/2+\ell}y^{2\ell-1}}$$

whenever $\lambda \ge \lambda_0$, $y \ge y_0$.

This comes from a calculation I will sketch in a moment. First I will show how the Proposition follows. Each Φ_n satisfies the above conditions with $\lambda = |2\pi n|/M \ge 2\pi/M$ since $\Phi_0 = 0$. So take $\lambda_0 = 2\pi/M$ Then for y_0 as above and $y \ge y_0$, and any ℓ,

$$|\Phi(x+iy)| \le \sum|\Phi_n(y)|$$

$$\le \frac{C_\ell}{y^{2\ell-1}} \sum_{n\ne 0} \frac{\|D^\ell_{2\pi|n|/M}\Phi_n\|}{(2\pi|n|)^{1/2+\ell}}$$

$$\le \frac{C_\ell}{y^{2\ell-1}} \|\Delta^\ell\Phi\| \left[\sum_{n\ne 0}\left(\frac{1}{2\pi|n|}\right)^{1+2\ell}\right]^{1/2}$$

(last by Cauchy-Schwarz). Thus Φ is of rapid decrease on $y \ge 1$.

As for the derivatives of Φ, let A be the operator $y\partial/\partial x$, B be $y\partial/\partial y$. Then A and B form a Lie algebra with $[B,A] = A$ and

$$\Delta = A^2 + B^2 - B .$$

As a consequence every operator of degree $\le k$ in the universal enveloping algebra of this algebra can be written as a sum of elements in one of the forms $A^m\Delta^n$ or $BA^m\Delta^n$ with these also of degree $\le k$. Now if Φ satisfies the hypotheses of the Proposition, so do all $\Delta^n\Phi$ as well. Since

$$-\langle\Delta\Phi,\Phi\rangle = \|A\Phi\|^2 + \|B\Phi\|^2$$

both $A\Phi$ and $B\Phi$ will lie in $\mathbb{L}^2$. Since $A\Phi$ is, so is $\partial\Phi/\partial x$, and since Δ and $\partial/\partial x$ commute, so is each $\partial^n\Phi/\partial x^n$. Since this is rapidly vanishing, so is $A^n\Phi$. But also from the remarks above, $B\Phi$ will satisfy the same conditions as Φ (consider $\Delta^n B\Phi = B\Delta^n\Phi - [B,\Delta^n]\Phi$). Hence every $A^m B^n\Phi$ will be of rapid decrease.

It remains to prove 2.5. Consider

$$\phi = Ce^{-\lambda y} - \frac{1}{2\lambda}\int_1^y e^{-\lambda|y-x|}\psi(x)x^{-2}dx$$

where C is given by (2.10). Applying Cauchy-Schwarz we see that

$$(2.11)\qquad |c| \le \|\psi\|\cdot\frac{1}{2\lambda}\left(\int_1^\infty e^{-2\lambda y}y^{-2}dy\right)^{1/2}$$

$$\le \|\psi\|\cdot\frac{1}{2\lambda}\cdot\frac{1}{\sqrt{2\lambda}}\quad .$$

As for the integral, decompose it into two parts. By lemma 2.4,

$$(2.12)\qquad \left|\int_1^y e^{-\lambda(y-x)}\psi(x)x^{-2}dx\right| \le \|\psi\|\left(\int_1^y e^{-2\lambda(y-x)}x^{-2}dy\right)^{1/2}$$

$$\le \frac{C'}{\lambda^{1/2}y}\|\psi\|\ .$$

Also, more easily from Cauchy-Schwarz,

$$(2.13)\qquad \left|\int_y^\infty e^{-\lambda(x-y)}x^{-2}\psi(x)dx\right| \le \|\psi\|\frac{1}{\lambda^{1/2}y}\quad .$$

Combining (2.11)-(2.13) we see that there exists C_1 and y_0 such that

$$|\phi(y)| \le \frac{C_1\|D\phi\|}{\lambda^{3/2}y}$$

whenever $\lambda \ge \lambda_0$, $y \ge y_0$. The rest of the proof is a similar calculation, proceeding by induction from ℓ to $\ell+1$.

This concludes the proof of Theorem 2.2. Incidentally, the Schwartz spaces may be given the obvious Fréchet norms, and the closed graph theorem (see, for example, Trèves [1967], p. 165 and p. 173) says that the decomposition is continuous. Thus dual to the decomposition above we have one of tempered distributions: if $A(X)$ is the continuous dual of $S(X)$ then $A(X) = A_{cusp}(X) \oplus A_{Eis}(X)$, where (1) A_{cusp} is the dual of S_{cusp} and the annihilator of S_{Eis}; (2) A_{Eis} is the dual of S_{Eis} and the annihilator of S_{cusp}. I do not know how to prove it in an elementary way, and will not attempt a proof here, but it is true that S_{cusp} is dense in A_{cusp}. This has as a consequence that the Eisenstein series E_ϕ, as ϕ ranges over $C_c^\infty(0,\infty)$, are dense in S_{Eis} (which does not quite seem to be a trivial fact).

I might add that I do not know whether or not $C_c^\infty(X)$ possesses a decomposition into cuspidal and Eisenstein components.

The whole of the above theory does, however, extend to Schwartz forms, and also to arbitrary arithmetic $\Gamma \subseteq SL_2(\mathbb{Q})$. Combining this with results of §1, we see that the cohomology of X comes in two pieces: (1) that of the cuspidal currents $C^\bullet_{cusp}$ and (2) that of the Eisenstein currents $C^\bullet_{Eis}$. The first component is represented uniquely by the harmonic cuspidal automorphic forms on X. It is the second that we are specifically concerned with here.

3. Harmonic analysis of $\mathbb{L}^2(X)$

The Eisenstein series defined earlier converged for simple reasons. Now something classical:

3.1. Lemma. *Suppose* $\sigma > 1$. *The series*

$$E_\sigma(z) = \sum_{\Gamma_\infty \backslash \Gamma} y(\gamma(z))^\sigma$$

converges uniformly on compact subsets of $\mathcal{H}$.

This is well known, but I will include a proof (which generalizes easily to higher dimensions). Assume that Γ has no torsion elements. This is true of some subgroup of finite index if not of Γ itself, so it is no serious restriction. Let U be a small (non-Euclidean) disc around i, invariant under $K = SO(2)$. Then for $g \in SL_2(\mathbb{R})$, $z = g(i)$, the set $U_z = gU$ depends only on z.

For $u \in U$, the function $(c,d) \mapsto |cu + d|^2$ is a positive definite quadratic form. Hence for a fixed u the ratio

$|cu+d|^2/(c^2+d^2)$ lies in a fixed neighborhood of 1 for all $(c,d) \neq (0,0)$. Since U is relatively compact, there exists $\lambda > 1$ with

$$\lambda^{-1} < \frac{|cu+d|^2}{c^2+d^2} < \lambda$$

for all $(c,d) \neq (0,0)$, $u \in U$. Also, of course, one may find $K > 1$ such that

$$K^{-1} < y(u) < K$$

for all $u \in U$. From these remarks it is straightforward to prove that

$$K^{-1}\lambda^{-1} < \frac{y(u)}{y(z)} < K\lambda$$

for all $u \in U_z$.

Now for a given z choose U small enough so that the sets $\{\gamma U_z | \gamma \in \Gamma\}$ are all disjoint. Choose Y so $y \geq Y$ on U_z. Then for $\gamma \in \Gamma - \Gamma_\infty$, $y \leq Y^{-1}$ on γU_z. If Γ_∞ is generated by $\begin{pmatrix} \pm 1 & M \\ 0 & \pm 1 \end{pmatrix}$ then some Γ_∞-translate of γU_z will lie in the band $|x| \leq M + Y^{-1}$. So

$$E_\sigma(z) = y(z)^\sigma + \sum_{\Gamma_\infty \backslash \Gamma - \Gamma_\infty} y(\gamma(z))^\sigma$$

and

$$|E_\sigma - y^\sigma| \leq (\text{meas } U)^{-1}(K\lambda)^{2\sigma} \iint_{\substack{y \leq Y^{-1} \\ |x| \leq M+Y^{-1}}} y^{\sigma-2}\,dx\,dy$$

which is finite if $\sigma > 1$.

A similar argument and the geometrical observations in §2 will prove further:

3.2. Proposition. <u>For any</u> $s \in \mathbb{C}$ <u>with</u> $\mathrm{Re}(s) > 1$, <u>the series</u>

$$E_s(z) = \sum_{\Gamma_\infty\backslash\Gamma} y(\gamma(z))^s$$

converges to a real-analytic function on $X = \Gamma\backslash H$, with the properties

(1) In any region $y \geq Y$, $E_s(z) - y^s = O(y^{-(s-1)})$:

(2) In the neighborhood of any cusp of Γ not equivalent to $i\infty$, $E_s = O(y^{s-1})$;

(3) $\Delta E_s = s(s-1)E_s$.

The second property refers to the inside of any circle in H just touching the real axis at P.

The property (3) follows from the identity

$$\Delta y^s = s(s-1)y^s$$

together with the fact that Γ and Δ commute.

If Y is a large positive number and χ_Y is the characteristic function of $[Y,\infty)$ then let $E_{Y,s}$ be the Eisenstein series

$$E_{Y,s}(z) = \sum_{\Gamma_\infty\backslash\Gamma} \chi_Y(y(\gamma(z)))y(\gamma(z))^s$$

For any given z, only one term is non-zero. And in the region $y \geq Y_*$ it agrees with $\chi_Y \cdot y^s$ for large enough Y_*. One immediate consequence of 3.2 is that $E_s - E_{Y,s}$ always lies in $\mathbb{L}^2(X)$. This - as Colin de Verdière [1981] points out - and the equation $\Delta E_s = s(s-1)E_s$ are the characteristic properties of E_s. Indeed, he shows that it is almost trivial to deduce from this that E_s possesses an analytic continuation into the region $Re(s) > 1/2$, $s \notin (1/2, 1]$. With only a little more work he goes on to imitate the classical theory of Sturm-Liouville operators on infinite intervals and to prove:

3.3. Theorem. The Eisenstein series continues meromorphically to all of $\mathbb{C}$. The constant term of E_s at the cusp $i\infty$ is of the form $y^s + c_\infty(s)y^{1-s}$, at a cusp P not equivalent to $i\infty$ of the form

$c_P(s)y^{s-1}$, <u>where the</u> $c_P(s)$ <u>have poles in the region</u> $\mathrm{Re}(s) \geq 1/2$ <u>at most on</u> $(1/2, 1]$.

I am not sure of the history of this result. There are many expositions, such as Kubota [1973], but the paper of Colin de Verdière is extraordinarily lucid, and recommended highly. Similar techniques probably extend to what might be called the rank-one situation for more general rational reductive groups (see Langlands [1966]). I will take this result as given, without further explanation, and proceed from it.

I will again assume for convenience from now on that Γ has but one cusp. Let $c(s) = c_\infty(s)$. The function E_{1-s} has the same Δ-eigenvalue as E_s and its constant term is $y^{1-s} + c(1-s)y^s$. Hence $c(1-s)E_s = E_{1-s}$, and $c(s)c(1-s) = 1$. The function E_s is real for real s, so $c(\overline{s}) = \overline{c(s)}$. Hence for $\mathrm{Re}(s) = 1/2$, $|c(s)| = 1$. (It is this sort of thing that is responsible for the lack of poles of $c(s)$ on $1/2 + i\mathbb{R}$ in general.)

The first step in the further analysis of E_s is a formula for the inner product of two <u>truncated</u> Eisenstein series. These already play a role in proving 3.3. What are they? Let Y be a large positive number. Then the <u>truncation</u> $T^Y E_s$ of E_s at Y in $y \geq Y$ is the difference between E_s and $[E_{Y,s} + c(s)E_{Y,1-s}]$. Roughly speaking, in other words, it is equal to E_s in the inside of X but equal to E_s less its constant term when $y \geq Y$. A slight modification of Proposition 2.3 shows that $T^Y E_s$ and all its derivatives are rapidly decreasing as $y \to \infty$. In particular, $T^Y E_s \in \mathbb{L}^2(X)$. Another way to define $T^Y E_s$:

$$\text{let } M_s(y) = \begin{cases} -c(s)y^{1-s} & y \geq Y \\ y^s & y < Y . \end{cases}$$

Then $T^Y E_s$ is the Eisenstein series $E_{M(s)}$:

$$T^Y E_s(z) = \sum_{\Gamma_\infty \backslash \Gamma} M_s(y(\gamma(z))),$$

when this series converges, and its meromorphic continuation when not.

3.4. Proposition. _For_ $s, t \in \mathbf{C}$ _with_ $s-t \neq 0$, $s+t-1 \neq 0$,

$$\langle T^Y E_s, T^Y E_t \rangle = \frac{Y^{s+t-1} - c(s)c(t)Y^{1-s-t}}{s+t-1}$$

$$- \frac{c(s)Y^{-s+t} - c(t)Y^{s-t}}{s-t} .$$

Proof. This is a variant of Green's Theorem, since both E_s and E_t are eigenfunctions of Δ. But the simplest way to prove it is to use a generalization of the calculation (2.2):

$$\begin{aligned} \langle T^Y E_s, T^Y E_t \rangle &= \langle E_s, T^Y E_t \rangle \\ &= \langle E_s, E_{M_t} \rangle \\ &= \langle E_{s,0}, M_t \rangle . \end{aligned}$$

This last is a sum of two integrals:

$$(3.1) \qquad \int_0^Y [y^s + c(s)y^{1-s}]y^{t-2}dy$$

and

$$(3.2) \qquad -\int_Y^\infty [y^s + c(s)y^{1-s}]c(t)y^{1-t-2}dy .$$

These make sense as long as $\mathrm{Re}(s+t) > 1$, $\mathrm{Re}(t) > \mathrm{Re}(s)$. The formula in (3.4) is an easy calculation.

3.5. Corollary. When $t = 1-s$ but $t \neq s$, with $\mathrm{Re}(s) = 1/2$

$$\langle T^Y E_s, T^Y E_t \rangle = 2 \log Y - c'(s)c(1-s) - \frac{c(s)Y^{1-2s} - c(1-s)Y^{2s-1}}{1-2s} .$$

Proof. Take a limit in 3.4. Setting $t = \bar{s}$ in these:

3.6. Corollary. For $s \notin \mathbb{R}$, with $\mathrm{Re}(s-1/2) = \sigma$ and $\tau = \mathrm{Im}(s) \neq 0$,

$$\|T^Y E_s\|^2 = \frac{1}{2\sigma} [Y^{-2\sigma} - |c(s)|^2 Y^{2\sigma}] - \frac{1}{2i\tau} [c(s)Y^{-2i\tau} - \overline{c(s)}Y^{2i\tau}]$$

when $\sigma > 0$ and

$$= [2 \log Y - c'(s)\overline{c(s)}] - \frac{1}{2i\tau} [c(s)Y^{-2i\tau} - \overline{c(s)}Y^{2i\tau}]$$

when $\sigma = 0$.

We will not need the case $\tau = 0$.

The left-hand side is always positive, of course. This translates to:

$$|\tau c + i\sigma X| \leq |sX|$$

or

$$|c + \frac{i\sigma X}{\tau}| \leq |X| \sqrt{1+\frac{\sigma^2}{\tau^2}} .$$

The first implies that in any region $0 < \sigma \leq \sigma_0$, $|\tau| \leq \tau_0$, τc is bounded. The second implies that in a region $0 < \sigma \leq \sigma_0$, $|\tau| \geq \tau_0 > 0$, c is bounded. Looking back at 3.6, this gives:

3.7. Proposition. (a) <u>The function</u> E_s <u>has no poles in the region</u> $\mathrm{Re}(s) \geq 1/2$ <u>except on</u> $(1/2, 1]$;

(b) <u>Any poles on</u> $(1/2, 1]$ <u>are simple</u>;

(c) <u>In any region</u> $1/2 \leq \sigma \leq \sigma_0$, $|\tau| \geq \tau_0 > 0$, c <u>is bounded</u>;

(d) <u>In any region</u> $1/2 < \sigma \leq \sigma_0$, $|\tau| \geq \tau_0 > 0$, <u>there exists a constant</u> C <u>with</u>

$$\|T^Y E_s\| \leq C/\sqrt{\sigma} .$$

The general principle behind (a) is that the behavior of $c(s)$ controls that of E_s, but in practice this is often a subtle matter.

If σ is one of the poles of $c(s)$ in $(1/2, 1]$, then $\mathrm{Res}_\sigma E_s$ will be an eigenfunction of Δ with eigenvalue $\sigma(\sigma-1)$, and its constant term at $i\infty$ will be $(\mathrm{Res}_\sigma\, c(s))y^{1-\sigma}$. This is square-integrable in the neighborhood of $i\infty$, and in fact $E^*_\sigma = \mathrm{Res}_\sigma E_s$ is square-integrable on all of X.

There is always a pole at $s = 1$, and E^*_1 is a constant function. If Γ is arithmetic, or as we shall see later if it has only one cusp, then this is the only pole in $(1/2, 1]$. In the arithmetic case $c(s)$ may be evaluated explicitly in terms of Dirichlet L-functions (see for example p.46 of Kubota[1973]). In this case also the estimate 3.7(d) may be improved, but the best possible estimate is not yet known, and seems rather deep. These examples also show that where $\mathrm{Re}(s) < 1/2$ both $c(s)$ and E_s might be very badly behaved. This is not so important because of the functional equation for E_s.

Our next step will be to find the decomposition of $L^2_{\mathrm{Eis}}(X)$ in terms of Eisenstein series. I follow Langlands [1966].

First I must recall some results of harmonic analysis on the multiplicative group of positive real numbers. Let ϕ be in $C^\infty_c(0,\infty)$. Its multiplicative Fourier transform is entire:

$$(3.3) \qquad \Phi(s) = \int_0^\infty x^{-s-1}\phi(x)\,dx \qquad (s \in \mathbb{C}) .$$

A theorem of Paley-Wiener characterizes completely such functions, but all we need to know is that $\Phi(\sigma+it)$, considered as a function of τ, lies in the Schwartz space, uniformly in σ. The function ϕ is computed in terms of Φ by the formula

$$(3.4) \qquad \phi(y) = \frac{1}{2\pi i}\int_{\sigma-i\infty}^{\sigma+i\infty} \Phi(s)y^s\,ds$$

where σ is any real number. Given two functions $\phi, \psi \in C_c^\infty(0,\infty)$ with Fourier transforms Φ, Ψ, the product again lies in $C_c^\infty(0,\infty)$ and has Fourier transform $(2\pi i)^{-1}\Phi * \Psi$ (as can be seen by applying the inverse transform to the latter). In particular

$$(3.5) \qquad \int_0^\infty \phi(x)\psi(x)x^{-1}dx = \frac{1}{2\pi i}\,\Phi * \Psi(0)$$

$$= \frac{1}{2\pi i}\int_{\sigma-i\infty}^{\sigma+i\infty} \Phi(t)\Psi(-t)\,dt .$$

I should remark here that the convolution I use here is defined by an integral over a <u>vertical</u> line. These formulae extend to a more general situation: suppose $\phi(x)$ is not assumed to vanish at $x = 0$, but only to be bounded on left half-lines. Then $\Phi(s)$ may be defined only for s with $\mathrm{Re}(s) < 0$. All the above formulae still hold, however, as long as $\sigma < 0$. Equivalently, if ψ lies in this situation, one must assume $\sigma > 0$ in (3.5).

Continue to suppose $\phi \in C_c^\infty(0,\infty)$. Then at least formally it follows from 3.4 that

$$E_\phi = \frac{1}{2\pi i}\int_{\sigma-i\infty}^{\sigma+i\infty} \Phi(s)E_s\,ds .$$

This is easily justified for $\sigma > 1$ (where E_s behaves nicely). Taking constant terms, we have (for $\sigma > 1$)

$$E_{\phi,0}(y) = \frac{1}{2\pi i}\int_{\sigma-i\infty}^{\sigma+i\infty} \Phi(s)\,[y^s + c(s)y^{1-s}]ds .$$

This can be rewritten as

$$\frac{1}{2\pi i}\int_{\sigma-i\infty}^{\sigma+i\infty} [\Phi(s) + \check{c}(1-s)\Phi(1-s)]y^s ds$$

where now $\sigma < 0$. In other words, <u>the Fourier transform of</u> $E_{\phi,0}$ <u>is analytic for</u> $\mathrm{Re}(s) < 0$ <u>and equal to</u> $[\Phi(s) + c(1-s)\Phi(1-s)]$ <u>in that region</u>.

Now choose $\phi, \psi \in C_c^\infty(0,\infty)$ and consider the inner product $\langle E_\phi, E_\psi\rangle$. According to the calculation in (2.2),

$$\langle E_\phi , E_\psi\rangle = M\int_0^\infty \phi(y)E_{\psi,0}(y)y^{-2}dy$$

$$= M\int_0^\infty \phi(y)\,[E_{\psi,0}(y)y^{-1}]y^{-1}dy$$

which by (3.5) and an elementary shift

$$= \frac{M}{2\pi i}\int_{\sigma-i\infty}^{\sigma+i\infty} \Phi(s)\,[c(s)\Psi(s) + \Psi(1-s)]ds$$

with $\sigma > 1$. This contour, according to Theorem 3.3, may be moved to $\sigma = 1/2$. As this is done, a number of residues of $c(s)$ are encountered, say at $s_1, s_2, \ldots, s_n$. Let $c^*(s_i) = \mathrm{Res}_{s_i} c(s)$. Then the inner product is the sum of

(3.6) $$\sum Mc^*(s_i)\Phi(s_i)\Psi(s_i)$$

and

$$\frac{M}{2\pi i}\int_{\sigma=1/2}\Phi(s)[\Psi(1-s) + c(s)\Psi(s)]ds \ .$$

Since $c(s)c(1-s) = 1$, this integral is also

$$\frac{M}{4\pi i}\int_{\sigma=1/2}[\Phi(s) + c(1-s)\Phi(1-s)][\Psi(1-s) + c(s)\Psi(s)]ds$$

$$= \frac{M}{4\pi i}\int_{\sigma=1/2}c(1-s)[\Phi(1-s) + c(s)\Phi(s)][\Psi(1-s) + c(s)\Psi(s)]ds \ .$$

If $\psi = \overline{\phi}$, these equations give

$$\text{(3.7)} \qquad \|E_\phi\|^2 = \sum[Mc^*(s_i)]|\Phi(s_i)|^2 + \frac{M}{4\pi}\int_{-\infty}^{\infty}|\Phi(\tfrac{1}{2} - i\tau) + c(\tfrac{1}{2} + i\tau)\Phi(\tfrac{1}{2} + i\tau)|^2 d\tau \ .$$

How is this to be interpreted? Given any $f \in C_c^\infty(X)$ define its Fourier-Eisenstein transform $F(s)$ by the formula

$$\text{(3.8)} \qquad F(s) = \langle f, E_s\rangle \ .$$

It will be meromorphic in s, and satisfy the functional equation $F(s) = c(s)F(1-s)$ since $E_s = c(s)E_{1-s}$. Its residue at a pole s_i is $\langle f, E^*(s_i)\rangle = F^*(s_i)$. If $f = E_\phi$, then

$$F_\phi(s) = M[\Phi(1-s) + c(s)\Phi(s)]$$

and

$$F_\phi^*(s_i) = Mc^*(s_i)\Phi(s_i) \ .$$

All in all, (3.7) thus may be read that if $f = E_\phi$ then

(3.9) $$\|f\|^2 = \sum [Mc^*(s_i)]^{-1} |F_\phi^*(s_i)|^2 + \frac{1}{4\pi M}\int_{-\infty}^{\infty} |F_\phi(\tfrac{1}{2} + i\tau)|^2 d\tau,$$

while (3.6) may be read as

(3.10) $$\langle E_\phi , E_\psi\rangle = \sum \Phi(s_i) F_\psi^*(s_i) + \frac{1}{2\pi i}\int_{\sigma=1/2} \Phi(s) F_\psi(s)\, ds \quad .$$

Since $|c(s)| = 1$, multiplication by $c(s)$ is an isometry of $L^2(1/2 + i\mathbb{R})$ with itself, and the map $\Phi \mapsto \Lambda\Phi$ where

$$\Lambda\Phi(s) = \frac{1}{2}[\Phi(s) + c(s)\Phi(1-s)] ,$$

is a projection onto the closed subspace of F with $F(s) = c(s)F(1-s)$. Now $\mathbb{L}^2_{Eis}$ is by definition the closure of the E_ϕ with $\phi \in C_c^\infty(0,\infty)$ and $\mathbb{L}^2(1/2 + i\mathbb{R})$ is the closure of the functions $\Phi(s)$ which are Fourier transforms of such ϕ, so that the functions $\Lambda\Phi$ are dense in $\Lambda\mathbb{L}^2(1/2 + i\mathbb{R})$. The Fourier-Eisenstein transform $F(s)$ of E_ϕ is $(\Lambda\Phi)(1-s)$. Therefore Equation (3.8) implies

3.8 Theorem. _The map_

$$E_\phi \mapsto \{F^*(s_i),\ F(\tfrac{1}{2} + i\tau)\}$$

where F _is the Fourier-Eisenstein transform of_ E_ϕ , _extends to an isometry of_ $\mathbb{L}^2_{Eis}(X)$ _with_

$$\mathbb{C}^n \oplus \Lambda\mathbb{L}^2(\tfrac{1}{2} + i\mathbb{R})$$

where the norm on the latter is given by

$$\sum [Mc^*(s_i)]^{-1} |F^*(s_i)|^2 + \frac{1}{4\pi M} \int_{-\infty}^{\infty} |F(\tfrac{1}{2} + i\tau)|^2 d\tau \ .$$

The isomorphism given by this theorem is difficult to work with directly. However, it is easy to see that there is a very useful characterization of the Fourier-Eisenstein transform of $f \in \mathbb{L}^2_{Eis}(X)$. Say $\phi \in C_c^\infty$ with multiplicative Fourier transform Φ, and suppose $\{F^*(s_i), F(1/2 + i\tau)\}$ is the transform of f. Then from the definition it follows (see Equation (3.10)) that

$$\text{(3.9)} \qquad \langle f, E_\phi \rangle = \sum \Phi(s_i) F^*(s_i) + \frac{1}{2\pi i} \int_{\sigma=1/2} \Phi(s) F(s)\, ds \ .$$

Furthermore, the transform of f is determined by the condition that this holds for all $\phi \in C_c^\infty(0,\infty)$.

Even with this criterion, however, it seems not quite trivial to answer the following question: suppose $f \in C_c^\infty(X)$. The $F(s) = \langle f, E_s \rangle$ is meromorphic in s, with poles only where the poles of $c(s)$ are. Is the restriction of F to $1/2 + i\mathbb{R}$ equal to the continuous part of its $\mathbb{L}^2$ Fourier-Eisenstein transform? This, and something slightly better, will be dealt with in the next section.

The inverse of the isomorphism above is not hard to describe explicitly. Suppose $F \in \mathbb{L}^2(1/2 + i\mathbb{R})$, and consider for $T > 0$ the function

$$f_T = \frac{1}{4\pi M i} \int_{1/2-iT}^{1/2+iT} F(1-s) E_s\, ds \ .$$

It is certainly a C^∞ function on X. Thus

$$\langle f_T\, ,\, E_\phi\rangle = \frac{1}{4\pi Mi}\int_{1/2-iT}^{1/2+iT} F(1-s)\langle E_\phi\, ,\, E_s\rangle ds$$

$$= \frac{1}{4\pi Mi}\int_{1/2-iT}^{1/2+iT} MF(1-s)[\Phi(1-s) + c(s)\Phi(s)]ds$$

$$= \frac{1}{4\pi i}\int_{1/2-iT}^{1/2+iT} \Phi(s)[F(s) + c(s)F(1-s)]ds$$

$$= \frac{1}{2\pi i}\int_{1/2-iT}^{1/2+iT} \Phi(s)\Lambda F(s)ds\ .$$

Hence, applying the criterion above, we see that the transform of f_T is $\Lambda F(s)$, truncated at $1/2 \pm iT$. Therefore as $T \to \infty$ the function f_T converges in $L^2(X)$ to a function f whose transform is ΛF.

One final remark: the formula (3.8) requires $c^*(s_i)$ to be positive at all s_i. However, we know that the poles of $c(s)$ are simple, so if $n \geq 2$ the signs of the $c^*(s_i)$ must alternate. Hence there can be only one pole.

4. A Paley-Wiener Theorem

What is to be done now is characterize the Fourier-Eisenstein transforms of functions in $S(X)$ - or, rather S_{Eis}.

Let $f \in S(X)$. Define

$$F(s) = \langle f, E_s\rangle\ .$$

Because f vanishes rapidly at $i\infty$ while E_s is of moderate growth, this is well defined whenever E_s is, and in fact it is meromorphic as a function of s. It has simple poles in the region $\mathrm{Re}(s) \geq 1/2$, at most where the poles of $c(s)$ are. Let $F^*(s_i)$ be its residue at s_i, for $s_i \in (1/2, 1]$. (I am ignoring my last remark in §3.)

4.1 Lemma. *If $f \in S_{Eis}(X)$ then the $F^*(s_i)$ together with the restriction of F to $1/2 + i\mathbb{R}$ comprise the $\mathbb{L}^2$ transform of f.*

Proof. According to a remark made at the end of §2, I can find a sequence of $\phi \in C_c^\infty(0,\infty)$ with the E_ϕ converging to f in $S(X)$. Thus $F_\phi(s) \to F(s)$ for all $s \in \mathbb{C}$, where $F_\phi(s) = \langle E_\phi , E_s \rangle$. Similarly for the residues. This convergence is uniform on compact subsets in $\mathbb{C}$. But also by definition $F_\phi(s)$ converges in $\mathbb{L}^2(1/2 + i\mathbb{R})$ to the continuous $\mathbb{L}^2$-transform of f, and the $F_\phi^*(s_i)$ converge to the discrete part of the transform. It is easy to deduce the claim.

In view of this result, which says that the $\mathbb{L}^2$-transform of $f \in S_{Eis}$ is determined by the meromorphic function $F(s)$, I will call $F(s)$ the Fourier-Eisenstein transform of f, for every $f \in S(X)$. Of course if $f \notin S_{Eis}(X)$ then one cannot hope to recover f from F, but we will see that at least one can recover the component of f in $S_{Eis}(X)$.

If $f \in S(X)$ then so is every $\Delta^n f$. The transform of $\Delta^n f$ is $s^n(s-1)^n F(s)$, if F is the transform of f.

4.2. Theorem. *Suppose $f \in S(X)$ and for each $n \geq 0$ let $F_n(s)$ be the Fourier-Eisenstein transform of $\Delta^n f$. Then for every $n \geq 0$:*

(1) *$F_n(s)$ is meromorphic in all of $\mathbb{C}$;*

(2) $F_n(s) = c(s)F_n(1-s)$;

(3) *The poles of $F_n(s)$ in the region $\mathrm{Re}(s) \geq 1/2$ are among the poles of $c(s)$, and are simple;*

(4) *The restriction of $F_n(s)$ to $\mathrm{Re}(s) = 1/2$ is square-integrable;*

(5) *In any region $|\tau| \geq \tau_0 > 0,\ 0 < \sigma \leq \sigma_0$, $F_n(s) = O(1/\sqrt{\sigma})$*

Conversely, suppose given $f(s)$ such that all the $F_n(s) = s^n(s-1)^n F_0$

<u>satisfy these conditions.</u> <u>Then</u> $F(s)$ <u>is the Fourier-Eisenstein transform of some</u> $f \in S(X)$.

In condition (5), $\sigma = \mathrm{Re}(s) - 1/2$, $\tau = \mathrm{Im}(s)$.

The proof that every $F_n(s)$ satisfies (1)-(5) is simple, given what has been done already. The only slightly tricky point is (5). For this, write

$$E_s = T^Y E_s + R_s$$

where Y is a large positive number, $T^Y E_s$ is the truncation of E_s constructed in §3, and the remainder R_s will be null outside the region $y \geq Y$, and $y^s + c(s)y^{1-s}$ inside it. Then

$$F(s) = \langle E_s , f\rangle = \langle T^Y E_s , f\rangle + \langle R_s , f\rangle .$$

By 3.7(d), $\|T^Y E_s\| = O(1/\sqrt{\sigma})$ in the region $0 < \sigma \leq \sigma_0$, $|\tau| \geq \tau_0 > 0$, so that by Cauchy-Schwarz it only remains to show that $\langle R_s , f\rangle = O(1/\sqrt{\sigma})$ there. Because $c(s)$ is bounded in that region and R_s is invariant under translations $z \mapsto z+x$, it suffices to show that the integral

$$\int_Y^\infty y^{s-2} f_0(y)\,dy$$

is bounded in regions $|\sigma| \leq \sigma_0$. Since f is rapidly decreasing at $i\infty$, this is clear.

The proof of the converse is more difficult.

Given $F(s) = F_0(s)$ satisfying the conditions of the theorem, the candidate for f is more or less clear from remarks at the end of §3: define it to be the sum of

$$\lim_{T\to\infty} \frac{1}{4\pi M i} \int_{1/2-iT}^{1/2+iT} F(1-s) E_s\, ds$$

and

$$\sum [Mc^*(s_i)]^{-1} F^*(s_i) E^*_{s_i} \quad .$$

At the beginning, all we know is that f and indeed all the $\Delta^n f$ lie in $\mathbb{L}^2(X)$, and their $\mathbb{L}^2$-transforms are given by F. What we want to prove is that $f \in S(X)$. This is simply a condition on the behaviour of f near the cusp $i\infty$. Now near $i\infty$ the function f may be written as $f_0 + (f-f_0)$. Proposition 2.3 may be applied to see that $(f-f_0)$ and all its derivatives vanish rapidly at $i\infty$, so that all that remains is to show the same for the constant term f_0.

Note that so far we have used almost none of the properties (1)-(5).

What is f_0? It is the sum of

$$\lim_{T\to\infty} \frac{1}{4\pi i} \int_{1/2-iT}^{1/2+iT} F(1-s)\,[y^s + c(s)y^{1-s}]\,ds \tag{4.1}$$

and

$$\sum F^*(s_i) y^{1-s_i} \quad . \tag{4.2}$$

Note that the first is a function, call it $\phi(y)$, such that $\phi(y)y^{1/2}$ lies in $L^2(0,\infty)$ (with respect to the multiplicatively invariant measure dy/y). Because of the functional equation (2) for $F(s)$, the integral is the same as

$$\lim_{T\to\infty} \frac{1}{2\pi i} \int_{1/2-iT}^{1/2+iT} F(s)y^{1-s}\,ds \quad .$$

The aim now is to move the line of integration to the right. This is possible precisely because of condition (5) applied, say to $F_2(s)$, since $1/\sqrt{\sigma}$ is integrable on any interval $(0,\varepsilon)$. Therefore in fact we can keep moving the line of integration as far to the right as we want, except that we pick up some residues. These amount to

$$- \sum F^*(s_i) y^{1-s_i} ,$$

and of course cancel out the summand (4.2). So all in all, the constant term of f is equal to

$$\frac{1}{2\pi i} \int_{\sigma - i\infty}^{\sigma + i\infty} F(s) y^{1-s} d|s| = O(y^{1-\sigma}).$$

This is true for all the F_n as well, so that f_0 and all the $\Delta^n f_0$ are rapidly vanishing at the cusp. But on constant terms Δ acts as $y^2 \partial/\partial y^2 = (y\partial/\partial y)^2 - (y\partial/\partial y)$, so that it is easy to see all the derivatives $(y\partial/\partial y)^n f_0$ vanish rapidly also. This is enough to get the same condition on the $(\partial/\partial y)^n f_0$.

I have talked only about functions, not forms of higher degree, and have assumed Γ has only one cusp in the last part of §3 and so far in §4, but similar techniques will apply without these restrictions.

Note that by starting with $f \in S(X)$ and reconstructing a function from $F(s)$, we obtain the Eisenstein component of f. This gives a new, more explicit, proof of the decomposition $S = S_{cusp} \oplus S_{Eis}$, but of course the key technical result is the same in both proofs.

At any rate, the consequence we will need in analyzing cohomology is this:

4.3. Corollary. <u>On the Eisenstein component of</u> $\Omega_S^{\bullet}$ <u>each operator</u> Δ^n <u>acts injectively, with closed image and finite-dimensional cokernel</u>.

This is a simple deduction from the generalization of 4.2, since Δ becomes multiplication by $s(s-1)$ on the Fourier-Eisenstein transform.

4.4. Theorem. <u>The operator</u> Δ <u>is surjective on the Eisenstein component of the tempered currents</u>.

This follows from 4.3 by duality.

It is either of 4.3 or 4.4 which is the basic contribution of analysis to the Hodge theory we are looking for.

5. Applications to cohomology

There will be two steps in which the Paley-Wiener theorem will contribute to analyzing cohomology. The first is:

5.1. Theorem. The cohomology of the complex made up of the Eisenstein component of tempered currents is the same as that of its subcomplex of currents annihilated by some power of the Laplacian.

This is the variant of the Hodge theorem I was referring to earlier. Note that it does not identify the cohomology with a subspace of the currents - it does not find unique representatives for the cohomology classes, but only allows a kind of reduction of the calculation. We will see later to what extent this is useful.

Proof. Let $C^{\bullet}[\Delta]$ be the subspace of tempered currents annihilated by some power of Δ. I claim that Δ induces an isomorphism of the quotient $C^{\bullet}/C^{\bullet}[\Delta]$ with itself. Injectivity is an immediate consequence of the definition of $C^{\bullet}[\Delta]$. Surjectivity follows from Theorem 4.4.

Consider the short exact sequence

$$0 \to C^{\bullet}[\Delta] \to C^{\bullet} \to C^{\bullet}/C^{\bullet}[\Delta] \to 0.$$

The de Rham differential induces a differential on each of these complexes (note that Δ and d commute). The result we want will follow from the fact that the cohomology of $C^{\bullet}/C^{\bullet}[\Delta]$ must be trivial. This is a consequence of what has just been proved, together with the equation

$$\Delta = d\delta + \delta d$$

or

$$Id = d(\Delta^{-1}\delta) + (\Delta^{-1}\delta)d$$

on $C^{\cdot}/C^{\cdot}[\Delta]$, which says that $\Delta^{-1}\delta$ is a homotopy operator on this complex.

Note that from an analyst's point of view this proof is a bit peculiar, since the spaces in the short exact sequence we used are not topological vector spaces. The space $C^{\cdot}[\Delta]$ is in some sense an algebraic object.

Only in some weak sense, however, because the Paley-Wiener Theorem must be used again to identify it. Once again to illustrate the situation I will look at functions only. How can I describe the tempered distributions annihilated by some Δ^n? The answer is in terms of Eisenstein series. Note that since $(\Delta - s(s-1))^n(y^s \log^{n-1} y) = 0$, so also is $d^{n-1}/ds^{n-1}(E_s)$ annihilated by $(\Delta - s(s-1))^n$. To obtain those annihilated by Δ^n, we must set $s = 0$, $s = 1$. There are two problems, however: there is a pole of E_s at $s = 1$, and $s = 0$ lies in the "forbidden region" $Re(s) < 1/2$. Neither of these is a serious difficulty, as we will see.

Suppose $P_1, \ldots, P_n$ are the distinct cusps of X, and say the corresponding M-numbers are $M_1, \ldots, M_n$. What I mean by this is that if N_i is the subgroup of unipotent elements fixing P_i, then M_i is the index of $\Gamma \cap N_i$ in $SL_2(\mathbb{Z}) \cap N_i$. Then it is easy to see that if $E_s^{(1)}, \ldots, E_s^{(n)}$ are the Eisenstein series associated to these cusps, the linear combination $\sum a_i E_s^{(i)}$ will have residue at $s = 1$ equal to the constant

$$\sum a_i M_i c^{(i)*}(1)$$

where $c^{(i)*}(1)$ is the residue of the corresponding c-function at $s = 1$. In particular, this linear combination will have no pole at $s = 1$ if and only if $\sum a_i M_i c^{(i)*}(1) = 0$. Call any such linear combina-

of the $E_s^{(i)}$ admissible. One way to obtain functions annihilated by powers of Δ^n is to consider derivatives in s of such admissible combinations, evaluated at $s = 1$. Another way is by considering derivatives in s of any of the $(s-1)E_s^{(i)}$, again evaluated at $s = 1$. The second consequence of the Paley-Wiener Theorem we need is that all tempered distributions in A_{Eis} annihilated by some power of Δ are obtained as linear combinations of these two types. It is pretty clear, in fact, that for the second type, only one cusp need be taken into account.

Something similar holds for differential forms constructed from Eisenstein series.

I will not prove these assertions, except for a few remarks. The first is that the subspace of A_{Eis} annihilated by Δ^n is the dual of the quotient $S_{Eis}/\Delta^n S_{Eis}$, which one can see from Theorem 4.2 is clearly related to derivatives of Eisenstein series, and moreover is finite-dimensional. The second is that the point $s = 0$ plays no role because the functional equation for E_s says that the Eisenstein series and their derivatives obtained from the region $Re(s) < 1/2$ are obtained already from the region $Re(s) \geq 1/2$.

We come now to the final question: how does one compute the cohomology of this rather large complex made up of Eisenstein series and their residues and derivatives? One could in fact sort out the situation without too much trouble in the context we are dealing with, but it is here perhaps that it is most helpful to use representation theory.

The construction of Eisenstein series may be carried out for all weights of automorphic forms, not just functions or differential forms. One gets them all, in the region $Re(s) > 1$, by considering the series

$$\sum_{\Gamma_\infty \backslash \Gamma} y(\gamma(z))^s \varepsilon(cz+d)^n$$

where $\varepsilon(x) = x/|x|$ for $x \in \mathbb{C}^{\times}$, and (c,d) make up the bottom row of $\gamma \in \Gamma$, as well as the analogous series for all the cusps. One gets all automorphic forms in the Eisenstein component by meromorphically continuing these, at least as far as $\mathrm{Re}(s) = 1/2$, and taking residues and s-derivatives. In terms of representation theory the Eisenstein series themselves correspond to embeddings of principal series representations into the dual of the Schwartz space of $\Gamma\backslash SL_2(\mathbb{R})$, which may be defined much as $S(X) = S(\Gamma\backslash SL_2(\mathbb{R}))^K$ was. By a well known construction, the relative Lie algebra cohomology of these principal series (with respect to the pair $(\mathfrak{sl}_2, \mathfrak{so}_2)$) maps into the cohomology of X. Taking derivatives with respect to s amounts to mapping into the tempered distributions certain representations of G induced from non-irreducible representations of the subgroup $\begin{pmatrix} * & * \\ & * \end{pmatrix}$, related to the principal series. A variant of Shapiro's Lemma enables one to see easily what the cohomology amounts to.

I will be a little more precise. It is well known that by lifting forms on $\Gamma\backslash\mathcal{H}$ to $\Gamma\backslash SL_2(\mathbb{R})$, one obtains an isomorphism of the de Rham complex of X with $C^{\bullet}(\mathfrak{g},\mathfrak{k}, C^{\infty}(\Gamma\backslash G))$, where now $G = SL_2(\mathbb{R})$, $K = SO(2)$, and $\mathfrak{g}$ and $\mathfrak{k}$ are their Lie algebras. If we consider Schwartz forms, we obtain $S(\Gamma\backslash G)$ instead of $C^{\infty}(\Gamma\backslash G)$. And if we start with tempered currents on X, we obtain the space $A(\Gamma\backslash G)$ of tempered distributions on $\Gamma\backslash G$. The space $S(\Gamma\backslash G)$ possesses a G-stable decomposition $S_{cusp} \oplus S_{Eis}$, and so does $A(\Gamma\backslash G)$. The results at the beginning of this § translate to the statement that the Eisenstein component of the cohomology of X is the same as the $(\mathfrak{g},\mathfrak{k})$-cohomology of the subspace of elements of $A_{Eis}(\Gamma\backslash G)$ annihilated by some power of the Casimir operator C in the universal enveloping algebra of $\mathfrak{g}$.

Let B be the subgroup of elements

$$\begin{pmatrix} a & x \\ 0 & a^{-1} \end{pmatrix} \qquad (a \in \mathbb{R}^{\times},\ x \in \mathbb{R})$$

in $SL_2(\mathbb{R})$. To $s \in \mathbb{C}$ is associated a character χ_s of B:

$$\begin{pmatrix} a & x \\ 0 & a^{-1} \end{pmatrix} \mapsto |a|^s .$$

The principal series representation of G associated to this is the representation π_s induced by this, the space of all C^∞ functions $f\colon\ G \to \mathbb{C}$ such that

$$f(bg) = \chi_s(b)f(g)$$

for $b \in B$, $g \in G$. The Eisenstein series E_s gives rise, when analytic at s, to an embedding $\pi_s \to A_{Eis}(\Gamma\backslash G)$. When it has a pole at s then the normalized Eisenstein series $(t-s)E_t$, evaluated at $t = s$, gives rise to a map $\pi_s \to A_{Eis}(\Gamma\backslash G)$ which will have a non-trivial kernel.

Associated also to s is a sequence of finite-dimensional representations $\chi_s^{(1)}, \chi_s^{(2)}, \ldots$ of B, obtained by letting B act by the right regular representation on the spaces spanned by $|a|^s$, $|a|^s\log|a|, \ldots, |a|^s\log^{n-1}|a|$. Let $|a|^s[\log|a|]$ be the union of all these spaces, of infinite dimension. Then since $|a|^s\log|a|$ is the derivative of $|a|^s$ with respect to s, the derivatives of normalized Eisenstein series give rise to maps from $\pi_{s,\log} = \mathrm{Ind}(|a|^s[\log|a|]\,|B,G)$ to $A_{Eis}(\Gamma\backslash G)$.

This construction can be carried out for all cusps as well. As a consequence of the Paley-Wiener theorem one can see that the space of all tempered Eisenstein distributions annihilated by some power of C is as a G-module isomorphic to the direct sum of $(n-1)$ copies of $\pi_{1,\log}$ and one copy of $\pi_{0,\log}$ (n = number of cusps). (The latter comes from the pole of E_s at 1.)

By Shapiro's Lemma,

$$H^{\cdot}(\mathfrak{g},\mathfrak{k},\pi_{s,\log}) \cong H^{\cdot}(\mathfrak{b},|a|^{s}[\log|a|]).$$

Hochschild-Serre give a spectral sequence converging to this with E_2-term

$$H^{\cdot}(\mathfrak{a},H^{\cdot}(\mathfrak{n},\mathbb{C}) \otimes |a|^{s}[\log|a|]).$$

Here $\mathfrak{b} = \mathfrak{a}+\mathfrak{n}$, $\mathfrak{n}$ the Lie algebra of unipotents, $\mathfrak{a}$ that of diagonal matrices. This is equal to 0 unless $s = 0$ or 1, and then, by an interesting calculation, gives

$$H^{m}(\mathfrak{g},\mathfrak{k},\pi_{0,\log}) = \begin{cases} \mathbb{C} & m = 0 \\ 0 & m \neq 0 \end{cases}$$

$$H^{m}(\mathfrak{g},\mathfrak{k},\pi_{1,\log}) = \begin{cases} \mathbb{C} & m = 1 \\ 0 & m \neq 1 \end{cases} .$$

Finally, the Eisenstein component of cohomology turns out to be what it has to be:

$$H^{0}_{Eis} = \mathbb{C}$$

$$H^{1}_{Eis} = \mathbb{C}^{n-1}$$

$$H^{2}_{Eis} = 0 .$$

References

Y. Colin de Verdière, Une nouvelle démonstration du prolongement méromorphe des séries d'Eisenstein, C.R. Acad. Sci. Paris, t. 293 (1981), 361-363.

J. Dixmier and P. Malliavin, Factorisation de fonctions et de vecteurs indéfiniment différentiables, Bull. Sc. Math. 102(1978), 305-330.

R. Godement, Topologie algébrique et théorie des faisceaux, Hermann, Paris, 1958.

P. Griffiths and W. Schmid, Recent developments in Hodge Theory, in Discrete Subgroups of Lie Groups and Applications to Moduli, Oxford Press, Bombay, 1975, 31-127.

G. Harder, On the cohomology of discrete arithmetically defined groups, in Discrete Subgroups of Lie Groups and Applications to Moduli, Oxford Press, Bombay, 1975, 129-160.

G. Harder, Cohomology of $SL_2(\mathcal{O})$, in Lie groups and their representations, I.M. Gelfand ed, Halsted Press, New York, 1975, 139-150.

T. Kubota, Elementary Theory of Eisenstein Series, Halsted Press, New York, 1973.

R.P. Langlands, Eisenstein series, in Proc. Symp. Pure Math IX, A.M.S., Providence, 1966.

G. de Rham, Variétés différentiables, Hermann, Paris, 1960.

F. Trèves, Topological Vector Spaces, Distributions and Kernels, Academic Press, New York, 1967.

AUTOMORPHIC FORMS AND L-FUNCTIONS FOR THE UNITARY GROUP*)

Stephen Gelbart
Department of Mathematics
Cornell University
Ithaca, New York 14853/USA

and

Ilya Piatetski-Shapiro
Departments of Mathematics
Yale University, New Haven, CT. 06520/USA
Tel Aviv University, Ramat-Aviv, Israel

Introduction.

Our purpose is to define and analyze L-functions attached to automorphic cusp forms on the unitary group $G = U_{2,1}$ and a six-dimensional representation

$$\rho: {}^LG \to GL_6(\mathbb{C})$$

of its L-group.

The motivation for this work is three fold.

Firstly, we use these L-functions to analyze the lifting of cusp forms from $U_{1,1}$ to $U_{2,1}$; here the model for our work is Waldspurger's L-function theoretic characterization of the image of Shimura's map for modular forms of half-integral weight (cf. [Wald]).

A second motivation comes from the need to relate the poles of the L-functions for G to integrals of cusp forms over cycles coming from $U_{1,1}$. The prototype here is the recent proof of Tate's conjecture for Hilbert modular surfaces due to Harder, Langlands, and Rapaport.

Thirdly, we view this work as a special contribution to the general program of constructing local L and ε factors of Langlands type for representations of arbitrary reductive groups. In [PS1], such a program was sketched generalizing classical methods of Hecke, Rankin-Selberg, and Shimura. Related developments are discussed in [Jacquet], [Novod], [PS2], and [PS3]. For the unitary group $U_{2,1}$, the present paper extends the developments initiated in [PS3].

---

*) Notes based on lectures by S.G. at the University of Maryland Special Year on Lie Group Representations, 1982-83.

Finally, we mention the recent works of [Kudla 1,2]; here the lifting from $U_{1,1}$ to $U_{2,1}$ is described using classical theta-series and an Euler product of degree 6 is defined following [Shintani]. Complementary recent works include [Kottwitz], [Rogawski], and [Flicker], whose combined efforts produce a deep analysis of automorphic forms on the unitary group by means of the Selberg trace formula.

We are grateful to R.P. Langlands for explaining the connections with Tate's conjecture, and H. Jacquet for perfecting the local-global "L-function machine" which we appeal to so frequently.

TABLE OF CONTENTS

Notation

(i) F is a field (sometimes local, sometimes an $\mathbb{A}$-field),
E is a quadratic extension of F with Galois involution $z \to \bar{z}$.

(ii) V is a 3-dimensional vector space over E, with basis $\{\ell_{-1},\ell_0,\ell_1\}$.
(,) is a Hermitian form on E, with matrix

$$\begin{bmatrix} 0 & 0 & 1 \\ 0 & 1 & 0 \\ 1 & 0 & 0 \end{bmatrix}$$

with respect to $\{\ell_{-1},\ell_0,\ell_1\}$.

(iii) $G = U_{2,1} = U(V)$ is the unitary group for the form $(,)_V$.
P = parabolic subgroup stabilizing the isotropic line through ℓ_{-1}

= MN with

$$M = \left\{ \begin{bmatrix} \delta & 0 & 0 \\ 0 & \beta & 0 \\ 0 & 0 & \bar{\delta}^{-1} \end{bmatrix} \quad : \delta \in E^\times,\ \beta \in E^1 = \{z : z\bar{z} = 1\} \right\}$$

and unipotent radical

$$N = \left\{ \begin{bmatrix} 1 & b & z \\ 0 & 1 & -\bar{b} \\ 0 & 0 & 1 \end{bmatrix} : z,b \in E,\ z+\bar{z} = -b\bar{b} \right\} .$$

The center of N is

$$Z = \left\{ \begin{bmatrix} 1 & 0 & z \\ 0 & 1 & 0 \\ 0 & 0 & 1 \end{bmatrix} : \bar{z} = -z \right\}$$

§1. Whittaker Models (Ordinary and Generalized)

Some kind of Whittaker model is needed in order to introduce L-functions on G.

Fix F local (not of characteristic two), and suppose (π, H_π) is an irreducible admissible representation of G. Naively, we should look for functionals on H_π which transform under N according to a one-dimensional representation. However, since such functionals need not exist in general, and since there are irreducible representations of N which are not 1-dimensional, it is natural to pursue a more general approach.

(1.1) Recall N is the maximal unipotent subgroup of G and E is a quadratic extension of F.

We fix, once and for all, an element i in E such that $\bar{i} = -i$, so $\mathrm{Im}(z) = (z-\bar{z})/2i$. Regarding E as a 2-dimensional symplectic space over F with skew-form $\langle z_1, z_2 \rangle = \mathrm{Im}((z_1, z_2))$ we have

$$N = \left\{ \begin{bmatrix} 1 & b & z \\ 0 & 1 & -\bar{b} \\ 0 & 0 & 1 \end{bmatrix} : z, b \in E,\ z+\bar{z} = -b\bar{b} \right\} \approx H(E),$$

the Heisenberg group attached to E over F. In particular, N is non-abelian, with commutator subgroup

$$[N,N] = \left\{ \begin{bmatrix} 1 & 0 & z \\ 0 & 1 & 0 \\ 0 & 0 & 1 \end{bmatrix} : z = -\bar{z} \right\} = Z,$$

the center of N. The maximal abelian subgroup of N is

$$N' = \left\{ \begin{bmatrix} 1 & b & z \\ 0 & 1 & -b \\ 0 & 0 & 1 \end{bmatrix} \in N : b \in F \right\}$$

(1.2) The irreducible representations of the Heisenberg group, and hence those of N, are well known:

(i) σ is 1-dimensional.

In this case, σ must be trivial on

$$Z = [N,N]$$

and define a character of N/Z. So

$$N/Z \approx \left\{ \begin{bmatrix} 1 & a & 0 \\ & 1 & -\bar{a} \\ & & 1 \end{bmatrix} \right\} \approx E$$

implies σ corresponds to a character of E, i.e.,

$$\sigma = \psi_N\left(\begin{bmatrix}1 & a & z\\ 0 & 1 & -\bar{a}\\ 0 & 0 & 1\end{bmatrix}\right) = \psi(\operatorname{Im} a)$$

with ψ a character of F.

(ii) σ is infinite-dimensional.

In this case (by the Stone-von Neumann uniqueness theorem), σ is completely determined by its "central" character. In particular, if

$$\sigma\left(\begin{bmatrix}1 & 0 & z\\ 0 & 1 & 0\\ 0 & 0 & 1\end{bmatrix}\right) = \psi(\operatorname{Im} z)I$$

for some (additive) character ψ of F, then

$$\sigma = \rho_\psi = \operatorname{Ind}_{N'}^{N} \psi_{N'} ,$$

with $\psi_{N'}$ the character of (the maximal abelian subgroup) N' obtained by trivially extending ψ from Z to N'.

(1.3) Definition. By a (generalized) Whittaker functional for (π, H_π) we understand an N-map from H_π to some irreducible representation (σ, L_σ) of N (possibly infinite dimensional).

(1.4) Remark. The torus

$$T = \left\{\begin{bmatrix}\delta & 0 & 0\\ 0 & 1 & 0\\ 0 & 0 & \bar{\delta}^{-1}\end{bmatrix} : \delta \in E^x\right\}$$

acts by conjugation on N, taking

$$\begin{bmatrix}1 & b & z\\ 0 & 1 & -\bar{b}\\ 0 & 0 & 1\end{bmatrix} \quad \text{to} \quad \begin{bmatrix}1 & \delta b & \delta\bar{\delta}z\\ 0 & 1 & -(\overline{\delta b})\\ 0 & 0 & 1\end{bmatrix}$$

So if ψ_N denotes the 1-dimensional representation of N corresponding to the fixed character ψ of F as in (1.2)(i), Pontrygin duality for $E \approx N/Z$ implies that any other 1-dimensional representation is trivial or of the form

$$\psi_N^\delta(n) = \psi_N\left(\begin{bmatrix}\delta & & \\ & 1 & \\ & & \bar{\delta}^{-1}\end{bmatrix} n \begin{bmatrix}\delta & & \\ & 1 & \\ & & \bar{\delta}^{-1}\end{bmatrix}^{-1}\right)$$

for some $\delta \in E^x$.

(1.5) If σ is a one-dimensional representation of N of the form ψ_N, a given irreducible admissible representation (π,H_π) need not possess a non-trivial ψ_N-Whittaker functional $\mathcal{L}$. However, if it does, then by (1.4) it possesses a σ-Whittaker functional for any one-dimensional representation ψ_N^δ, given by the formula

$$\mathcal{L}^\delta(v) = \mathcal{L}\left(\pi\cdot\left(\begin{bmatrix} \delta & & \\ & 1 & \\ & & \bar{\delta}^{-1}\end{bmatrix}\right)v\right) \quad , \ v \in H_\pi$$

In this case, we call (π,H_π) non-degenerate. By a well-known Theorem of Shalika and Gelfand-Kazhdan (cf. [Shal]), the space of such σ-Whittaker functionals is one-dimensional. In particular, the corresponding Whittaker models

$$\mathbb{D}(\pi,\psi) = \{W(g) = \mathcal{L}(\pi(g)v) : v \in H_\pi\}$$

are unique.

(1.6) In general, (π,H_π) is not non-degenerate, examples being provided by the Weil representations discussed in §6. Thus it is necessary to consider σ-Whittaker models for infinite dimensional σ as well. Such σ, however, are completely determined by their central character ψ_Z, so it is convenient to work with a slight thickening of N. More precisely, consider the stablizer R in P of the central character ψ_Z of Z. Because $\begin{bmatrix} \delta & \beta & \\ & & \bar{\delta}^{-1}\end{bmatrix}$ conjugates

$$\begin{bmatrix} 1 & 0 & z \\ 0 & 1 & 0 \\ 0 & 0 & 1\end{bmatrix} \text{ to } \begin{bmatrix} 1 & 0 & \delta\bar{\delta}z \\ 0 & 1 & 0 \\ 0 & 0 & 1\end{bmatrix},$$

$$R = \left\{\begin{bmatrix} \delta & * & * \\ 0 & \beta & * \\ 0 & & \delta\end{bmatrix} \in P: \quad \delta, \beta \in E^1\right\} \approx (E^1 \times E^1) \ltimes N.$$

In particular, each irreducible infinite dimensional representation ρ_ψ of N extends to a like representation ρ_ψ^α of R with α a character of $E^1 \times E^1$.

Theorem. (Existence and Uniqueness of Generalized Whittaker Models; [PS3]). Any (π,H_π) possesses a ρ_ψ^α-Whittaker functional for some choice of ρ_ψ^α; moreover, the space of such functionals is at most one dimensional.

We shall discuss this result in more detail in the global context of §7.

§2. Some Fourier Expansions and Hypercuspidality

Now F is an $\mathbb{A}$-field not of characteristic 2, and π is an automorphic cuspidal representation of $G_{\mathbb{A}}$ which we suppose realized in some subspace of cusp forms H_π in $L_0^2(G_F\backslash G_{\mathbb{A}})$. To attach an L-function to π, it is useful to take forms f in H_π and examine their Fourier coefficients along the maximal unipotent subgroup N. When such coefficients are non-zero, π is non-degenerate, and we are led back to the local Whittaker models $\mathbb{W}(\pi_v,\psi_v)$ of (1.5); in this case, we can (and eventually do) introduce L-functions using Jacquet's generalization of the "Rankin-Selberg method".

On the other hand, if these Fourier coefficients represent zero, then π is hypercuspidal; in this case, looking at Fourier expansions along Z will bring us back to the generalized Whittaker models of (1.6), and ultimately allow us to introduce an L-function for π using the so-called "Shimura method".

Henceforth, let us fix a non-trivial character ψ of $F\backslash\mathbb{A}$, and define characters ψ_N and ψ_Z of $N = N_{\mathbb{A}}$ and $Z = Z_{\mathbb{A}}$ by

$$\psi_N\left(\begin{bmatrix}1 & a & z\\ 0 & 1 & -\bar{a}\\ 0 & 0 & 0\end{bmatrix}\right) = \psi(\mathrm{Im}\, a)$$

and

$$\psi_Z\left(\begin{bmatrix}1 & 0 & z\\ 0 & 1 & 0\\ 0 & 0 & 1\end{bmatrix}\right) = \psi(\mathrm{Im}\, z).$$

(2.1). Fix f in H_π. To obtain a Fourier expansion of f "along N", we introduce the familiar ψ-th coefficient

$$W_f^\psi(g) = \int_{N_F\backslash N_{\mathbb{A}}} f(ng)\overline{\psi_N(n)}\,dn\ .$$

The transitivity of $T_A = \left\{\begin{bmatrix}\delta & & \\ & 1 & \\ & & \bar{\delta}^{-1}\end{bmatrix}\right\}$ acting on $N_{\mathbb{A}}\backslash Z_{\mathbb{A}}$ implies - as in the local theory - that

$$\begin{aligned} W_f^{\psi^\delta}(g) &= \int_{N_F\backslash N_{\mathbb{A}}} f(ng)\overline{\psi_N^\delta(n)}\,dn \\ &= \int_{N_F\backslash N_{\mathbb{A}}} f(ng)\psi_N\left(\begin{bmatrix}\delta & & \\ & 1 & \\ & & \bar{\delta}^{-1}\end{bmatrix} n \begin{bmatrix}\delta & & \\ & 1 & \\ & & \bar{\delta}^{-1}\end{bmatrix}^{-1}\right) dn\end{aligned}$$

$$= W_f^{\psi}\left(\begin{bmatrix} \delta & & \\ & 1 & \\ & & \bar{\delta}^{-1} \end{bmatrix} g\right).$$

In other words, knowing W_f^{ψ} determines $W_f^{\psi^{\delta}}$ for all ψ^{δ}, $\delta \in E^{\times}$.

However, though $N_F\backslash N_{\mathbb{A}}$ is compact, it is not abelian; to obtain a nice Fourier expansion, we must bring into play the compact abelian group $N_F Z_{\mathbb{A}}\backslash N_{\mathbb{A}}$.

(2.2). We compute

$$W_f^{\psi}(g) = \int_{N_F\backslash N_{\mathbb{A}}} f(ng)\overline{\psi_N(n)}dn$$

$$= \int_{N_F Z_{\mathbb{A}}\backslash N_{\mathbb{A}}} \left(\int_{Z_F\backslash Z_{\mathbb{A}}} f(nzg)dz\right)\overline{\psi_N(n)}dn$$

$$= \int_{N_F Z_{\mathbb{A}}\backslash N_{\mathbb{A}}} f_{00}(ng)\overline{\psi_N(n)}dn$$

with

(2.2.1) $$f_{00}(g) = \int_{Z_F\backslash Z_{\mathbb{A}}} f(zg)dz$$

the constant term (in the Fourier expansion) of $f(zg)$ along Z.

Fix g in $G_{\mathbb{A}}$. As a function on the compact abelian group $N_F Z_{\mathbb{A}}\backslash N_{\mathbb{A}}$, $f_{00}(ng)$ has Fourier expansion

(2.2.2) $$f_{00}(g) = \sum_{\delta\in E^{\times}} W_f^{\psi^{\delta}}(g) + \int_{N_F Z_{\mathbb{A}}\backslash N_{\mathbb{A}}} f_{00}(n'g)dn'.$$

Indeed, the last paragraph says precisely that $W_f^{\psi}(g)$ is the ψ_N-th Fourier coefficient of $f_{00}(ng)$ along $Z\backslash N \approx E$. Moreover, the constant term is actually zero since f cuspidal implies

$$\int_{N_F Z_{\mathbb{A}}\backslash N_{\mathbb{A}}} f_{00}(n'g)dn' = \int_{N_F Z_{\mathbb{A}}\backslash N_{\mathbb{A}}} \left(\int_{Z_F\backslash Z_{\mathbb{A}}} f(zn'g)dz\right)dn'$$

$$= \int_{N_F\backslash N_{\mathbb{A}}} f(ng)dn = 0.$$

(2.3) Let $\mathfrak{w}(\pi,\psi)$ denote the space of ψ-th Fourier coefficients $W_f^{\psi}(g)$, $f \in H_{\pi}$.

Proposition. The vanishing or nonvanishing of $\mathfrak{w}(\pi,\psi)$ is independent of ψ; in particular, $\mathfrak{w}(\pi,\psi) \equiv \{0\} \Longleftrightarrow$

$$f_{00}(g) = 0 \ \forall f \in H_{\pi}.$$

Proof. According to (2.2) and (2.1)

$$f_{00}(g) = \sum_{\delta\varepsilon E^{x}} W_f^{\psi^{\delta}}(g) \tag{2.3.1}$$

$$= \sum_{\delta\varepsilon E^{x}} W_f^{\psi}\left(\begin{bmatrix}\delta & & \\ & 1 & \\ & & \bar{\delta}^{-1}\end{bmatrix} g\right),$$

with

$$W_f^{\psi}(g) = \int_{N_F Z_{\mathbb{A}}\backslash N_{\mathbb{A}}} f_{00}(ng)\overline{\psi_N(n)}dn.$$

(2.4) Definition. We call (π,H_π) hypercuspidal if $f \varepsilon H_\pi$ implies $f_{00} \equiv 0$.

Proposition 2.4. Let $L^2_{0,1}$ denote the orthocomplement in L^2_0 of all hypercusp forms. Then:

(i) $L^2_{0,1}$ has multiplicity 1,

(ii) each $(\pi,H_\pi) \subset L^2_{0,1}$ is non-degenerate, and

(iii) for any f in $H_\pi \subset L^2_{0,1}$, the constant term

$$f_{00}(g) = \sum_{\delta\varepsilon E^{x}} W_f^{\psi^{\delta}}(g)$$

completely determines f.

Proof. We start with (iii). Suppose f and f' in H_π are such that $f_{00} = f'_{00}$. Then $(f-f')_{00} = 0$ implies $f-f' = 0$ (by the hypothesis $H_\pi \subset L^2_{0,1}$). This proves (iii). To prove (i) and (ii), suppose there exists $H'_\pi \subset L^2_{0,1}$ such that the right regular representation restricted to H'_π again realizes π. If $f \varepsilon H_\pi$ and $f' \varepsilon H'_\pi$, then

$$f_{00}(g) = \sum_{\delta\varepsilon E^{x}} W_f^{\delta}\left(\begin{bmatrix}\delta & & \\ & 1 & \\ & & \bar{\delta}^{-1}\end{bmatrix} g\right), \tag{2.4.1}$$

and

$$f'_{00}(g) = \sum_{\delta\varepsilon E^{x}} W_{f'}^{\psi}\left(\begin{bmatrix}\delta & & \\ & 1 & \\ & & \bar{\delta}^{-1}\end{bmatrix} g\right).$$

Note that each W_f^{ψ} (or $W_{f'}^{\psi}$) satisfies the condition $W_f^{\psi}(ng) = \psi_N(n)W_f^{\psi}(g)$, $n \varepsilon N$, i.e., the spaces $\{W_f^{\psi}\}$ and $\{W_{f'}^{\psi}\}$ afford Whittaker models for π. But by (2.3) these spaces are non-zero (which proves (ii)) and by the uniqueness of Whittaker models quoted in (1.5), these spaces coincide. Thus by (2.4.1), the spaces $\{f_{00}\}$ and $\{f'_{00}\}$ coincide; by (iii) the spaces $H_\pi = \{f\}$ and $H'_\pi = \{f'\}$

also coincide, thereby proving (i).

(2.5) Remarks.

(i) It is conjectured (cf. [Flicker]) that multiplicity 1 holds for the entire space of cusp forms; however, at the present time, we can prove this only for $L^2_{0,1}$.

(ii) Hypercuspforms do exist; again, examples are provided by the Weil representation discussed in §6.

(iii) Although $\mathfrak{w}(\pi,\psi) \neq \{0\}$ implies π non-degenerate (in the sense that an abstract ψ-Whittaker functional exists), the converse is not clear. Indeed, the work of [Wald] indicates that characterizing the non-vanishing of a space of Fourier coefficients is a delicate matter.

§3. L-functions à la Rankin-Selberg-Jacquet

We are now ready to attach (global) L-functions to (non-degenerate) cuspidal representations π of $G_{\mathbb{A}}$. The method used goes back to [Rankin] and [Selberg] who used it to analytically continue the convolution of Dirichlet series corresponding to classical holomorphic modular forms. The reformulation of their construction in the language of representation theory was carried out in detail by [Jacquet], for $GL_2 \times GL_2$, and by [PS 1] in general (but without details or explicit computation). In this Section (and the next), we carry out the construction for $G = U_{2,1}$.

(3.1). Recall V is a 3-dimensional vector space over E/F and $(\ ,\ )_V$ is a Hermitian form on V whose matrix with respect to the basis $\{\ell_{-1},\ell_0,\ell_1\}$ is $\begin{bmatrix} 0 & 0 & 1 \\ 0 & 1 & 0 \\ 1 & 0 & 0 \end{bmatrix}$.

Let $H \subset G$ denote the stabilizer of the anisotropic line $\{\ell_0\}$. Then H also preserves the orthocomplement of $\{\ell_0\}$, namely

$$W = \{\ell_{-1},\ell_1\}.$$

Regarding W as a 2-dimensional Hermitian space, (whose matrix with respect to the basis $\{\ell_{-1},\ell_1\}$ is $\begin{bmatrix} 0 & 1 \\ 1 & 0 \end{bmatrix}$), we have $H \approx U(W) \approx U_{1,1}$ via the embedding

$$\begin{bmatrix} * & * \\ * & * \end{bmatrix} \to \begin{bmatrix} * & 0 & * \\ 0 & 1 & 0 \\ * & 0 & * \end{bmatrix} .$$

Let B denote the standard maximal parabolic (Borel) subgroup

$$\left\{ \begin{bmatrix} \alpha & 0 & \beta \\ 0 & 1 & 0 \\ 0 & 0 & \bar{\alpha}^{-1} \end{bmatrix} \right\} \quad \text{of } H$$

so that $B \approx T \ltimes Z$, with T the torus

$$T = \left\{ \begin{bmatrix} \alpha & & \\ & 1 & \\ & & \bar{\alpha}^{-1} \end{bmatrix} : \alpha \in E^\times \right\}$$

and Z the center of N.

(3.2). Given an automorphic cuspidal representation (π,H_π) of $G_{\mathbb{A}}$, and $f \in H_\pi$, we shall analyze a global zeta-integral of the form

(3.2.1) $$L^{\mu}(f,F,s) = \int_{H_F\backslash H_{\mathbb{A}}} f(h)E^{\mu}(h,F,s)dh.$$

First we need to describe the (as yet) undefined terms μ, F, E, etc.

Fixing a (not necessarily unitary) character μ of the idele class group $E^{\times}\backslash\mathbb{A}_E^{\times}$ of $E^{\times}$, and $s \in \mathbb{C}$, define a character ω_{μ}^{s} of the Borel subgroup B by the formula

(3.2.2) $$\omega_{\mu}^{s}\left(\begin{bmatrix}\alpha & 0 & \beta \\ 0 & 1 & 0 \\ 0 & 0 & \bar{\alpha}^{-1}\end{bmatrix}\right) = \mu(\alpha)|\alpha|_E^{s}, \quad \alpha \in \mathbb{A}_E^{\times} .$$

Fixing an arbitrary Schwartz-Bruhat function Φ in the space $\mathcal{S}(W_{\mathbb{A}_E})$, set

(3.2.3) $$F_{\Phi}(h) = \int_{\mathbb{A}_E^{\times}} (h\cdot\Phi)(t\ell_{-1})\mu(t)|t|_E^{s}d^{\times}t,$$

where $(h\cdot\Phi)(w) = \Phi(h^{-1}\cdot w)$, and $h\cdot w$ denotes the natural action of $H_{\mathbb{A}}$ on $W_{\mathbb{A}} \subset V_{\mathbb{A}}$; as usual, this integral converges for Re(s) sufficiently large, and continues meromorphically to define a function of h on $H_{\mathbb{A}}$ for all s in $\mathbb{C}$. Note

$$F(bh) = \omega_{\mu}^{s}(b)F(h) \quad \text{for} \quad b \in B_{\mathbb{A}},\ h \in H_{\mathbb{A}}.$$

Finally, the Eisenstein series $E^{\mu}(h,F,s)$ is defined by the familiar series

(3.2.4) $$E^{\mu}(h,F,s) = \sum_{\gamma\in B_F\backslash H_F} F(\gamma h);$$

it converges initially only for Re(s) large, but the Selberg-Langlands theory of Eisenstein series implies that the function it defines continues meromorphically in s and defines an automorphic form on $H_{\mathbb{A}}$.

(3.3) Remarks.

(i) Because E(h,F,s) is a automorphic form on H, and the restriction of the cusp form f from $G_{\mathbb{A}}$ to $H_{\mathbb{A}}$ is still rapidly decreasing, the integral defining $L^{\mu}(f,F,s)$ (cf. (3.2.1)) is convergent. The resulting function of s - the global zeta-function of f - has poles which can arise only from poles of E(h,F,s).

(ii) The function E(h,F,s) is essentially the familiar GL_2-Eisenstein series discussed (for example) in [Jacquet]. Indeed, SL_2 is isomorphic to a subgroup of H, namely $SU_{2,1}$, and $H \approx SU_{2,1} \rtimes U_1$, where $U_1 \approx \{\left[\begin{smallmatrix}\alpha & 0 \\ 0 & 1\end{smallmatrix}\right]:\alpha \in E^1\}$ and $\left[\begin{smallmatrix}\alpha & 0 \\ 0 & 1\end{smallmatrix}\right]$ acts on $SU_{2,1}$ by conjugation. Thus the restriction of E(h,F,s) to $SU_{2,1}$ is the familiar Eisenstein series on $SL_2(\mathbb{A})$ with functional equation and non-trivial residue given by the constant function (arising from the pole at s = 1). At the

$U_{2,1}$ level, these residues become proportional to $\mu(\det h)$, and the functional equation relates $E(h,F,s)$ to a "partially Fourier transformed" Eisenstein series at 1-s.

(3.4). From the theory above, we conclude that $L^{\mu}(f,F,s)$ is meromorphic in $\mathbb{C}$ with functional equation relating values at s and 1-s; more significantly for the sequel, the only possible residues of $L^{\mu}(f,F,s)$ are proportional to

(3.4.1)
$$\int_{H_F\backslash H_{\mathbb{A}}} f(h)\mu(\det h)dh.$$

In particular, if this last integral vanishes, then the zeta-function $L^{\mu}(f,F,s)$ is entire.

Regarding $H_F\backslash H_{\mathbb{A}}$ as an (algebraic) cycle in $G_F\backslash G_{\mathbb{A}}$, we (ultimately) obtain the following statement: The existence of a pole for $L^{\mu}(f,F,s)$ (and ultimately the L-function $L(s,\pi,\mu)$) is related to the non-vanishing integral of f in H_{π} (suitably tensored with μ) over the cycle coming from $U_{1,1}$.

We shall return to these considerations in §5. For the moment, we content ourselves with a factorization of $L^{\mu}(f,F,s)$ into local integrals.

(3.5). Proposition.

$$L^{\mu}(f,F,s) = \int_{Z_{\mathbb{A}}\backslash H_{\mathbb{A}}} W_f^{\psi}(h)F(h)dh.$$

Proof. From the definition of the series E,

$$L^{\mu}(f,F,s) = \int_{H_F\backslash H_{\mathbb{A}}} \sum_{B_F\backslash H_F} f(\gamma h)F(\gamma h)dh$$

$$= \int_{B_F\backslash H_{\mathbb{A}}} f(h)F(\text{[illegible]}$$

Recall our subgroups

$$Z = \left\{\begin{bmatrix}1 & 0 & z\\ 0 & 1 & 0\\ 0 & 0 & 1\end{bmatrix}\right\} \subset H = \left\{\begin{bmatrix}* & 0 & *\\ 0 & 1 & 0\\ * & 0 & *\end{bmatrix}\right.$$

Since $F(h)$ is invariant by $Z_{\mathbb{A}}$ (cf. (3.2.[illegible]

$$L^{\mu}(f,F,s) = \int_{B_FZ_{\mathbb{A}}\backslash H_{\mathbb{A}}} F(h)\ \{ \int_{Z_F\backslash\text{[illegible]}}$$

$$= \int_{B_FZ_{\mathbb{A}}\backslash H_{\mathbb{A}}} F(h)f_{00}(h\text{[illegible]}$$

Now recall the Fourier expansion

$$f_{00}(h) = \sum_{\delta\varepsilon E^{x}} W_f^{\psi}\left(\begin{bmatrix} \delta & & \\ & 1 & \\ & & \bar{\delta}^{-1} \end{bmatrix} h\right) ;$$

cf. (2.2.1). Since

$$B_F = \left\{\begin{bmatrix} \delta & 0 & z \\ 0 & 1 & 0 \\ 0 & 0 & \bar{-}^{-1} \end{bmatrix} : \delta \varepsilon E^{x},\ z \varepsilon E\right\}$$

we have

$$Z_{\mathbb{A}}\backslash B_F Z_{\mathbb{A}} \approx \left\{\begin{bmatrix} \delta & & \\ & 1 & \\ & & \bar{\delta}^{-1} \end{bmatrix} : \delta \varepsilon E^{x}\right\}$$

and therefore

$$L^{\mu}(f,F,s) = \int_{B_F Z_{\mathbb{A}}\backslash H_{\mathbb{A}}} \left\{\sum_{Z_{\mathbb{A}}\backslash B_F Z_{\mathbb{A}}} W_f^{\psi}(bh)\right\} F(h)dh$$

$$= \int_{Z_{\mathbb{A}}\backslash H_{\mathbb{A}}} W_f^{\psi}(h)F(h)dh,$$

as was to be shown.

<u>Remarks</u>. (i) We defined $L^{\mu}(f,F,s)$ for any $f \varepsilon H_{\pi}$ without assuming H_{π} orthogonal to all hypercuspforms. However, this last proposition shows that $L^{\mu}(f,F,s)$ is identically 0 if $\mathfrak{W}(\pi,\psi) \equiv \{0\}$, i.e., if f is a hypercuspform. This is why the Rankin-Selberg method fails for arbitrary π, and why (in §9)) we need to use Shimura's method.

(ii) The significance of this Proposition is that it allows us to factor $L^{\mu}(f,F,s)$ into local zeta-integrals, one for each place v of F. Note that whenevery v splits in E, i.e., whenever $E \otimes_F F_v$ splits as the direct sum of two fields E_{w_1} and E_{w_2} (isomorphic to F_v), we have

$$G_v = G_{F_v} \approx GL_3(F_v).$$

<u>(3.6) Proposition</u>. Suppose $\Phi = \Pi\Phi_w$ in $\mathcal{S}(W_{\mathbb{A}_E})$, the product being taken over all the primes w of E. Assuming f is not hypercuspidal, we have

$$L^{\mu}(f,F,s) = \prod_v \left(\int_{Z_v\backslash H_v} W_v(h)F_v(h)dh\right)$$

where v is an arbitrary place <u>of F</u>, W_v belongs to the local Whittaker space $\mathfrak{W}(\pi_v,\psi_v)$, and

$$F_v(h_v) = \prod_{w|v} \int_{E_w^\times} (h_v \cdot \Phi_w)(t\ell_{-1})\mu_w(t)|t|_w^s d^\times t .$$

Proof. Since the domain of integration factors, we need only check that the integrand does also. By uniqueness of Whittaker models, $W(h)$ in $\mathfrak{W}(\pi,\psi)$ with $\psi = \prod_v \psi_v$ implies $W(h) = \prod W_v(h_v)$, with each $W_v \in \mathfrak{W}(\pi_v,\psi_v)$. Also, from the definition of $F_\Phi(h)$, it follows

$$F_\Phi(h) = \prod_v \Big(\prod_{w|v} \int_{E_w^\times} (h_i \cdot \Phi_w)(t\ell_{-1})\mu_w(t)|t|_w^s d^\times t\Big),$$

as claimed.

(3.7). When v splits in E, each integral

$$\int_{E_{w_i} \approx F_v} (h_v \cdot \Phi)(t\ell_{-1})\mu_{w_i}(t)|t|_{w_i}^s d^\times t$$

reduces to the distribution $z(\alpha^s\mu, h\cdot\Phi)$ treated in §14 of [Jacquet]. Indeed, in this case, $H_v \approx U_{2,1} \approx GL_2(F_v)$, and ℓ_{-1} corresponds to the vector $(0,1)$ in F_v^2 fixed by the unipotent subgroup of $GL_2(F_v)$ (under the action $(x\ y)\cdot\begin{pmatrix} a & b \\ c & d\end{pmatrix}$).

In this case, our local zeta-integral takes the form

$$\int_{GL_2} W_v(h) z(\alpha^s\mu_{w_1}, h\cdot\Phi_{w_1}) z(\alpha^s\mu_{w_2}, h\cdot\Phi_{w_2}) dh$$

where $W_v(h)$ is a Whittaker function on GL_3 restricted to GL_2 (via the embedding $\begin{bmatrix} * & * \\ * & * \end{bmatrix} \to \begin{bmatrix} * & 0 & * \\ 0 & 1 & 0 \\ * & 0 & * \end{bmatrix}$), and w_1, w_2 are the two primes of E which divide v.

§4. Local L-functions (non-degenerate π).

(4.1). Suppose E (quadratic) over F is local and π is an irreducible admissible representation of G_F. Because of Proposition (3.6), it is natural to define local zeta-integrals (of Rankin-Selberg-Jacquet type) by

$$L^{\mu}(W,F_{\Phi},s) = \int_{Z\backslash H} W(h)F(h)dh$$

where $W \in \mathcal{W}(\pi,\psi_N)$, μ is a character of $E^{\times}$, Φ is a Schwartz-Bruhat function on the two-dimensional E-vector space $W = \{\ell_{-1},\ell_1\}$, and

$$F^{\mu}(h) = \int_{E^{\times}} (h\Phi)(t\ell_{-1})\mu(t)|t|_E^s d^{\times}t \ .$$

Note

$$F(bh) = \mu(\alpha)|\alpha|_E^s F(h)$$

for all $h \in H$, and

$$b = \begin{bmatrix} \alpha & 0 & \beta \\ 0 & 1 & 0 \\ 0 & 0 & \bar{\alpha}^{-1} \end{bmatrix} \in B \ .$$

Remark. There are other kinds of local integrals to consider, namely those which arise from the splitting primes for E; cf. (3.7). Since these zeta integrals involve the more familiar groups GL_3 and GL_2, we shall concentrate on the unitary integrals instead (dealing with the splitting primes only parenthetically).

(4.2). Suitably modifying the program in §14 of [Jacquet] one can obtain the following:

Proposition. For each W, Φ, and μ as above, the local zeta-integrals $L^{\mu}(W,F_{\Phi},s)$ converge for $\mathrm{Re}(s) \gg 0$ and define rational functions of q^{-s} satisfying the following conditions:

(i) The sub-vector space of $\mathbb{C}(q^{-s})$ spanned by $L^{\mu}(W,F_{\Phi},s)$ is in fact a fractional ideal of the ring $\mathbb{C}[q^{-s},q^{s}]$ generated by some polynomial Q_0 in $\mathbb{C}[q^{-s}]$ which is independent of W and Φ;

(ii) There is a rational function of q^{-s}, denoted $\gamma(s)$, and a "contragredient" L-function $\tilde{L}^{\mu^{-1}}(W,F_{\hat{\Phi}},s)$, such that for all W and Φ (and $\hat{\Phi}$ a special kind of "partial Fourier transform"),

$$\tilde{L}^{\mu^{-1}}(W,F_{\hat{\Phi}},1-s) = \gamma(s)L^{\mu}(W,F_{\Phi},s).$$

(4.3). Remarks. (i) If we demand that $Q_0(0) = 1$, then $L(s,\pi,\mu) = Q_0(q^{-s})^{-1}$ is the unique Euler factor such that

$$\frac{L^{\mu}(W,\Phi,s)}{L(s,\pi,\mu)}$$

is entire (actually polynomial in q^s and q^{-s}) for all W and Φ, and equal to 1 for appropriately chosen W and Φ. A similar statement holds for $L(s,\tilde{\pi},\mu)$ and $\tilde{L}^{\mu^{-1}}(W,\Phi,s)$. As usual, we regard $L(s,\pi,\mu)$ as the normalized g.c.d. of the zeta-functions $L^{\mu}(W,\Phi,s)$, and as the local component of a (soon to be defined) global L-function $L(s,\pi,\mu)$.

(ii) If we let

$$\varepsilon(s,\pi,\psi) = \frac{L(1-s,\tilde{\pi},\mu^{-1})\gamma(s)}{L(s,\pi,\mu)},$$

then ε is the monomial factor relating

$$\frac{\tilde{L}^{\mu^{-1}}(W,\hat{\Phi},1-s)}{L(s,\tilde{\pi},\mu^{-1})} \quad \text{and} \quad \frac{L^{\mu}(W,\Phi,s)}{L(s,\pi)}.$$

(iii) Throughout this Section, we are implicitly dealing only with non-archimedean fields; the case of $\mathbb{R}$ (or $\mathbb{C}$) is an unfortunately thorny yet unavoidable reality.

(4.4). Unramified computations.

In the next few sections we shall compute $L(s,\pi,\mu)$ when everything in sight is unramified.

Thus we suppose F is a local non-archimedean field of odd characteristic, and E is an unramified quadratic extension of F. Let O_F (resp. O_E) denote the ring of integers of F (resp. O_F), $\tilde{\omega}$ (resp. $\tilde{\omega}_E$) a generator of the prime ideal $\mathfrak{p}$ (resp. $\mathfrak{p}_E$) of O_F (resp. O_E), and ψ a character of F trivial on O_F but not on $\mathfrak{p}^{-1}O_F$.

Let K denote the standard maximal compact subgroup of G_F consisting of "integral" matrices (entries in O_E, determinant in O_E^{x}). Because E is unramified over F, we have

$$G = NAK,$$

where

$$A = \left\{ \begin{bmatrix} t & 0 & 0 \\ 0 & 1 & 0 \\ 0 & 0 & t^{-1} \end{bmatrix} : t \in F^{x} \right\}$$

is the maximal F-split torus of G_F.

(4.5). Class -1 Whittaker functions.

Suppose π is a class 1 (with respect to K) irreducible admissible representation of G_F. Then π is of the form

$$\pi = \pi(\nu) = \mathrm{Ind}_P^G \nu^*$$

where ν is an unramified (quasi-) character of $E^\times$,

$$P = MN = \left\{ \begin{bmatrix} \delta & * & * \\ 0 & \beta & * \\ 0 & 0 & \bar{\delta}^{-1} \end{bmatrix} \right\} ,$$

and ν^* is defined on $M = P/N$ by

$$\nu^*\left(\begin{bmatrix} \delta & & \\ & \beta & \\ & & \bar{\delta}^{-1} \end{bmatrix} \right) = \nu(\delta).$$

From [Cas Sh] we know $\pi(\nu)$ is non-degenerate; moreover, the K-fixed Whittaker function W in $\mathfrak{W}(\pi(\nu),\psi_N)$-normalized by the condition $W(k) \equiv 1$, is uniquely determined by the following formulas:

(i) $W(nak) = \psi_N(n)W(a)$ for all $n \in N$, $a \in A$, and $k \in K$;

(ii) $W\left(\begin{bmatrix} \delta & & \\ & 1 & \\ & & \delta^{-1} \end{bmatrix} \right) = 0$ if $|\delta|_F > 1$; and

(iii) for all $n \geq 0$,

$$W\left(\begin{bmatrix} \tilde{\omega}^n & 0 & 0 \\ 0 & 1 & 0 \\ 0 & 0 & \tilde{\omega}^{-n} \end{bmatrix} \right) = |\tilde{\omega}|_F^{2n} \frac{\nu(\tilde{\omega})^{n+1} - \nu(\tilde{\omega})^{-(n+1)}}{\nu(\tilde{\omega}) - \nu(\tilde{\omega})^{-1}} ,$$

(cf. Theorem 5.4 of [Cas Sh]).

<u>(4.6)</u>. We compute $L^\mu(W,F_\Phi,s)$ with μ an unramified character of $E^\times$, $W(g)$ as in (4.5), and Φ the characteristic function of the O_E-module in $E\ell_{-1} \oplus E\ell_1$ generated by ℓ_{-1} and ℓ_1.

Let $K_H = K \cap H$. Since $Z = N \cap H$, we have $H = ZAK_H$, with corresponding integration formula

$$\int_{Z\backslash H} f'(h)dh = \int_{K_H}\left(\int_{F^\times} f'\left(\begin{bmatrix} a & & \\ & 1 & \\ & & a^{-1} \end{bmatrix} k \right) |a|^{-2} d^\times a \right) dk .$$

Here f' is a function of $Z\backslash H$, and Haar measure $d^\times a$ on $F^\times$ is normalized so that $m(O_F^\times) = 1$.

Note $k \in K_H$ implies

$$F_\Phi\left(\begin{bmatrix} a & & \\ & 1 & \\ & & a^{-1}\end{bmatrix}k\right) = \mu(a)|a|_E^s F^\mu(k)$$

$$= \mu(a)|a|_E^s \int_{E^\times} (k\cdot\Phi)((t\ell_{-1}))\mu(t)|t|_E^s d^\times t$$

$$= \mu(a)|a|_E^s F_\Phi(1)$$

since $k\cdot\Phi(t\ell_{-1}) = \Phi(t\ell_{-1})$ for our unramified choice of Φ. Thus we have

$$L^\mu(W,F_\Phi,s) = F_\Phi(1)\int_{F^\times} W\left(\begin{bmatrix} a & & \\ & 1 & \\ & & a^{-1}\end{bmatrix}\right)\mu(a)|a|_F^{2s-2}d^\times a .$$

with

$$F_\Phi(1) = \int_{E^\times} \Phi(t\ell_{-1})\mu(t)|t|_E^s d^\times t$$

$$= \int_{E^\times} 1_{O_E}(t)\mu(t)|t|_E^s d^\times t$$

$$= \frac{1}{1-\mu(\tilde\omega_E)|\tilde\omega_E|^s} = L_E(s,\mu).$$

Here $L_E(s,\mu)$ is the local Hecke-Tate factor attached to the quasi-character μ of $E^\times$; cf. [Goldstein], §8.1.

(4.7). It remains to compute the integral of our class 1 Whittaker function. From the formula (4.5)(iii), we have

$$\int_{F^\times} W\left(\begin{bmatrix} a & & \\ & 1 & \\ & & a^{-1}\end{bmatrix}\right) \mu(a)|a|^{2s-2}d^\times a$$

$$= \frac{1}{\nu(\tilde\omega)-\nu(\tilde\omega)^{-1}} \sum_{n=0}^{\infty} |\tilde\omega^n|^2\mu(\tilde\omega^n)|\tilde\omega^n|^{2s-2}\{\nu(\tilde\omega)^{n+1}-\nu(\tilde\omega)^{-(n+1)}\}$$

$$= \sum_{n=0}^{\infty} \mu(\tilde\omega^n)|\tilde\omega^n|^{2s} \sum_{i+j=n} \nu(\tilde\omega)^i\nu(\tilde\omega^{-1})^j$$

$$= (\sum_{n=0}^{\infty} \mu(\tilde\omega^n)|\tilde\omega^n|^{2s}\nu(\tilde\omega)^n)(\sum_{m=0}^{\infty} \mu(\tilde\omega^m)|\tilde\omega^m|^{2s}\nu(\tilde\omega^{-1})^m)$$

$$= \left(\frac{1}{1-\mu(\tilde{\omega})\nu(\tilde{\omega})|\tilde{\omega}|^{2s}}\right)\left(\frac{1}{1-\mu(\tilde{\omega})\nu^{-1}(\tilde{\omega})|\tilde{\omega}|^{2s}}\right)$$

$$= L_F(2s,\mu\nu)L_F(2s,\mu\nu^{-1}) .$$

Summing up,

$$L^{\mu}(W,F_{\Phi},s) = L(s,\pi,\mu)$$

$$= L_E(s,\mu)L(2s,\mu\nu)L(2s,\mu\nu^{-1}).$$

Here μ is regarded as a character of $E^{\times}$ in the L-factor $L_E(s,\mu)$; and (by restriction - along with ν) as a character of $F^{\times}$ in the remaining two L-factors. Altogether,

$$L(s,\pi,\mu) = Q_0(q^{-s})^{-1}$$

with Q_0 a polynomial <u>of degree 6</u> in $q^{-s} = |\tilde{\omega}_F|^s$.

<u>(4.8)</u>. To which conjugacy class and representation of the L-group ${}^LF$ does $L(s,\pi,\mu)$ correspond?

Recall ${}^LG$ is the semi-direct product

$${}^LG = GL_3(\mathbb{C}) \rtimes W_{E/F}$$

where $W_{E/F}$ fits into the exact sequence

$$1 \to E^{\times} \to W_{E/F} \to Gal(E/F) \to 1$$

and τ in $W_{E/F}$ acts on $GL_3(\mathbb{C})$ through its projection onto $Gal(E/F)$. In particular, in the present context, $Gal(E/F) = \{I, \text{Frobenius } Fr\}$, with Fr taking

$$g(\text{in } GL_3(\mathbb{C})) \to \begin{bmatrix} 0 & 0 & 1 \\ 0 & -1 & 0 \\ 1 & 0 & 0 \end{bmatrix} {}^tg^{-1} \begin{bmatrix} 0 & 0 & 1 \\ 0 & -1 & 0 \\ 1 & 0 & 0 \end{bmatrix} .$$

Now let ρ_0 denote the standard representation of $GL_3(\mathbb{C})$ and set

$$\rho = \mathrm{Ind}_{GL_3(\mathbb{C})\times E^{\times}}^{{}^LG}(\rho_0) .$$

Since $W_{E/F}$ acts non-trivially on ρ_0 (taking it to its "twisted" contragredient), ρ is an <u>irreducible</u> six-dimensional representation of ${}^LG$ whose restriction to ${}^LG^0 = GL_3(\mathbb{C})$ is the direct sum of $\rho_0(g)$ and $\tilde{\rho}_0(g) = \rho_0\left(\begin{bmatrix} & & 1 \\ & -1 & \\ 1 & & \end{bmatrix} {}^tg^{-1} \begin{bmatrix} & & 1 \\ & -1 & \\ 1 & & \end{bmatrix}\right)$.

<u>Proposition</u>. Let t_ν denote the conjugacy class in ${}^LG$ determined (via the Satake isomorphism) by the unramified representation $\pi(\nu)$ of G_F. Then

$$L(s,\pi,1) = \{\det[I-\rho(t_\nu|\tilde{\omega}|_F^s]\}^{-1},$$

the (local) L-function Langlands attaches to the data $\{t_\nu,\rho\}$

<u>Proof</u>. The conjugacy class t_ν determined by $\pi(\nu)$ is

$$\begin{bmatrix} \nu(\tilde{\omega}) & 0 & 0 \\ 0 & 1 & 0 \\ 0 & 0 & 1 \end{bmatrix} \rtimes Fr \in {}^LG ,$$

so

$$\rho(t_\nu) = \left[\begin{array}{ccc|ccc} & & & 1 & & \\ & & & & 1 & \\ & & & & & \nu^{-1}(\tilde{\omega}) \\ \hline \nu(\tilde{\omega}) & & & & & \\ & 1 & & & & \\ & & 1 & & & \end{array}\right]$$

A straightforward calculation with determinants then shows

$$\det[I-\rho(t_\nu)|\tilde{\omega}_F|^s]^{-1} = L_E(s,1)L(2s,\nu)L(2s,\nu^{-1}),$$

as claimed.

<u>(4.9)</u> Finally, we relate $L(s,\pi,\mu)$ to the standard L-functions on GL_3 over E. Still working locally, let

$$G' = Res_F^E(G_F) = Res_F^E(GL_3).$$

Then $G_F' = GL_3(E)$,

$${}^LG_F' = GL_3(\mathbb{C}) \times GL_3(\mathbb{C}) \rtimes W_{E/F}$$

and

$${}^L(GL_3)_E = GL_3(\mathbb{C}) \times W_{E/E} .$$

Now consider the representation $\pi^\nu \otimes \mu$ of $G_F' = GL_3(E)$ induced from the unramified character

$$\begin{bmatrix} a_1 & * & * \\ 0 & a_2 & * \\ 0 & 0 & a_3 \end{bmatrix} \mapsto \mu\nu(a_1)\mu(a_2)\mu\nu^{-1}(a_3),$$

of the standard Borel subgroup of $GL_3(E)$. The conjugacy class in ${}^L(GL_3)_E$ corresponding to $\pi^\nu \otimes \mu$ is

$$\begin{bmatrix} \mu\nu(\tilde{\omega}) & 0 & 0 \\ 0 & \mu(\tilde{\omega}) & 0 \\ 0 & 0 & \mu\nu^{-1}(\tilde{\omega}) \end{bmatrix} \times \mathrm{Fr}.$$

Therefore, the following L-factors coincide:

(i) $L(s,\pi^\nu \otimes \mu,\rho_0)$, as an L-factor for GL_3 <u>over E</u>, and

(ii) $L(s,\pi(\nu),\mu)$, our L-factor attached to G over F.

Note that π^ν is clearly the <u>base change lift</u> of $\pi(\nu)$ on G to G'.

§5. Global Base Change Lifting (non-degenerate π).

Suppose π is an automorphic cuspidal representation of $G_{\mathbb{A}}$ which is not hypercuspidal. We claim (and sketch a proof of the fact) that π lifts (via "base change") to an automorphic representation Π of

$$G' = \mathrm{Res}_F^E\, G = \mathrm{Res}_F^E\, GL_3;$$

moreover, this lift will be cuspidal if and only if $L(s,\pi,\mu)$ has no pole.

(5.1) Suppose $\pi = \otimes \pi_v$, and $L(s,\pi,\mu) = \Pi\, L(s,\pi_v,\mu_v)$, with each $L(s,\pi_v,\mu_v,)$ as defined by the local theory of (4.3).

Applying the local-global L-function machine first devised by [Tate], generalized by [Jacquet-Langlands], and well-oiled by [Jacquet], [God-Jac], etc., we can conclude $L(s,\pi,\mu)$ satisfies the following properties:

(i) it is meromorphic in s, its only possible poles proportional to some

$$\int_{H_F\backslash H_{\mathbb{A}}} f(h)\, \mu(\det h) dh, \ f \in H_\pi; \text{ and}$$

(ii) it satisfies a functional equation of the form

$$\tilde{L}(1-s,\tilde{\pi},\mu^{-1}) = \varepsilon(s,\pi,\mu)\, L(s,\pi,\mu).$$

(5.2) By completing the analysis of §4, we get, for almost every place v, that

$$L(s,\pi_v,\mu_v) = \prod_{w|v} L(s,\Pi_w \otimes \mu_w)$$

for some irreducible Π_w on $GL_3(E_v)$. With some additional local information (mostly about the ε-factors $\varepsilon(s,\pi(v),\mu,\psi)$), we can apply the converse theorem of [JPSSh] to conclude that there exists a (global) automorphic representation Π of $GL_3(\mathbb{A}_E)$ such that

(1) $\Pi_w = \pi^{\nu_w}$ whenever $E_v = E \otimes_F F_v$ is an unramified quadratic extension of F_v, with $w|v$, and $\pi_v = \pi(\nu_w)$;

(2) Π is cuspidal if and only if

$$\int_{H_F\backslash H_{\mathbb{A}}} f(h)\, \mu(\det h)\, dh = 0$$

for all μ and $f \in H_\pi$. The detailed reasoning is exactly analogous to what Jacquet proves for (quadratic) base-change lifting for GL_2 in [Jacquet].

(5.3) Remark. The lifting $\pi \to \Pi$ exists for hypercuspidal π too, and will follow from our general construction of $L(s,\pi)$ using Shimura's method (§9). For a trace-formula proof of this lifting (for arbitrary cuspidal π), see [Flicker].

(5.4) Summing up, the following are equivalent.

(A) $L(s,\pi,\mu)$ has a pole for some μ;

(B) π base-change lifts to an automorphic (as opposed to automorphic cuspidal) representation Π of GL_3 over E;

(C) for some μ, and f in H_π,

$$\int_{H_F \backslash H_{\mathbb{A}}} f(h) \quad \mu(\det h)dh \neq 0$$

What remains to be done is to extend this theory to hypercuspidal π (as much as is possible) and to relate the existence of a pole for $L(s,\pi)$ (condition (A)) to π being in the image of a lifting of automorphic forms from $U_{1,1}$ to $U_{2,1}$.

For all this, the Weil representation, which we shall now describe, plays a crucial role.

§6. Weil Representation

(6.1) First suppose F is a local field not of characteristic 2, E is quadratic over F, and V is over 3-dimensional (isotropic) Hermitian space over E.

Regarding V as a 6-dimensional space over F, we equip it with the F-valued skew-symmetric form $\text{Im}(\ ,\ )_V$. Then V_F is a symplectic space, and we let $Sp(V_F)$ denote its symplectic group.

(6.2) Fixing an additive character ψ of F, there is defined a unique representation ω_ψ of $Sp(V_F)$, called the Weil representation (attached to V_F and ψ); cf. [Rallis]. Actually, unless $F = \mathbb{C}$, ω_ψ only defines an ordinary representation of the (unique) two-fold metaplectic cover of $Sp(V_F)$, denoted $\tilde{Sp}(V_F)$; on $Sp(V_F)$ itself, ω_ψ defines a multiplier representation with non-trivial co-cycle.

(6.3) To describe ω_ψ explicitly, it is convenient to choose a "complete polarization" for V_F; cf. [Howe] and [Ge]. Thus, if V_1 and V_2 are maximal totally isotropic subspaces of V_F such that $V_F = V_1 \oplus V_2$, we may describe $\omega_\psi(g)$ through the action of familiar operators in the Schwartz-Bruhat space $\mathcal{S}(V_1)$; this gives a "Schrodinger realization" of ω_ψ. In particular, if $g = g_\beta$ is an element of $Sp(V_F)$ which acts as the identity in V_2, then

$$\omega_\psi(g_\beta)\ \Phi(X_1) = \psi(\tfrac{1}{2}\,\text{Im}(X_1, X_1\beta))\ \Phi(X_1),$$

with X_β the projection of $g \cdot X_1$ onto V_2; other formulas for ω_ψ are compiled in [Rallis] and [Ho PS 2].

(6.4) Suppose we now regard $SU(V)$ as a subgroup of $Sp(V_F)$. (Since each g in $U(V)$ preserves $(\ ,\ )_V$, it certainly preserves the symplectic form $\text{Im}(\ ,\ )_V$.) Then the restriction of ω_ψ to SU can be made into an ordinary representation, and extended to an ordinary representation of $U(V)$ (which we again denote by ω_ψ).

Remarks. (i) That ω_ψ on SU_n can be made into an ordinary representation was first spelled out by Kazhdan, following remarks of Sah; cf. [Kazhdan]. That this representation can be extended to one of $U(V)$ follows from a careful analysis of the restriction of ω_ψ to a certain parabolic subgroup of $U(V)$; cf. §7 for a few more details.

(ii) Since $SU(V)$ is its own commutator subgroup, there is only one way to make ω_ψ on SU into an ordinary representation. On the other hand, the extension of ω_ψ on SU up to $U(V)$ is not unique; because

$$U(V)/SU(V) \simeq U(1) \simeq E^1,$$

the extension to $U(V)$ is unique only up to twisting by $\mu(\det g)$, μ a character of E^1.

(6.5) Let ω_ψ denote any one of the ordinary Weil representation of $U(V)$ just introduced. The center of $U(V)$ consists of scalar operators αI, $\alpha \in E^1$. Thus ω_ψ decomposes according to the characters χ of $E^1 \approx U(1)$.

Proposition. Let ω_ψ^χ denote the subrepresentation of ω_ψ which transforms according to the central character χ. Then:

(i) ω_ψ^χ is an irreducible admissible representation of $G_F = U_{2,1}(F)$;

(ii) if χ is "unramified relative to ψ", ω_ψ^χ is equivalent to the non-tempered Langlands quotients of some reducible principal series representation "at $s = 1/2$"; otherwise, ω_ψ^χ is supercuspidal (which for an archimedean place we shall take to mean "discrete series").

Remarks (i) By χ "unramified relative to ψ" we mean χ occurs in the restriction of the Weil representation r_ψ of $SL_2(F)$ restricted to $U(1)$(embedded as a torus); in particular, if χ,ψ and E are all unramified (which is the case "almost everywhere"), then χ occurs in $r_\psi|_{E^1}$, and ω_ψ^χ is class 1.

(ii) These facts are due to Howe and Piatetski-Shapiro, who proved them in order to construct cuspidal representations $\Pi = \otimes\, \omega_{\psi_v}^{\chi_v}$ contradicting the generalized Ramanujan conjecture; the published version of their work - for Sp_4 in place of $U_{2,1}$, appears in [Ho PS1].

(6.6) As the remarks above already suggest, the construction of ω_ψ^χ is significant globally as well as locally.

Suppose F is an $\mathbb{A}$-field not of characteristic 2, $\psi = \Pi\psi_v$ is a non-trivial character of $F\backslash\mathbb{A}$, and $\chi = \prod_w \chi_w$ is a character of $E^\times|E^1_{\mathbb{A}}$. For any prime v of F, let $E_v = E \otimes_F F_v$, $V_v = V_F \otimes_v F_v$, etc.

Suppose first that v splits in E, so $E_v \approx F_v \oplus F_v$, and $U_v = G_v \approx GL_3(F_v)$ embeds "diagonally" in $Sp(V_v) = Sp_6(F_v)$. Then it is easy to prove directly that $\omega_{\psi_v}^{\chi_v}|_{SU}$ can be made into an ordinary representation and extended up to one of U_v.

Otherwise, if E_v remains a field, we use (6.4) to define $\omega_{\psi_v}^{\chi_v}$, and then construct the product

$$\omega_\psi^\chi = \otimes\, \omega_{\psi_v}^{\chi_v}$$

The result is an irreducible unitary representation of $G_{\mathbb{A}} = \Pi\, G_v$ which we shall now describe in more detail.

(6.7) First we intertwine ω_ψ^χ with the space of automorphic forms on $G_{\mathbb{A}}$ via the imbedding

$$\Phi \text{ in } \mathcal{S}(V_1)_{\mathbb{A}}) \to \Theta_\phi(g) = \sum_{\xi \in V_1} (\omega_\psi^\chi(g)\Phi)(\xi)$$

The fact that $\Theta_\Phi(g)$ is left G_F invariant follows from the well-known fact (cf. [Weil] or [Ho]) that the map

$$\Phi \to \sum_{\xi \in V_1} \Phi(\xi)$$

defines the (unique up multiples) functional on $\mathcal{S}((V_1)_{\mathbb{A}})$ (the space of smooth vectors for ω_ψ^χ) which is invariant for $\omega_\psi^\chi(Sp(V_F))$.

(6.8) The question remains, when is ω_ψ^χ actually cuspidal?

Using an explicit Schrodinger realization of ω_ψ^χ we can compute the constant coefficients of $\Theta_\Phi(g)$ along the unipotent subgroup N and its center Z. The startling result is that

$$(\Theta_\Phi)_0(g) = (\Theta_\Phi)_{00}(g)$$

Therefore the following statements are equivalent:

(i) ω_ψ^χ is cuspidal;

(ii) at least one local component $\omega_{\psi_v}^{\chi_v}$ of ω_ψ^χ is "supercuspidal";

(iii) the Z-constant terms $(\Theta_\Phi)_{00}$ vanish identically;

(iv) ω_ψ^χ is hypercuspidal.

Similar local computations show that when E_v is a field, $\omega_{\psi_v}^{\chi_v}$ cannot possess a non-trivial ψ_N-Whittaker functional. Indeed, what is actually shown is that such a $\omega_{\psi_v}^{\chi_v}$ is "locally hypercuspidal" in the sense that it supports no Z-invariant functionals and (for $G = U_{2,1}$) this is equivalent to the non-existence of ψ_N-Whittaker functionals.

In conclusion, the Weil representation gives us examples of local degenerate representations which are also globally hypercuspidal. Other examples of hypercusp forms on $G_{\mathbb{A}}$ are those whose $G_{\mathbb{R}}$-components are holomorphic discrete series respresentations of $U_{2,1}$. In the Sections which follow, we shall discuss (generalized) Whittaker models and L-functions for such cusp forms.

§7. More Fourier Expansions and Hypercuspidality

In §2, we examined Fourier expansions for an arbitrary cusp form f on $G_{\mathbb{A}}$. This led to the Whittaker models $\mathcal{W}(\pi,\psi_N)$, hence to a dead end in the case of hypercuspforms.

Given an arbitrary (cuspidal) automorphic representation (π,H_π) of $G_{\mathbb{A}}$, we want to produce a generalized Whittaker model for π by looking at Fourier expansions of f in H_π along Z (instead of N).

(7.1). Recall

$$Z = \left\{ \begin{bmatrix} 1 & 0 & z \\ 0 & 1 & 0 \\ 0 & 0 & 1 \end{bmatrix} : \bar{z} = -z \right\},$$

so an arbitrary character of Z is of the form

$$\psi_Z\left(\begin{bmatrix} 1 & 0 & z \\ & 1 & 0 \\ & & 1 \end{bmatrix}\right) = \psi(\operatorname{Im} z),$$

with ψ a character of F.

If we define the ψ-th coefficient of f along Z to be

$$f_\psi(g) = \int_{Z_F\backslash Z_{\mathbb{A}}} f(zg)\overline{\psi_Z(z)}dz,$$

then harmonic analysis on the compact abelian group $Z_F\backslash Z_{\mathbb{A}}$ gives

$$f(g) = f_{00}(g) + \sum_{\text{all } \psi \neq 1} f_\psi(g).$$

Remark. Contrary to the situation encountered for the Fourier coefficients W_f^ψ in (2.1), knowing f_ψ for one fixed ψ_0 does not (in general) determine the remaining f_ψ. Indeed, from (1.4) we recall that the diagonal element $\delta = \begin{bmatrix} \delta & & \\ & 1 & \\ & & \bar{\delta}^{-1} \end{bmatrix}$, $\delta \in E^x$, conjugates $\begin{bmatrix} 1 & 0 & z \\ & 1 & 0 \\ & & 1 \end{bmatrix}$ to $\begin{bmatrix} 1 & 0 & \delta\bar{\delta}z \\ & 1 & 0 \\ & & 1 \end{bmatrix}$; therefore

$$f_{\psi^{N\delta}}(g) = f_\psi\left(\begin{bmatrix} \delta & & \\ & 1 & \\ & & \bar{\delta}^{-1} \end{bmatrix} g\right),$$

but no other relations between coefficients are obtained.

(7.2) Choose ψ_0 such that

$$f_{\psi_0}(g) \neq 0,\ f \in H_\pi.$$

Such a ψ_0 must exist, since otherwise we would have $f \equiv f_{00}$, which is clearly impossible.

Let $\mathcal{L}_{\psi_0}$ denote the space of restrictions of the functions f_{ψ_0} to $R_{\mathbb{A}}$. (Recall R is the stabilizer in P of the character $(\psi_0)_Z$.) Since $\gamma \in R_F$ implies

$$f_{\psi_0}(\gamma r) = \int_{Z_F\backslash Z_{\mathbb{A}}} f(z\gamma r)\overline{\psi_0(z)}dz$$

$$= \int_{Z_F\backslash Z_{\mathbb{A}}} f(\gamma\ \ zr)\overline{\psi_0(z)}dz = f_{\psi_0}(r)$$

for all $r \in R_{\mathbb{A}}$, it follows $\mathcal{L}_{\psi_0}$ is a subspace of $L^2(R_F\backslash R_{\mathbb{A}})$ invariant on the right by $R_{\mathbb{A}}$.

Let σ denote any irreducible automorphic representation of $R_{\mathbb{A}}$ which occurs in the $R_{\mathbb{A}}$-module $\mathcal{L}_{\psi_0}$, say in the subspace L_σ. Then the map from (π, H_π) to the projection of $\mathcal{L}_{\psi_0}$ on L_σ is a non-trivial σ-Whittaker map for π in the sense of (1.6).

(7.3). Example. Suppose π is the (hypercuspidal) representation $\omega^X_{\psi_0}$, and $f = \theta_\Phi = \sum_{\xi\varepsilon V_1} \omega^X_\psi(g)\Phi(\xi)$. In this case,

$$f_\psi(g) \equiv 0$$

unless $\psi = \psi_0^{N\delta}$ for some $\delta \varepsilon E^X$. In other words, π (or $f = \theta_\Phi$) is distinguished in the sense that only one (non-trivial) orbit of Fourier coefficients along Z is non-zero, and therefore

$$f(g) = \sum_{\delta\varepsilon E^X} f_{\psi_0}\left(\begin{bmatrix} \delta & 0 & 0 \\ 0 & 1 & 0 \\ 0 & 0 & \bar{\delta}^{-1} \end{bmatrix} g\right).$$

In general, such a simple Fourier expansion is not possible.

For $\omega^X_{\psi_0}$, the resulting $R_{\mathbb{A}}$-module $\mathcal{L}_{\psi_0}$ is of particular interest. Indeed in this case, ω^X_ψ further "distinguishes" itself by the property that $\mathcal{L}_{\psi_0}$ is already irreducible as an $R_{\mathbb{A}}$-module. (For a general definition of what is meant by a "distinguished representation" of G_F, see the last Section of [PS 1].

Remark. Exactly which irreducible representation of $R_{\mathbb{A}}$ is realized by $\mathcal{L}_{\psi_0}$ depends on which extension of $\omega_\psi|_{SU}$ to $U_{2,1}$ is chosen. However, by varying ψ_0 and the extension chosen for ω_{ψ_0}, all irreducible representations $\rho^\alpha_{\psi_0}$ of $R_{\mathbb{A}}$ are eventually obtained. This results from a careful analysis of the representation theory of the Heisenberg group N and its (almost) trivial extension R; cf.§(1.6).

(7.4). For an arbitrary (possibly hypercuspidal) cuspidal representation (π,H_π), the fact that π is not in general distinguished makes it difficult to define an Euler product for π using only the Fourier expansion of f in H_π. However, the idea of Shimura's method is that by integrating such an f against a distinguished form such as θ_Φ, all but one orbit of Fourier coefficients will end up disappearing, and an Euler product will become possible after all. Before explaining this in detail in §9, we pause to define the lifting from $U_{1,1}$ to $U_{2,1}$ using Weil's representation.

§8. Lifting from $U_{1,1}$ to $U_{2,1}$;

Our lifting of forms from $U_{1,1}$ to $U_{2,1}$ is a particular example of the duality correspondence which results from R. Howe's theory of dual reductive pairs ([Ho]; cf. [Ge]).

(8.1) The dual reductive pair is (H,G), where $G \approx U_{2,1}$ is the unitary group of the Hermitian space V spanned by the vectors $\ell_{-1}, \ell_0, \ell_1$, and $H \approx U_{1,1}$ is the unitary group of the two-dimensional W spanned by the isotropic vectors $w_1 = \ell_{-1}$, $w_2 = \ell_1$. The matrices of these hermitian forms are, respectively, $\begin{bmatrix} & & 1 \\ & 1 & \\ 1 & & \end{bmatrix}$ and $\begin{bmatrix} & 1 \\ 1 & \end{bmatrix}$.

To embed (H,G) in a symplectic group we form the product $W \otimes V$ and consider the symplectic space $(W \otimes V)_F$ obtained by taking the imaginary part of the Hermitian form $(\ ,\ )_W (\ ,\ )_V$. Thus $X_F = (W \otimes V)_F$ is 12 dimensional over F, and we have $(H,G) \subset Sp_{12}(F) = Sp(X_F)$, a dual reductive pair in the sense of [Ho].

(8.2) Given the character $\psi = \Pi\psi_v$ on $F\backslash\mathbb{A}$, we recall the (projective) Weil representation ω_ψ of $Sp(X)$ discussed in §6. (This earlier discussion was for Sp_6, but the theory works equally well for Sp_{2n}, let alone Sp_{12}). By ω_ψ we also denote the restriction of the Weil representation to SU, made into an ordinary representation there and extended to an ordinary representation of $U(W \otimes V)$. In this global context, ω_ψ is determined on $U(W \otimes V)_{\mathbb{A}}$ only up to twisting by a character of $E^1_{\mathbb{A}}$ composed with the determinant. Therefore, we choose one such extension ω_ψ, and fix it once and for all.

(8.3) Suppose (τ, V_τ) is an automorphic cuspidal representation of H in the space of cusp forms V_τ on $H_{\mathbb{A}}$. For each ϕ in V_τ, we define its lifting to $G_{\mathbb{A}}$ by

$$f_\phi(g) = \int_{H_F\backslash H_{\mathbb{A}}} \Theta_\psi^\Phi(h,g)\, \phi(h)dh$$

Here Θ_ψ^Φ is the theta-kernel

(8.3.1) $$\Theta_\psi(h,g) = \sum_{X \in (X_1)_F} \omega_\psi(h,g)\, \Phi(X),$$

with $X_F = X_1 \oplus X_2$ a complete polarization of X, and $\Phi \in \mathcal{S}((X_1)_{\mathbb{A}})$.

(8.4) The polarization used is

$$X_1 = (\ell_1 \otimes W) \oplus (\ell_0 \otimes E_{W_1}),$$

$$X_2 = (\ell_{-1} \otimes W) \oplus (\ell_0 \otimes E_{W_2}).$$

Here W is regarded as a 4-dimensional vector space over F, and

$$X_1 = \{\ell_1 \otimes w + \ell_0 \times tw_1\} \approx \{w,t\},$$

with $w \in W$ and $t \in E$. Note that

$$Z = \left\{\begin{bmatrix} 1 & 0 & z \\ & 1 & 0 \\ & & 1 \end{bmatrix}\right\}$$

leaves ℓ_0 and ℓ_{-1} alone, hence acts as the identity on X_2. This means (cf. §(6.2) and [Rallis]) that the formula for ω_ψ is especially simple. In particular, for $X = (w,t)$,

$$\text{(8.4.1)} \qquad \omega_\psi\left(h, \begin{bmatrix} 1 & 0 & z \\ & 1 & 0 \\ & & 1 \end{bmatrix} g\right) \Phi(X) = \psi\left(\frac{\operatorname{Im} \bar{z}(w,w)_W}{2}\right) \omega_\psi(g,h)\, \Phi(X).$$

<u>(8.5)</u> Using (8.4.1), and a similar (yet more complicated) formula for $\omega_\psi(g,nh)\,\Phi(X)$, $n \in N$, we can prove the following:

<u>Proposition</u>. Given (τ, V_τ) as before, assume ψ such that $\mathfrak{W}(\tau,\psi)$ exists. Then:

(i) The space of ψ-lifts of τ, namely

$$\Theta(\tau,\psi) = \{f_\phi^\psi(g) : \phi \in V_\tau\},$$

is non-zero, and generates an irreducible automorphic representation $\pi(\tau)$ of $G_{\mathbb{A}}$;

(ii) $\pi(\tau)$ is cuspidal if and only if τ is orthogonal to all Θ-series on $U_{1,1}$ derived from the pair $(U_1, U_{1,1})$;

(iii) $\pi(\tau) \subset L^2_{0,1}$ the orthocomplement of the space of hypercusp forms; in fact,

$$\text{(8.5.1)} \qquad W_f^\psi(g) = \int_{U_{\mathbb{A}}\backslash H_{\mathbb{A}}} W_\phi^\psi(h)\ \omega_\psi(h,g)\ \Phi(w_{2,1})\,dh,$$

where $U = H \cap Z = \left\{\begin{bmatrix} 1 & 0 & u \\ & 1 & 0 \\ & & 1 \end{bmatrix}\right\}$ is the maximal unipotent subgroup of H, and

$$\text{(8.5.2)} \qquad W_\phi^\psi(h) = \int_{U_F\backslash U_{\mathbb{A}}} \phi(uh)\ \overline{\psi(u)}\ du$$

is the ψ-Fourier coefficient of ϕ along U.

<u>(8.6)</u> <u>Remarks.</u>

(i) By $\mathfrak{W}(\tau,\psi)$ we mean the Whittaker space of ψ-Fourier coefficients of ϕ in V_τ as defined by (8.5.2); for at least one orbit of characters ψ (under conjugation action of the torus T on U) this space must be non-zero.

(ii) By theta-series on $U_{1,1}$ derived from U_1 we mean precisely the functions

$$\Theta(h) = \sum_{t\in E} \omega_\psi(1,h)\ \Phi(0,t).$$

Restricted to $SU_{1,1} \approx SL_2$, these are precisely (generalizations of) the classical "theta-series with grossencharacter" constructed by Hecke and Maass, and then Shalika-Tanaka ([Shal-Tan]). Indeed, the restriction of ω_ψ to $SU \subset H$ is just the usual Weil representation r_ψ of SL_2 in $\mathcal{S}(F)$.

(iii) The formula (8.5.1) for W_f^ψ not only shows that f_{00} (and hence $\mathcal{W}(\pi,\psi)$) is non-zero, it also affords a purely local definition of the lifting $\tau \to \pi(\tau)$. Indeed, suppose τ_v is an irreducible admissible representation of H_v, and ψ_v on F_v is such that the Whittaker model $\mathcal{W}(\tau_v,\psi_v)$ exists. Then define the lift of τ_v to be the space of (Whittaker) functions

(8.6.1) $$W_v(g) = \int_{U_v\backslash H_v} W_{\tau_v}(h_v)\ \omega_{\psi_v}(h,g)\ \Phi(w_2,1)dh$$

where $g \in G_v$, $W_{\tau_v} \in \mathcal{W}(\tau_v,\psi_v)$, and Φ is a Schwartz-Bruhat function.

We note that if $\tau = \tau(\nu)$, i.e., the class 1 representation of $(U_{1,1})_v$ induced from the character

$$\begin{bmatrix} \alpha & * \\ & \bar{\alpha}^{-1} \end{bmatrix} \to \nu(\alpha),$$

then $\pi(\tau)$ is the induced representation $\pi(\nu) = \mathrm{Ind}_P^G\ \nu^*$ described in §4.

(8.7) It remains to characterize the image of the lift $\tau \to \pi(\tau)$, i.e., to characterize those $\pi \subset L^2_{0,1}$ which are of the form $\pi(\tau)$ for some cuspidal τ on $H_{\mathbb{A}}$.

Theorem. Suppose (π,H_π) is an automorphic cuspidal representation of $G_{\mathbb{A}}$, with $H_\pi \subset L^2_{0,1}$. Then π is of the form $\pi(\tau)$ for some (τ,V_τ) cuspidal on $H_{\mathbb{A}}$ if for some μ,

(8.7.1) $$\int_{H_F\backslash H_{\mathbb{A}}} f(h)\ \mu(\det h)dh \neq 0$$

for some f in H_π.

Proof. The idea is to define a lifting

$$f \to \phi_f$$

from cusp forms on G to cusp froms on H such that $f = f_{\phi_f}$ as soon as f

satisfies condition (8.7.1). To make this lift adjoint to $\phi \to f$, we define

$$\phi_f(h) = \int_{G_F \backslash G_{\mathbb{A}}} \overline{\Theta_\psi(h,g)}\, f(g)dg;$$

with $\Theta_\psi(h,g)$ the theta-kernel defined by (8.3.1). This time, however, we choose the simpler polarization

$$W \otimes V = (w_1 \otimes V) \oplus (w_2 \otimes V).$$

Computing the constant term of $\phi = \phi_f$, we find that

$$\phi_0(h) = \int_{N_{\mathbb{A}} \backslash G_{\mathbb{A}}} \omega_\psi(h,g)\Big(\int_{N_F \backslash N_{\mathbb{A}}} f(ng)dn\Big)dg = 0,$$

since f itself is a cusp form.

In other words, the lift to $H_{\mathbb{A}}$ of any cusp form on $G_{\mathbb{A}}$ is automatically cuspidal on $H_{\mathbb{A}}$. To show this cusp form is non-zero, we compute

$$W_\phi^\psi(1) = \int_{H_{\mathbb{A}} \backslash G_{\mathbb{A}}} \Big(\int_{H_F \backslash H_{\mathbb{A}}} \omega_\psi(1,g)\, \Phi(h\ell_0)\, f(hg)\, dh\Big)dg$$

Choosing Φ such that $\Phi(v) \neq 0$ only if (v,v) is close to 1, we get

$$W_\phi^\psi(1) = \int_{\substack{\text{compact}\\ \text{set}}} \omega_\psi(1,g)\, \Phi(\ell_0)\, \Big(\int_{H_F \backslash H_{\mathbb{A}}} f(hg)dh\Big)dg \neq 0$$

<u>if</u>

$$\int_{H_F \backslash H_{\mathbb{A}}} f^g(h)dh \neq 0$$

In other words, ϕ_f is a <u>non-zero</u> cuspform, and possesses a non-trivial ψ-Whittaker model, as soon as (8.7.1) holds with $\mu = 1$; hence it may be lifted back to $f = f_{\phi_f}$ on $U_{2,1}$. More generally, if (8.7.1) holds, we can argue similarly with $\mu \cdot \omega_\psi$ in place of ω_ψ.

<u>(8.8)</u> <u>Corollary</u>. Suppose $(\pi,H_\pi) \subset L^2_{0,1}$. Then $\pi = \pi(\tau)$ (for some (τ,V_τ) cuspidal on $H_{\mathbb{A}}$) if <u>and only if</u> $L(s,\pi,\mu)$ has a pole for some μ.

One direction follows immediately from Theorem 8.7 combined with (5.4). For the reverse direction we need to argue more indirectly, analyzing the poles of $L(s,\pi,\mu)$ in terms of its explicit Euler product expansion; details are given in [PS 4] for the analogous dual reductive pair $(PGSp_4, SL_2)$.

<u>(8.9)</u> <u>Remark</u>. This pretty picture mostly falls apart - or becomes meaningless - when Whittaker models fail to exist. For example, if (π,H_π) is hypercuspidal, then $L(s,\pi)$ is not yet even defined. Moreover, if f in H_π is hypercuspidal, then the integral

$$\int_{H_F \backslash H_{\mathbb{A}}} f(h)dh$$

is <u>automatically</u> zero, since $f_{00} = 0$ means precisely that the restriction of f to $H_{\mathbb{A}}$ is cuspidal (and hence orthogonal to the space of constant functions).

What we shall attempt to do in the next Section is to define $L(s,\pi)$ for arbitrary cuspidal π and relate the existence of a pole for $L(s,\pi)$ to π being in the image of a lifting from U_2 or to an f in H_π satisfying some kind of analogue of the condition $\int f\, dh \neq 0$.

§9. L-functions of Shimura-type

This method was developed for classical modular forms of half-integral weight by [Shimura], then in representation-theoretic terms (but again for modular forms of half-integral weight) by us in [Ge PS]. In the generality of arbitrary reductive groups, Shimura's method was developed by Piatetski-Shapiro in [PS 1], initial results for $U_{2,1}$ being worked out by him in lectures at Yale in 1977-78 ([PS 3]).

Here, we shall give more details for the case $G = U_{2,1}$. The idea is to define global zeta functions

(9.0) $$L^{\mu}(f,F,s) = \int_{G_F \backslash G_{\mathbb{A}}} f(g)\overline{\Theta(g)}E(g,F,s)dg,$$

with Θ (resp. E) a theta (resp. Eisenstein) series <u>on $G_{\mathbb{A}}$</u>, and then proceed as in §4 to define local factors $L(s,\pi)$, etc.

<u>(9.1)</u> Recall $G = MNK$ with

$$M = \left\{ \begin{bmatrix} \delta & & \\ & \beta & \\ & & \bar{\delta}^{-1} \end{bmatrix} \delta \in E^{\times},\ \beta \in E^{1} \right\}.$$

To define an Eisenstein series on $G_{\mathbb{A}}$ we fix a character μ of $E^{\times}\backslash \mathbb{A}_E^{\times}$; then we set

$$\mu_s\left(\begin{bmatrix} \delta & * & * \\ 0 & \beta & * \\ 0 & 0 & \bar{\delta}^{-1} \end{bmatrix}\right) = \mu(\delta)|\delta|_E^{s},$$

$$F(g) = F(g,\Phi,s) = \int_{E^{\times}} (g\cdot\Phi)(t\ell_{-1})\mu(t)|t|_E^{s}dt,$$

with Φ a Schwartz-Bruhat function on $V_{\mathbb{A}}$, and

$$E^{\mu}(g,F,s) = \sum_{P_F \backslash G_F} F(\gamma g).$$

As usual, the series $E^{\mu}(g,F,s)$ converges for $\mathrm{Re}(s)$ large, and by Selberg-Langlands theory continues meromorphically in s to define a (slowly increasing) automorphic form on $G_{\mathbb{A}}$. Note that $F(g)$ is so defined that for all $b \in P_{\mathbb{A}}$,

$$F(bg) = \mu_s(b)F(g).$$

Finally, we define the theta-series $\Theta(g)$ by

$$\Theta(g) = \sum_{\xi \in V_1} \omega_{\psi}^{\chi}(g)\phi(\xi).$$

Here ϕ is a Schwartz-Bruhat function on the adelization of the polarizing subspace V_1 of V. Which Weil representation ω_ψ^χ we choose will depend on the nature of our given (π,H_π), as we shall presently explain.

(9.2). Suppose (π,H_π) is an arbitrary automorphic cuspidal (possibly hypercuspidal) representation of $G_{\mathbb{A}}$, and $f \in H_\pi$. In defining $L^\mu(f,F,s)$ as in (9.0), we must define $\Theta(g)$ as follows.

Recall each automorphic representation σ of $R_{\mathbb{A}} = \left\{\begin{bmatrix} \delta & * & * \\ & \beta & * \\ & & \delta \end{bmatrix} : \delta,\beta \in E^1_{\mathbb{A}}\right\}$ is indexed by parameters ψ_0, $\{\mu_1,\mu_2\}$, where ψ_0 is a non-trivial character of $F\backslash\mathbb{A}$, and μ_1, μ_2 are idele characters of $E_F\backslash E^1_{\mathbb{A}}$. In particular,

$$\sigma = \mu_1\mu_2\,\rho_{\chi_0},$$

where ρ_{ψ_0} is the (unique) irreducible representation of (the Heisenberg group) $N_{\mathbb{A}}$ with central character ψ_0, and $\mu_1\mu_2$ is how we extend ρ_{ψ_0} to $(E^1_{\mathbb{A}} \times E^1_{\mathbb{A}}) \ltimes N_{\mathbb{A}}$.

Now given $f \in H_\pi$, choose ψ_0 and $\{\mu_1,\mu_2\}$ such that the space of ψ_0-Fourier coefficients

$$\{f_{\psi_0}(g) = \int f(zg)\overline{\psi_0(z)}dz\} \neq \{0\}$$

and such that the space of restrictions of these $f_{\psi_0}(g)$ to $R_{\mathbb{A}}$ constitutes a subspace of $L^2(R_F\backslash R_{\mathbb{A}})$ containing the representation $\sigma(\mu_1,\mu_2,\mu_0) = \mu_1\mu_2\rho_{\psi_0}$. We note that this choice of ψ_0 and $\{\mu_1,\mu_2\}$ is (in general) not unique.

Finally, pick the distinguished representation $\omega_{\psi_0}^\chi$ which belongs to this $\sigma(\mu_1,\mu_2,\psi_0)$. In other words (cf. Example (7.3)), we extend ω from SU to U so that the space of restrictions to $R_{\mathbb{A}}$ of the theta-functions

$$\Theta_\phi(g) = \sum \omega_\psi^\chi(g)\phi(\xi)$$

realizes precisely the representation $\mu_1\mu_2\rho_{\psi_0}$.

(9.3). Returning to our (now-defined) integral

$$L^\mu(f,F,s) = \int_{G_F\backslash G_{\mathbb{A}}} f(g)\overline{\Theta_\phi(g)}E(g,F,s)dg,$$

let us examine its analytic properties.

As with the Rankin-Selberg type integral, the poles of $L^\mu(f,F,s)$ come from those of $E(g,F,s)$ (since $f(g)$ rapidly increasing implies that the integrand is integrable). The residues generated by $E(g,F,s)$, though now non-trivial, are nevertheless known.

(9.4). Product Expansion for L^μ.

Specializing the arguments of [PS 1], §III, we shall derive a product expansion for the zeta-function $L^\mu(f,F,s)$.

Replacing $E(g,F,s)$ by the series defining it, we get

$$L^\mu(f,F,s) = \int_{P_F\backslash G_{\mathbb{A}}} f(g)\overline{\Theta_\phi(g)}F(g)dg.$$

From the fact that $\Theta_\phi(g)$ is distinguished, it follows from Example (7.3) that

$$\Theta_\phi(g) = \Theta_{00}(g) + \sum_{R_F\backslash P_F} \Theta_{\psi_0}(pg).$$

Thus, since $\Theta_{00} = \Theta_0$,

$$\begin{aligned} L^\mu(f,F,s) &= \int_{P_F\backslash G_{\mathbb{A}}} f(g)\overline{\Theta_0(g)}F(g)dh + \int_{P_F\backslash G_{\mathbb{A}}} \sum_{R_F\backslash P_F} \overline{\Theta_{\psi_0}(pg)}f(pg)F(pg)dg \\ &= I + II. \end{aligned}$$

Now Θ_0 and $F(g)$ are $N_{\mathbb{A}}$-invariant, and $f(g)$ is cuspidal. Therefore

$$\begin{aligned} I &= \int_{N_{\mathbb{A}}P_F\backslash G_{\mathbb{A}}} \left(\int_{N_F\backslash N_{\mathbb{A}}} f(ng)dn\right)\overline{\Theta_0(g)}F(g)dh \\ &= 0. \end{aligned}$$

On the other hand, since f and F are left P_F-invariant,

$$II = \int_{R_F\backslash G_{\mathbb{A}}} f(g)F(g)\overline{\Theta_{\psi_0}(g)}dg.$$

Thus it remains to find the Euler product expansion of II.

Recall that the character $\mu_s : P_{\mathbb{A}} \to \mathbb{C}^\times$ is such that for $r \in R_{\mathbb{A}}$,

$$F(rg) = \mu(\delta_r)F(g).$$

Therefore

$$L^\mu(f,F,s) = II = \int_{R_{\mathbb{A}}\backslash G_{\mathbb{A}}} \langle f(g),\Theta(g)\rangle_\mu F(g)dg,$$

with

$$\langle f(g),\Theta(g)\rangle_\mu = \int_{R_{\mathbb{A}}} \mu(\delta_r)f(rg)\overline{\Theta_{\psi_0}(rg)}dr,$$

and it remains to analyze $\langle\ ,\ \rangle_\mu$.

Note that for each fixed g, $f(rg)$ is a function on $R_F \backslash R_{\mathbb{A}}$ with Fourier expansion

$$f(rg) = \sum_{\psi \neq 1} f_\psi(rg) + f_{00}(rg),$$

(cf. (7.1)). Therefore, straightforward computation shows

$$\langle f(g), \Theta(g) \rangle_\mu = \int_{R_F Z_{\mathbb{A}} \backslash R_{\mathbb{A}}} \mu(\delta_r) f_{\psi_0}(rg) \overline{\Theta_{\psi_0}(rg)} dr,$$

i.e., $\langle f(g), \Theta(g) \rangle_\mu$ is a sesquilinear form on $H_\pi \times H_{\omega^\chi_{\psi_0}}$ which is completely determined by the appropriate projections onto $\sigma = \mu_1 \mu_2 \rho_{\psi_0}$. More precisely, let $\mathcal{L}^1$ (resp. $\mathcal{L}^2$) denote an $R_{\mathbb{A}}$-map from H_π (resp. $H_{\omega^\chi_{\psi_0}}$) to (σ, L_σ) in $L^2(R_F \backslash R_{\mathbb{A}})$. Then for each $g \in G_{\mathbb{A}}$

$$\langle f(g), \Theta(g) \rangle = \langle \mathcal{L}^1(\pi(g) f), \mathcal{L}^2(\omega^\chi_{\psi_0}(g)\Theta) \rangle \quad .$$

Now recall Thoerem (1.6) asserting the uniqueness of such local $R_{\mathbb{A}}$-maps. It follows that $\mathcal{L}^1$ and $\mathcal{L}^2$ are uniquely determined as the tensor product of local (generalized) Whittaker maps $\mathcal{L}^1_v$ and $\mathcal{L}^2_v$, i.e.,

$$\langle f(g), \Theta(g) \rangle = \Pi \langle \mathcal{L}^1_v(\pi_v(g) w^1_v), \mathcal{L}^2_v(\omega^{\chi_v}_{\psi_{0_v}}(g) w^2_v) \rangle$$

and finally, that

$$L^\mu(f,F,s) = \prod_v \int_{R_v \backslash G_v} W^{1,2}_v(g) F_v(g) dg$$

with

$$W^{1,2}_v(g_v) = \langle \mathcal{L}^1_v(\pi_v(g_v) w^1_v), \mathcal{L}^2_v(\omega^{\chi_v}_{\psi_{0_v}}(g_v) w^2_v) \rangle$$

and

$$F_v(g_v) = \prod_{w|v} \int_{E_w} (g_v \cdot \Phi_w)(t\ell_{-1}) \mu_w(t) |t|^s_w dt \ .$$

(9.5). A local theory can now be developed for the zeta-integrals $L^\mu(f,F,s)$ analogous to the one developed in §4 for the simpler integrals $L^\mu(W,F_\Phi,s)$; when $\pi_v = \pi(\nu)$ is unramified, the g.c.d. of the integrals $L^\mu(f,F,s)$ should coincide with the local factor $L(s,\pi(\nu),\mu)$ already computed using ordinary Whittaker models in §4.

Globally, the Euler product

$$L(s,\pi,\mu) = \prod L(s,\pi_v,\mu_v)$$

will be defined for arbitrary (possibly hypercuspidal) automorphic cuspidal representations π of $G_{\mathbb{A}}$, and will satisfy a functional equation coming from the local functional equations and the global Eisenstein series.

(9.6). What are the possible poles of $L(s,\pi,\mu)$, and how do we characterize their occurrence in terms of π?

The answer comes from examining the global zeta integral

$$L^{\mu}(f,F,s) = \int f(g)\overline{\Theta_{\phi}(g)}E(g,F,s)dg.$$

As remarked in (9.4), the poles of $L^{\mu}(f,F,s)$ arise from those of $E(g,F,s)$, and these are essentially known. More precisely, the poles of $E(g,F,s)$ correspond (by Langlands-Selberg theory) to points of reducibility of the global induced representations $\mathrm{Ind}(\mu|\ |^{s})$, and these are known; cf. [Keys] and [Flicker]. In particular, the residues of $E(g,F,s)$ at $s = 1/2$ generate a (non-tempered Langlands quotient) representation equivalent to the piece of the basic Weil representation ω_{ψ} on G indexed by (an automorphic) character of $U(1)$. Thus $L^{\mu}(f,F,s)$ (and hence $L(s,\pi,\mu)$) should have a pole precisely when

$$\int f(g)\overline{\Theta_{\phi}(g)}\Theta'(g)dg \neq 0. \tag{9.6.1}$$

(9.7). By basic functoriality properties of the duality correspondence for the reductive pairs $(U(1),U_{2,1})$ it should follow that the functions $\overline{\Theta_{\phi}(g)}\Theta'(g)$ generate a representation of $G = U_{2,1}$ coming (via the duality correspondence) from an automorphic representation of an appropriate unitary group H' in 2 variables. Thus condition (9.6.1) might be reformulated as the assertion that f itself is in the image of some automorphic representation τ of H'. i.e., the existence of a pole for $L(s,\pi,\mu)$ corresponds to π being in the image of a lift from some unitary group H'.

(9.8). Concluding Remarks.

To complete the analysis just sketched in this Section, a number of non-trivial computations must be completed. For example, it must be shown in general that the local Shimura-type factors $L(s,\pi_v)$ and $\varepsilon(s,\pi_v)$ are independent of the choice of generalized Whittaker models chosen to construct them, and that they equal the $L(s,\pi_v)$ and $\varepsilon(s,\pi_v)$ of §4 whenever π_v is non-degenerate.

Also, a detailed description of the unitary group H' obtained via (9.7) is crucial not only for comparisons with the results of [Flicker], but also for the search for a proper analogue of the condition (8.7.1) for hypercuspidal π. A complete theory, therefore, awaits future publication.

§10. Odds and Ends

(10.1) Suppose π is any automorphic cuspidal representation of $G_{\mathbb{A}}$ (not necessarily occuring in $L^2_{0,1}$, i.e., not necessarily non-degenerate). Applying the converse theorem for GL_3 (over the quadratic extension E, as in §5) makes it possible to prove that π lifts to an automorphic representation Π of $GL_3(\mathbb{A}_E)$, the base change lift of π; moreover, Π should be cuspidal if and only if π is not itself the lift of a cuspidal representation of some U_2(i.e., $L(s,\pi,\mu)$ is always entire).

(10.2) According to [Flicker], two cuspidal representations π and π' of $G_{\mathbb{A}}$ are L-indistinguishable, or in the same L-packet, if and only if they lift to one and the same Π on $GL_3(\mathbb{A}_E)$. More precisely, this equivalence is first defined in terms of a local lifting; by making certain assumptions about the "twisted" characters of local representations of GL_3, Flicker then describes these L-packets via the trace formula.

It is natural to ask if these L-packets can be described intrinsically, or in terms of the lifting from $U_{1,1}$ to $U_{2,1}$ discussed in these notes? Our conjecture is the following:

(a) Suppose π (locally) lifts from τ on $U_{1,1}$ via the duality correspondence $\Theta(\tau,\psi)$. Then the L-packet $\{\pi\}$ should consist of representations $\pi' = \Theta(\tau',\psi')$ where ψ' varies through the characters $\psi'(x) = \psi(ax)$ of F, and τ' varies through the L-packet $\{\tau\}$ of τ on $U_{1,1}$. We note that:

(i) $\pi' = \Theta(\tau',\psi')$ should be non-degenerate if and only if ψ' is such that the Whittaker model $\mathcal{W}(\tau',\psi')$ exists;

(ii) The theory of the Weil representation implies that the restriction of $\omega_{\psi'}$ to the unitary subgroup of $\widetilde{Sp}$ should depend only on ψ' modulo $N_{E/F}$, i.e. if $a \in N_{E/F}(E^X)$ and $\psi'(x) = \psi(ax)$, then the restrictions of ω_ψ and $\omega_{\psi'}$ to $U(V)$ should be equivalent. In particular, the size of the L-packet of $\{\pi\}$ is at most $2 = [F^X:N_{E/F}(E^X)]$ times the size of the packet $\{\tau\}$, an assertion which is consistent with [Flicker]. If π is a principal series representation, $\{\pi\}$ should consist of π alone.

(b) Suppose $\pi = \pi(\tau)$ is a discrete series representation and $\{\pi\}$ is the L-packet just described in (a). Working locally, and following the lead of [Wald 2], one might hope to realize the elements of $\{\pi\}$ alternatively by moving around the following (non-commutative) diagram:

$$\begin{array}{ccc} U_{1,1} & \xrightarrow{\Theta(\tau,\psi)\,=\,\pi(\tau)} & U_{2,1} \\ {\scriptstyle JL}\updownarrow & & \\ U_2 & \xrightarrow[\Theta(\tau',\psi)\,=\,\pi']{} & U_{2,1} \end{array}$$

The vertical arrows JL refer to (a minor modification of) the Jacquet-Langlands correspondence between discrete series representations τ of GL_2 (which is essentially $U_{1,1}$) and irreducible representations τ' of the multiplicative group of a quaternion algebra (which is essentially a compact unitary group U_2 in 2 variables). By varying τ through the L-packet $\{\tau\}$, the image representations $\Theta(\tau',\psi)$ should exhaust $\{\pi\}$, with ψ now fixed.

(c) Finally, suppose π is an arbitrary admissible representation of $G = U_{2,1}$ (not necessarily of the form $\Theta(\tau,\psi)$ for any τ on $U_{1,1}$). Then of course $\{\pi\}$ should consist of all π' sharing the same L and ε factors (à la §9) as π. In any global L-packet - except one containing a ω_ψ^χ, there should be at least one automorphic representation which has a Whittaker model, i.e., is non-degenerate.

(10.3) We close with some remarks about the condition

$$(8.7.1) \qquad \int_{H_F\backslash H_{\mathbb{A}}} f(h)\, \mu(\det h)dh \neq 0, \; f \in H_\pi,$$

characterizing the existence of a pole for the L-function $L(s,\pi,\mu)$. We have already remarked several times that such a condition is problematic when π is hypercuspidal; cf. Remark 8.9.

One way to get around this problem would be to replace π by a non-degenerate π' in the same L-packet as π. A more meaningful approach might also involve bringing into play the following recent (unpublished) result of Jacquet and Lai.

Let $\pi \leftrightarrow \pi'$ denote the Jacquet-Langlands correspondence alluded to in (10.2), and H_π(resp. $H_{\pi'}$) the corresponding irreducible space of cusp forms on GL_2 (resp. a quaternion group G') over a quadratic extension E of F. Let C(resp. C') denote the property that the integral of some f in H_π(resp. f' in $H_{\pi'}$) over the cycle coming from $GL_2(F)$ (resp. G' over F) does not vanish. Using the trace formula, Jacquet and Lai prove that H_π (usually) satisfies property C if and only if $H_{\pi'}$ satisfies property C'. Unfortunately, the press of a publication deadline prevents us from speculating further on the relevance of this beautiful result to (10.2)(b) and an analogue of our condition (8.7.1).

References

[Cas Sh] Casselman, W., and Shalika, J., "The unramified principal series of p-adic groups II: The Whittaker function", Compositio Math., Vol. 41, Fasc. 2 (1980), pp.207-231.

[Flicker] Flicker, Y., "L-packets and liftings for U(3)", preprint, Princeton University, 1982.

[Ge] Gelbart, S., "Examples of dual reductive pairs", in Proceedings of Symposia in Pure Mathematics, Vol. 33 (1979), part 1, pp.287-296, Amer. Math. Soc.

[Ge PS] Gelbart, S., and Piatetski-Shapiro, I., "On Shimura's correspondence for modular forms of half-integral weight", in Automorphic Forms, Representation Theory and Arithmetic (Bombay, 1979), pp.1-39, Tata Inst. Fund. Res. Studies in Math., 10, Tata Inst. Fund. Res., Bombay, 1981.

[God-Jac] Godement, R., and Jacquet, H., Zeta Functions of Simple Algebras, Springer Lecture Notes in Mathematics, Vol. 260, 1972.

[Goldstein] Goldstein, Larry, Analytic Number Theory, Prentice-Hall, 1971.

[Howe] Howe, R., "θ-series and automorphic forms", in Proceedings of Symposia in Pure Mathematics, Vol. 33 (1979), part 1, pp.275-286, Amer. Math. Soc.

[Ho PS 1] Howe, R., and Piatetski-Shapiro, I., "A counterexample to the "Generalized Ramanujan Conjecture" for (quasi-) split groups", in Proceedings of Symposia in Pure Mathematics, Vol. 33 (1979), part 1, pp.315-322, Amer. Math. Soc.

[Ho PS 2] ——————————, "Some examples of automorphic forms on SP_4", Duke Math. Journal, to appear.

[Jacquet] Jacquet, H., Automorphic Forms on GL(2): II, Springer Lecture Notes in Mathematics, No. 278, 1972.

[JL] Jacquet, H., and R. P. Langlands, Automorphic Forms on GL(2), Lecture Notes in Math., Vol. 114, Springer, 1970.

[J PS Sh] Jacquet, H., Piatetski-Shapiro, I., and Shalika, J., "Automorphic forms on GL(3): I and II", Annals of Math., 109 (1979), pp.169-258.

[Kazhdan] Kazhdan, D., "Some applications of the Weil representation", Journal d'Analyse Mathématique, Vol. 32 (1977), pp.235-248.

[Keys] Keys, D., Principal series representations of special unitary groups over local fields, preprint, U. of Utah, 1981.

[Kottwitz] Kottwitz, R., Unpublished notes.

[Kudla 1] Kudla, S., "On certain arithmetic automorphic forms for SU(1,q)", Inventiones Math. 52, pp.1-25 (1979).

[Kudla 2] ————, "On certain Euler products for SU(2,1)", Compositio Math., Vol. 42, Fasc. 3, 1981, pp.321-344.

[Langlands] Langlands, R. P., "Problems in the theory of automorphic forms", Lectures in Modern Analysis and Applications, Lecture Notes in Mathematics, Vol. 170, Springer, New York, 1970, pp.18-86.

[Novod] Novodvorsky, M., "Automorphic L-functions for the symplectic group GSp_4", in Proceedings of Symposia in Pure Mathematics, Vol. 33 (1979), part 2, pp.87-96, Amer. Math. Soc.

[PS 1] Piatetski-Shapiro, I., "Tate theory for reductive groups and distinguished representations", in Proceedings of the International Congress of Mathematicians, Helsinki 1978, pp.585-590.

[PS 2] ——————, "L-functions for GSp_4", preprint, 1981.

[PS 3] ——————, Lecture notes, Yale University, 1977-78.

[PS 4] ——————, "On the Saito-Kurokawa lifting", Inventiones Math., to appear.

[PS 5] ——————, and Soudry, D., "L and ε factors for GSp(4)", Jour. Fac. Science, Univ. Tokyo, Sec. 1A, Vol. 28, No.3 (Shintani Memorial Volume), pp.505-530, Feb. 1982.

[Rallis] Rallis, S., "Langlands' functoriality and the Weil representation", Amer. J. Math., Vol. 104, No. 3, 1982, pp.469-515.

[Rankin] Rankin, R., "Contributions to the theory of Ramanujan's function $\tau(n)$ and similar arithmetical functions", Proc. Cam. Phil. Soc., 1939.

[Rogawski] Rogawski, J., preprint, Yale Univ., 1981.

[Selberg] Selberg, A., Bemerkungen über eine Dirichletsche Reihe, die mit der Theorie der Modulformen naheverbunden ist, Ach. Math. Naturvid. 43 (1940), pp.47-50.

[Shal] Shalika, J., "The multiplicity one theorem for GL_n", Annals of Math. 100 (1974), pp.171-193.

[Shal-Tan] ——————, and Tanaka, S., "On an explicit construction of a certain class of automorphic forms", Amer. J. Math. 91 (1969), pp.1049-1076.

[Shimura] Shimura, G., "On modular forms of half-integral weight", Annals of Math. (2) 19 (1973), pp.440-481.

[Shintani] Shintani, T., "On automorphic forms on unitary groups of order 3", preprint 1979.

[Tate] Tate, J., "On Fourier analysis in number fields and Hecke's zeta-functions", in Algebraic Number Theory, Cassels, J. W. S. and A.Fröhlich, eds. Thompson Publishing Co., Washington, D.C., 1967, pp.305-347.

[Wald] Waldspurger, J.-L., "Correspondance de Shimura", J. Math. Pures et Appl. 59 (1980), pp.1-133.

[Wald 2] ——————, "Correspondances de Shimura et Quaternions", preprint, Paris, 1981.

[Weil] Weil, A., "Sur certains groupes d'operateurs unitaires, Acta Math. 111 (1964), 143-211.

ON THE RESIDUAL SPECTRUM OF GL(n)

Hervé Jacquet*
Columbia University
New York, NY 10027/USA

§1. Introduction.

(1.1) Let F be a number field, G or G_n the group $(GL(n)$ regarded as an F-group. Let ω be a character of module one of $F_{\mathbb{A}}^{\times}/F^{\times}$; we identify ω with a character of $Z(\mathbb{A})$, where $Z \simeq G_1$ is the center of G. Let then $L_2(\omega)$ denote the space of complex valued functions f on $G(\mathbb{A})$ such that:

(1) $f(\gamma z g) = \omega(z)\ f(g)$, all $\gamma \in G(F)$, $z \in Z(\mathbb{A})$;

(2) $\int |f|^2(g)\ dg < +\infty$, $g \in G(F)Z(\mathbb{A})\backslash G(\mathbb{A})$.

The group $G(\mathbb{A})$ operates by right shifts on $L^2(\omega)$. We denote by $L_d^{\ 2}(\omega)$ the closure of the space spanned by the closed invariant irreducible subspaces of $L^2(\omega)$. It is a standard fact there is a Hilbert space decomposition in irreducible representations:

(3) $L_d^{\ 2}(\omega) = \oplus n_\pi \pi$, with $0 < n_\pi < +\infty$.

We would like to enumerate all π in (3) and prove that $n_\pi = 1$ for all π. However, at the moment, our results are only partial. In §2, we exhibit representations occuring in (3), thereby extending a result of B. Speh. In §3, 4 we prove a partial result; hopefully the method, which uses in part a suggestion of I.N. Bernstein, can be extended to give a complete description of all π in (3) and to show that $n_\pi = 1$.

(1.2) We fix a maximal compact subgroup K_r or $K = \Pi K_v$ of $G_r(\mathbb{A})$ in the usual way. Let f be a K-finite automorphic form in $G(\mathbb{A})$ satisfying (1.1.1). We recall a necessary and sufficient condition for f to be square integrable, that is to satisfy (1.1.2).

Let Q be a parabolic subgroup of G defined over F, $Q = M_Q U_Q = MU$ a Levi-decomposition of Q. Then $G(\mathbb{A}) = Q(\mathbb{A})K$. We let A_M or A be the center of M, $X(P)$ the group $\mathrm{Hom}_F(P, GL(1))$ and set

(1) $\underline{a}_P = X(P) \otimes_{\mathbb{Z}} \mathbb{R}$,

(2) $\underline{a}_{P\mathbb{C}} = \underline{a}_P \otimes_{\mathbb{R}} \mathbb{C}$.

Note that $\underline{a}_P \simeq X(A_P) \otimes \mathbb{R}$. Every $\chi = \Sigma s_i \chi_i$ in $\underline{a}_{P\mathbb{C}}$ determines a morphism $P(\mathbb{A}) \mapsto \mathbb{C}^{\times}$:

---

* Partially supported by N.S.F. grant #MCS-8200551.

(3) $p \mapsto p^{\chi} = \prod_i |\chi_i(p)|^{s_i}$.

It is also convenient to introduce another morphism:

(4) $H_P : P(\mathbb{A}) \mapsto \underline{a}_P^{\vee}$

defined by

(5) $< H_P(p), \chi > = p^{\chi}, \chi \in \underline{a}_P$.

The kernel of H_P is noted P^0. We set $M^0 = P^0 \cap M(\mathbb{A})$ and we let A_+ be the set of $a \in A(\mathbb{A})$ such that $a_v = 1$ for v finite, and

(6) $\chi(a_v)$ = a positive number independent of v infinite for all $\chi \in X(A)$. Then:

(7) $M(\mathbb{A}) = A_+ M^0$.

We define Z_+ similarly.

Let f_Q be the constant term of f along Q:

(8) $f_Q(g) = \int f(ug)\, du,\ u \in U(F)\backslash U(\mathbb{A})$.

Then:

(9) $f_Q(g) = \sum_i < H_M(a), P_i > a^{\chi_i + \rho_Q} (\sum_j \varphi_{i,j}(m) f_{i,j}(k))$

if $g = uamk$, $u \in U(\mathbb{A})$, $a \in A_+$, $m \in M^0$, $k \in K$, where the χ_i are distinct elements of $\underline{a}_{P\mathbb{C}}$, P_i are polynomial functions on $\underline{a}_P^{\vee}$, $\varphi_{i,j}$ are automorphic forms on M^0, $f_{i,j}$ are K-finite functions. We may of course assume the sum $\sum_j$ to be not identically zero. Then the χ_i are called the <u>exponents</u> of f along Q. On the other hand if each $\varphi_{i,j}$ is orthogonal to all cusp-forms on M^0 then we may say that the <u>cuspidal component of</u> f <u>along</u> Q <u>is zero</u>.

We denote by $\Sigma(\underline{a}_P, P)$ the set of positive roots of P, by $\Delta(\underline{a}_P, P)$ the set of simple roots, by ρ_P the half sum of the positive roots (counting multiplicity). If α is a positive root we denote by $\alpha^{\vee}$ the corresponding co-root. If $Q \supset P$ then we can choose a Levi decomposition $M_Q U_Q$ of Q with $M_Q \supset M_P$. Then the restriction morphisms

$$X(Q) \to X(P),\ X(A_P) \to X(A_Q)$$

allow us to identify $\underline{a}_Q$ with a direct factor of $\underline{a}_P$:

$$\underline{a}_P = \underline{a}_Q \oplus \underline{a}_P^Q.$$

In particular there is a subset S of Δ such that $\underline{a}_Q$ is the set of $s \in \underline{a}_P$ such that

$$\alpha^{\vee}(s) = 0 \text{ for all } \alpha \in S.$$

Then $\underline{a}_P^Q$ is the linear span of the $\alpha \in S$. In particular, this applies to $Q = G$, $S = \Delta$. Then:

$$\underline{a}_P = \underline{a}_G \oplus \underline{a}_P^G .$$

We denote by $\mathcal{C}(\underline{a}_P)$ or $\mathcal{C}(\underline{a}_P^G)$ the positive Weyl chamber, that is the set of s in $\underline{a}_P^G$ such that

$$\alpha^\vee(s) > 0 \quad \text{for} \quad \alpha \in \Delta .$$

On the other hand we denote by $\mathcal{O}(a_P)$ or $\mathcal{O}(a_P^G)$ the set of s in $\underline{a}_P^G$ of the form

$$s = \sum_\Delta x_\alpha \quad \text{with} \quad x_\alpha > 0 .$$

Recall that $\mathcal{O}(a_P^G) \supset \mathcal{C}(a_P^G)$.

This being so let $\mathcal{P}$ be an association class of parabolic subgroups. We say that f is concentrated on $\mathcal{P}$ if the cuspidal component of f along any parabolic subgroup not in $\mathcal{P}$ is zero.

Proposition: Suppose that f is concentrated on $\mathcal{P}$. Then f is square integrable if and only if for any $P \in \mathcal{P}$ the exponents of f along P have a real part in $-\mathcal{O}(\underline{a}_P)$.

Example. Let f be the constant function equal to one. Then f is concentrated on the class of the subgroup B of upper triangular matrices. Furthermore $f_B = f$ and so:

$$f_B(g) = a^{-\rho_B + \rho_B} .$$

Since ρ_B is in $\mathcal{C}(\underline{a}_B)$ it is also in $\mathcal{O}(\underline{a}_B)$ and so f is square integrable!

2. Construction of some elements of L^2_d.

(2.1) We now produce some irreducible components of $L^2_d(\omega)$. In that, we merely extend the results of B. Speh. Let σ be a cuspidal automorphic representation of $G_r(\mathbb{A})$ with central character η. Denote by V the corresponding space of automorphic forms. We may arrange to have η trivial on Z_+. We let P be the parabolic subgroup of type $(r, r, \cdots, r)$ in $GL(n)$, $n = ra$. Then we take $M = M_P$ to be the group of matrices of the form:

$$(1) \quad m = \begin{pmatrix} m_1 & & & 0 \\ & m_2 & & \\ & & \ddots & \\ 0 & & & m_a \end{pmatrix}, \quad m_i \in GL(r) .$$

We may identify $\underline{a}_{P\mathbb{C}}$ with the set of a-tuples:

(2) $s = (s_1, s_2, \cdots, s_a)$

of complex numbers. Then

(3) $m^s = \prod_i |\det m_i|^{s_i}$.

For each s, denote by $\pi(s)$ the representation of $G(\mathbb{A})$ induced by the representation

(4) $m \mapsto \otimes\, \sigma(m_i)\, |\det m_i|^{s_i}$.

They form a fibre bundle of representations. We can construct sections as follows. Denote by $\mathcal{H}$ the space of functions f in $G(\mathbb{A})$ satisfying the following conditions:

(5) $f(u\gamma\, a\, g) = f(g)$

for $u \in U_P(\mathbb{A})$, $\gamma \in P(F)$, $a \in A_+$;

(6) f is K-finite and for each $k \in K$

$m \mapsto f(m\, k)$

belongs to the space $V^{\otimes a}$ of functions on

$M(\mathbb{A}) \simeq GL(r, \mathbb{A})^a$.

On the other hand extend H_P to a function on $G(\mathbb{A})$ still noted H_P which is K-invariant:

$H_P(umk) = H_P(m)$.

Then for each s, the representation of $G(\mathbb{A})$ (or the appropriate Hecke algebra) on the space of functions of the form

$g \mapsto f(g) \exp < H_P(g), s + \rho >, \quad f \in \mathcal{H}$

is equivalent to $\pi(s)$.

We form the Eisenstein series:

$E(g, s, f)$

$$= \sum_{P(F)\backslash G(F)} f(\gamma g) \exp < H_P(\gamma g), s + \rho >.$$

It converges absolutely for $\operatorname{Re} s \in \mathcal{C}(a_P) + \rho_P$ and extends to a meromorphic function of s. It is an automorphic form, concentrated on the association class of P; note that in this case any associate of P is actually conjugate to P. Furthermore the constant term of E along P is given by the formula:

$E_P(g, s, f) =$

$\sum_w [T(w, s) f](g) \exp < w s + \rho, H(g) > .$

The sum is over the Weyl group:

$W(A) = \text{Normalizer } (A)/\text{Centralizer } (A),$

a group isomorphic to $\mathfrak{S}_a$. For each w, $T(w, s)$ is an "intertwining" operator

$$\mathcal{H} \longrightarrow \mathcal{H}$$

defined by

$$[T(w, s) f](g) \exp < w s + \rho, H(g) >$$

$$= \int f(w^{-1} u g) \exp < s + \rho, H(w^{-1} u g) > du,$$

the integral over

$$U(F) \cap w U w^{-1}(F)\backslash U (\mathbb{A}) \cap w U w^{-1}(\mathbb{A}).$$

Remark that $\underline{a}_G$ is the set of a-tuples $(u, u, \cdots, u)$ and $\underline{a}_P^G$ is defined by $\Sigma s_i = 0$. We investigate the behavior of E near the point X defined by the equations:

$$\sum_i s_i = 0, \ \alpha^\vee(s) = 1 \ \text{ for all } \ \alpha \in \Delta (\underline{a}_P, P).$$

Here a positive root α maybe identified with a pair of integers (i, j) $1 \leq i < j \leq a$. Then

$$\alpha^\vee(s) = s_i - s_j .$$

(2.2) To that end we identify $T(w, s)$ to a tensor product

(1) $T(w, s) = \bigotimes_v T_v(w, s)$

of analogously defined local intertwining operators.

<u>Proposition</u>. <u>For each</u> v, $T_v(w, s)$ <u>is holomorphic near</u> X. <u>Furthermore, if</u> w_0 <u>is the longest element of</u> $W(A)$ <u>then the image of</u> $T_v(w_0, X)$ <u>is irreducible</u>.

If σ_v is tempered this is well known. In general we do not know that σ_v is tempered. However, we know the following: The representation σ_v is induced by unitary ones σ_i. In turn, each σ_i is either tempered or in a complementary series, in the sense that it is induced by a pair of representations of the form

$$m_1 \mapsto \xi(m) |\det m|^u, \ m_2 \mapsto \xi(m) |\det m|^{-u},$$

with ξ tempered and $0 < u < 1/2$. This information, together with the "functorial properties" of the intertwining operators, suffices to establish the proposition.

(2.3) Of course the infinite product (2.2.1) does not converge near X. To study its analytic properties we need to introduce an L-function. We choose a finite set S of places containing all infinite ones and so large that, for $v \notin S$, σ_v contains the unit representation of $K_r \cap GL(r, F_v)$. It corresponds to a certain semi-simple conjugacy class A_v in $GL(r, \mathbb{C})$. We set:

$$L_v(u) = \det(1 - q_v^{-u} A_v \otimes \bar{A}_v)^{-1}, \quad m_v(u) = \frac{L_v(u)}{L_v(1 + u)}$$

$$L^S(u) = \prod_{v \notin S} L_v(u)$$

$$m^S(u) = \frac{L^S(u)}{L^S(1 + u)} \quad .$$

Proposition. The product L^S, defined for $\mathrm{Re}\, u \gg 0$, extends to a meromorphic function of u, defined at least for $\mathrm{Re}\, u > 0$. It is non zero for $\mathrm{Re}\, u > 1$ and has a simple pole at $u = 1$.

(2.4) This being so we remark that if f_v, $v \in S$, is a $K_r \cap GL(r, F_v)$ fixed vector in the space of σ_v, then:

$$T_v(w, s)\, f_v = m_v(w, s)\, f_v,$$

where

$$m_v(w, s) = \prod_{\substack{\alpha>0 \\ w\alpha<0}} m_v(\alpha^\vee(s))$$

$$= \prod_{\substack{i<j \\ wi>wj}} m_v(s_i - s_j).$$

We are led to define N_v, for $v \notin S$, by

$$T_v(w, s) = N_v(w, s)\, m_v(w, s).$$

Note that $m_v(w, s)$ is holomorphic and non-zero at X. So $N_v(w, s)$ is holomorphic at X and the image of $N_v(w_0, X)$ irreducible. Then

$$T(w, s) = T_S(w, s)\, N^S(w, s)\, m^S(w, s)$$

where

$$T_S(w, s) = \bigotimes_{v \in S} T_v(w, s),$$

$$N_S(w, s) = \bigotimes_{v \notin S} N_v(w, s),$$

$$m^S(w, s) = \prod_{v \notin S} m_v(w, s)$$

$$= \prod_{\substack{i<j \\ wi>wj}} m^S(s_i - s_j) = \prod \frac{L^S(s_i - s_j)}{L^S(s_i - s_j + 1)} .$$

Both T_S and N^S are now holomorphic at X. As for the factor $m^S(w, s)$, its only singularities at X are the hyperplanes

$$\alpha^\vee(s) = 1 \quad \text{where } \alpha \text{ is simple and } w\alpha < 0.$$

Since the singularities of E are singularities of the $T(w, s)$ we see that the product

$$\prod_{\alpha \in \Delta} (\alpha^\vee(s) - 1)\, E(g, s, f)$$

is holomorphic at X. Its value at X is a certain automorphic form $H(g, f)$, concentrated on the class of P and whose constant term along P is given by

$$(\text{constant})\ [T_S(w_0, X) N^S(w_0, X) f](g) \exp \langle w_0 X + \rho, H(g) \rangle.$$

Since $w_0 X = - X$ and X is in $\mathcal{C}(\underline{a})$ hence in $\mathcal{O}(\underline{a})$ the Proposition recalled in §1 shows that $H(g, f)$ is square integrable. Furthermore the space spanned by the constant terms is irreducible under (the Hecke algebra of) $G(\mathbb{A})$. So the same is true of the space spanned by the $H(g, f)$. We have thus constructed an irreducible component of $L^2_d(\eta^a)$. Denote it by $B(\sigma, a)$. Of course $B(\sigma, a) = \sigma$ if $a = 1$.

(2.5) We would like to prove that the $B(\sigma, a)$'s exhaust $L^2_d(\omega)$. More precisely we would like to prove that

$$L^2_d(\omega) = \oplus B(\sigma, a),$$

the sum over all divisors a of n, all η such that $\eta^a = \omega$ and all cuspidal representations σ of $GL(\frac{n}{a}, \mathbb{A})$ with central character η.

We remark that we know that

$$B(\sigma, a) \simeq B(\sigma', a')$$

implies $a = a'$ and $\sigma = \sigma'$. However, we <u>have</u> <u>not</u> proved that the multiplicity of $B(\sigma, a)$ in $L^2_d(\omega)$ is one.

At any rate, we present only partial results in the next two sections.

§3. The scalar case: consequences of Langland's theory.

(3.1) We now specialize the previous discussion to the case where $r = 1$. Then P is the group of upper triangular matrices and σ is now a character of $F_{\mathbb{A}}^{\times}/F^{\times}$. We further specialize to the case where ω and σ are trivial and the ground field F is $\mathbb{Q}$. Finally we restrict our attention to K-invariant functions.

We complete and simplify our earlier notations as follows:

(1) $L(u) = \Pi L_v(u)$ (all v),

(2) $L_\infty(u) = \pi^{-s/2}\,\Gamma(\frac{u}{2})$,

(3) $L_p(u) = (1 - \frac{1}{p^u})^{-1}$,

(4) $m(u) = \dfrac{L(u)}{L(u+1)}$.

Recall that L has simple poles at 0 and 1 but no zero for $\mathrm{Re}\, u > 1$, for $\mathrm{Re}\, u < 0$, or for $0 < u < 1$. Furthermore:

(7) $L(u) = L(1-u)$.

Thus m has a simple pole at $u = 1$, a simple zero at $u = 0$ and no other real pole or zero. Furthermore:

(8) $m(0) = -1$.

We set

(9) $E(g, s) = \sum_{\gamma} \exp < s + \rho, H(\gamma g) >$.

Then:

(10) $E_U(g, s) = \sum_{w} m(w, s) \exp < ws + \rho, H(g) >$

where

(11) $m(w, s) = \Pi m(\alpha^\vee(s)),\ \alpha > 0,\ w\alpha < 0;$

$= \Pi m(s_i - s_j),\ i < j,\ wi > wj.$

(3.2) Following Langlands we construct a certain space of square integrable K-invariant functions on $G(\mathbb{Q})Z(\mathbb{A})\backslash G(\mathbb{A})$. Suppose Φ is a function on $A(\mathbb{A})$, invariant under $Z(\mathbb{A})$, $A(\mathbb{Q})$, and $K \cap A(\mathbb{A})$. Since

$A(\mathbb{A}) = A_+ \, A(\mathbb{Q}) \, K \cap A(\mathbb{A})$

Φ is really a function on $Z_+\backslash A_+ \simeq (\mathbb{R}_+^\times)^{n-1}$. We suppose that Φ is C^∞ of compact support and extend Φ to a K-invariant function on $G(\mathbb{A})$. The function:

(1) $\theta_\varphi(g) = \Sigma\Phi(\gamma g)$, $\gamma \in P(\mathbb{Q})\backslash G(\mathbb{Q})$

is square integrable. Here the function φ is the Mellin transform of Φ:

$$(2)\quad \varphi(s) = \int_{A(\mathbb{Q})Z(\mathbb{A})\backslash A(\mathbb{A})} \Phi(a)\, a^{-s-\rho}\, da$$

and θ_φ may be regarded as a "wave packet" of Eisenstein series:

$$(3)\quad \theta_\varphi(g) = \left(\frac{1}{2i\pi}\right)^{n-1} \int E(g, s)\, \varphi(s)\, ds;$$

the integral is for $\operatorname{Re} s = c \in \underline{a}_P^G$, with $c \in \mathcal{C}(\underline{a}_P) + \rho$.

There is a formula for the scalar product of two such functions:

$$(4)\quad (\theta_{\varphi_1}, \theta_{\varphi_2}) = \sum_w \left(\frac{1}{2i\pi}\right)^{n-1} \int m(w, s)\, \varphi_1(s)\, \bar{\varphi}_2(-w\bar{s})\, ds.$$

The closure $\mathcal{V}$ of the linear span of the θ_φ's is invariant under the action of the bi-K-invariant Hecke algebra $\mathcal{H}_0$. Using (4) as a starting point, Langlands obtains a complete description of the spectrum of $\mathcal{H}_0$ in $\mathcal{V}$.

We extract from his work the description of the discrete spectrum, that is, of the eigenspaces of $\mathcal{H}_0$ in $\mathcal{V}$. To begin with, for each s, the function

$$g \mapsto \exp < s + \rho, H(g) >$$

is an eigenvector of $\mathcal{H}_0$; the eigenvalue depends only in the orbit of s under W. Each eigenvalue of $\mathcal{H}_0$ in $\mathcal{V}$ corresponds to a certain orbit of W in $\underline{a}_P^G$. Let us emphasize that this orbit is <u>real</u>. The corresponding eigenspace is finite dimensional.

The vectors in the eigenspace are obtained as "multi-residues", as follows. If α is a root and c a real number we denote by $H(\alpha, c)$ the hyperplane of equation

$$\alpha^\vee(s) = c.$$

We consider only affine spaces of $\underline{a}_{P\,\mathbb{C}}^G$ which are intersections of hyperplanes of this form. We call them admissible. Suppose $H_1 \supset H_2$ are two admissible affine subspaces with H_2 of codimension one in H_1; suppose also that we have chosen a real unit vector of H_1 normal to H_2. Finally suppose that $F(s)$ is a meromorphic function on H_1, whose singularities are admissible hyperplanes. Then the residue of $F(s)$ along H_2 is defined. It is noted

$$\operatorname{Res}_{H_2} F$$

and it is a meromorphic function on H_2, whose singularities are admissible hyperplanes of H_2.

In general, suppose we have a sequence of admissible subspaces of $\underline{a}_{P\,\mathbb{C}}^{G}$:

$$\underline{a}_{P\,\mathbb{C}}^{G} = H_n \supset H_{n-1}, \supset \cdots \quad H_1 \supset H_0 = \{h\} ,$$

with H_i of codimension one in H_{i+1}; suppose also that we have chosen for each i a vector in H_{i+1}, normal to H_i. Then if F is a meromorphic function on $\underline{a}_{P\,\mathbb{C}}^{G}$ whose singularities are admissible hyperplanes we can define inductively F_i by

$$F_n = F, \quad F_i = \operatorname{Res}_{H_i} F_{i+1}.$$

For convenience let us call the previous data an admissible flag through h. Noting $\mathfrak{F}$ the flag we will set

$$F_0 = \operatorname{Res}_{\mathfrak{F}} F.$$

This being so suppose Ω is an orbit of W in $\underline{a}_P^G$ such that the corresponding eigenvalue occurs (discretely) in $\mathcal{V}$. Let $\mathcal{V}(\Omega)$ be the corresponding eigenspace. Then there are points h_i in Ω, for each i, admissible flags $\mathfrak{F}_{ij}$ through h_i and constants c_{ij} such that the functions

$$g \mapsto \sum_i \sum_j c_{ij} \operatorname{Res}_{\mathfrak{F}_{ij}}[E(g, s)\ \varphi\ (s)]$$

span $\mathcal{V}(\Omega)$.

Clearly the residues depend only on the derivatives of φ at h_i. By choosing an h_i, and taking a φ which vanishes at high order at all φ, we see that the function

$$g \mapsto \sum_j c_{ij} \operatorname{Res}_{\mathfrak{F}_{ij}}[E(g, s)\ \varphi\ (s)]$$

is square integrable. This simple consequence of Langland's extraordinary work will enable us to determine the possible Ω's.

(3.3) Now we describe the divisor of poles and zeroes of $E(g, s)$. Recall that the singularities of $E(g, s)$ are singularities of its constant term. Thus we see that the hyperplanes $H(\alpha, 1)$ are simple poles for $E(g, s)$. In general, any singular hyperplane H of $E(g, s)$ has the form $H(\alpha, c)$ for some $\alpha > 0$ and $c \in \mathbb{C}$. However, if H is real then $c = 1$.

For instance let X be the point of intersection of the hyperplanes

$H(\alpha, 1)$ with α simple.

Then the $H(\alpha, 1)$ are the only singular hyperplanes of $E(g, s)$ passing through X; the function

$$\prod_{\alpha \text{ simple}} (\alpha^{\vee}(s) - 1)\, E(g, s)$$

is holomoprhic at X; its value at X is a non zero constant function. Note that this value is also $\operatorname{Res}_{\mathcal{F}} E(g, s)$ where $\mathcal{F}$ is the flag determined by

$$F_i = F_{i+1} \cap H(\alpha_i, 1).$$

On the other hand, if $n = 2$, then

$$E_N(g, s) = \exp < s + \rho, H(g) > + \, m(w, s) \exp < ws + \rho, H(g) >.$$

If $s = 0$ then $m(w, 0) = -1$ and $s = ws$. Thus $E_N(g, 0) = 0$. It follows that $E(g, 0) = 0$. The inductive character of the Eisenstein construction shows then that $E(g, s)$ vanishes on any hyperplane $H(\alpha, 0)$ with α simple. If now α is any positive root there is a w such that $\beta = w\alpha$ is a positive simple root. Then

$$E(g, s) = E(g, t)\, m(w, s), \quad t = ws.$$

Since $wH(\alpha, 0) = H(\beta, 0)$ and $H(\alpha, 0)$ is not a singular hyperplane of $m(w, s)$ we conclude that $E(g, s)$ vanishes on $H(\alpha, 0)$.

(3.4) We are now ready to formulate the main theorem of this section.

Theorem. *Suppose Ω is an orbit of W in $\mathfrak{a}_P^G$ such that the corresponding eigenvalue of $\mathcal{H}_0$ occurs discretely in $\mathcal{V}$. Then Ω is the orbit of X and $\mathcal{V}(\Omega)$ consists of the constant functions.*

The proof will occupy the remainder of this section and the next section.

To begin with if Ω is the orbit of X then $\mathcal{V}(\Omega)$ contains the constant functions by the remarks of (3.3). Since the representation of $\mathcal{H}_0$ on the discrete spectrum is semi-simple this shows that $\mathcal{V}(\Omega)$ consists of all constant functions.

Next if Ω is such an orbit, then there is a point h in Ω, flags $\mathcal{F}_j$ through h and constants c_j such that for all φ

(1) $g \mapsto \sum_j c_j \operatorname{Res}_{\mathcal{F}_i} E(g, s)\, \varphi(s)$

is square integrable; furthermore there is at least one φ such that this is non zero.

Of course $\operatorname{Res}_{\mathcal{F}_j} E(g, s)\, \varphi(s)$ is zero unless the following happens: suppose $\mathcal{F}_j$ is the sequence

$$H_n \supset H_{n-1} \supset \cdots H_1 \supset H_0 = \{h\};$$

then for any i, there is at least one singular hyperplane H of $E(g, s)$ such that

$$H_i = H_{i+1} \cap H.$$

This hyperplane has the form $H(\alpha, c)$ for $\alpha > 0$ and appropriate c. It must be real, hence c must be one. In particular the $\alpha^{\vee}(h)$, for $\alpha > 0$, must be integers (that is h must be in the group generated by the weights).

In general if h is a point such that $\alpha^{\vee}(h)$ is an integer for all $\alpha > 0$ we will set

$$(2) \quad H_h(g, s) = E(g, s) \prod_{\alpha^{\vee}(h)=1} (\alpha^{\vee}(s) - 1) \left\{ \prod_{\alpha^{\vee}(h)=0} \alpha^{\vee}(s) \right\}^{-1}.$$

Then H_h is a meromorphic function, <u>holomoprhic</u> at h. Furthermore if $\mathfrak{F}$ is any real flag passing through h then

$$(3) \quad \operatorname{Res}_{\mathfrak{F}} E(g, s)\, \varphi(s)$$

$$= \sum_{|\alpha| \le m} c_\alpha D^\alpha (H_h(g, s)\, \varphi(s))|_{s=h}$$

where m is an appropriate integer. The sum is over all multi-indices α with $|\alpha| \le m$; of course if the residue is not zero then we may assume $c_\alpha \neq 0$ for at least one α with $|\alpha| = m$.

Now let us assume as before that there are flags $\mathfrak{F}_j$ through h and constants c_j such that for all φ

$$(4) \quad g \mapsto \sum_j c_j \operatorname{Res}_{\mathfrak{F}_j} E(g, s)\, \varphi(s)$$

is square integrable; assume further this is non zero for at least one φ. Then there is an $m \ge 0$ and constants c_α for $|\alpha| \le m$ such that (4) is actually equal to

$$(5) \quad \sum_{|\alpha| \le m} c_\alpha D^\alpha (H_h(g, s)\, \varphi(s))|_{s=h}.$$

So (5) is square integrable for all φ. Furthermore there is a β with $|\beta| = m$ and $c_\beta \neq 0$. If we choose a φ which vanishes at order m at h and is such that

$$D^\beta \varphi|_{s=h} = 1$$

$$D^\alpha \varphi|_{s=h} = 0 \quad \text{for } |\alpha| = m,\ \alpha \neq \beta,$$

then (5) reduces to

$$c_\beta H_h(g, h).$$

In particular $H_h(g, h)$ is square integrable.

(3.5) Thus we see that the theorem will be proved if we prove the following proposition:

Proposition. Suppose h is a point with $\alpha^\vee(h) \in \mathbb{Z}$ for all α. Suppose $H_h(g, h)$ is square integrable. Then h is conjugate to X.

(3.6) In order to establish (3.5) we will first derive a simple consequence of the functional equation of the Eisenstein series.

Proposition. Suppose h is a point with $\alpha^\vee(h) \in \mathbb{Z}$ for all α. Suppose $k = wh$. Then:

$$H_h(g, h) = c\, H_k(g, k)$$

with $c \neq 0$.

Proof. By induction on the length $\ell(w)$ of w it suffices to prove it when $\ell(w) = 1$, that is, when w is the reflection with respect to a simple root γ. We have then $w\gamma = -\gamma$ and $w\alpha > 0$ for any $\alpha > 0$, $\alpha \neq \gamma$. The functional equation reads:

$$E(g, s) = E(g, t)\, m(w, s), \text{ with } t = ws.$$

On the other hand:

$$E(g, s) = H_h(g, s) \prod_{\alpha^\vee(h)=0} \alpha^\vee(s) \left\{ \prod_{\alpha^\vee(h)=1} (\alpha^\vee(s) - 1) \right\}^{-1},$$

$$E(g, t) = H_k(g, t) \prod_{\beta^\vee(k)=0} \beta^\vee(t) \left\{ \prod_{\beta^\vee(k)=1} (\beta^\vee(t) - 1) \right\}^{-1}.$$

Suppose $\gamma(h) = 0$. Then $\gamma(k) = 0$ as well and $\gamma^\vee(s) = -\gamma^\vee(t)$, $t = ws$.

Thus:

$$\prod_{\alpha^\vee(h)=0} \alpha^\vee(s) = - \prod_{\beta^\vee(k)=0} \beta^\vee(t),$$

$$\prod_{\alpha^\vee(h)=1} (\alpha^\vee(s) - 1) \prod_{\beta^\vee(k)=1} (\beta^\vee(t) - 1),$$

and

$$H_h(g, s) = -\, m(w, s)\, H_k(g, s).$$

Since $m(w, h) = -1$ we have

$$H_h(g, h) = H_k(g, k).$$

Suppose $\gamma(h) = 1$. Then $\gamma(k) = -1$ and

$$\prod_{\alpha^\vee(h)=0} \alpha^\vee(s) = \prod_{\beta^\vee(k)=0} \beta^\vee(t);$$

but

$$\prod_{\alpha^\vee(h)=1} (\alpha^\vee(s)-1) = \prod_{\beta^\vee(k)=1} (\beta^\vee(t) - 1)(\gamma^\vee(s) - 1).$$

Thus

$$H_h(g, s) = H_k(g, s)\, m\,(w, s)(\gamma^\vee(s) - 1)$$

and

$$H_h(g, h) = H_k(g, k)\, c$$

with

$$c = m(w, s)(\gamma^\vee(s) - 1)|_{s=h}$$

$$= m(u)(u-1)|_{u=1} .$$

The case $\gamma^\vee(h) > 0$, $\gamma^\vee(h) \neq 1$ is easier:

$$H_h(g, h) = H_k(g, k)\, m(w, h).$$

The case $\gamma^\vee(h) < 0$ can be handled by interchanging h and k.

§4. The scalar case: Imitating Bernstein-Zelevinsky's argument.

(4.1) In order to establish Proposition (3.5) we will introduce the "residual Eisenstein series". Then, after taking in account Proposition (3.6), we will compute $H_h(g, h)$ as derivatives of these residual Eisenstein series at an appropriate point.

To that end, consider the set Γ of simple roots. Since a root is also a pair of integers we may think of Γ as a graph on the set $\{1, 2, \cdots, n\}$. For any c we denote by $H(\Gamma, c)$ the intersection

$$\cap\, H(\alpha, c),\ \alpha \in \Gamma.$$

If $c = 0$, then $H(\Gamma, 0)$ is the complexification of the space $\underline{a}_Q \subset \underline{a}_P$ attached to a parabolic subgroup $Q \supset R$. We have

$$\underline{a}_P = \underline{a}_Q \oplus \underline{a}_P^Q$$

where $\underline{a}_P^Q$ is the linear span of the α in Γ. Recall that the restrictions to $\underline{a}_Q^G$ of the $\alpha^\vee$, with $\alpha \in \Delta - \Gamma$, make up a basis of the dual of $\underline{a}_Q^G$. In particular the positive Weyl chamber $\mathcal{C}(\underline{a}_Q^G)$ is the cone of $\underline{a}_Q^G$ defined by

$\alpha^\vee(s) > 0$ for $\alpha \in \Delta - \Gamma$.

Furthermore we have a decomposition:

$$H(\Gamma, -1) = \underline{a}_{Q\mathbb{C}} + X(\Gamma)$$

where $X(\Gamma)$ denotes the normal translation in $H(\Gamma, -1)$. The coordinates of $X(\Gamma)$ are easily described: if Γ_i is a connected component of Γ with m elements then

$$\Gamma_i = \{j, j+1, \cdots, j+m-1\}$$

and the corresponding coordinates of $X(\Gamma)$ are

$$s_j = -\frac{m-1}{2}, \quad s_{j+1} = -\frac{m-1}{2} + 1, \cdots, s_{j+m-1} = \frac{m-1}{2}.$$

(4.2) Recall that the function m is holomorphic at -1 and in fact has a simple zero there. It follows that for any w, no singular hyperplane of the function $m(w, s)$ contains $H(\Gamma, -1)$. Thus we can define the restriction of $m(w, s)$ to $H(\Gamma, -1)$. It is a meromorphic function noted $m(w, s : \Gamma)$. Furthermore it is zero unless $w\gamma > 0$ for all $\gamma \in \Gamma$. Similarly we may define the restriction of $E(g, s)$ to $H(\Gamma, -1)$. It is noted $E(g, s : \Gamma)$. Furthermore

$$E_U(g, s : \Gamma) = \Sigma m(w, s : \Gamma) \exp \langle ws + \rho, H(g) \rangle,$$

the sum over all w such that $w\gamma > 0$ for all $\gamma \in \Gamma$.

In fact if $\Gamma = \Delta$ then $H(\Gamma, -1) = \{-X\}$ and

$$E(g, -X : \Gamma) = 1.$$

In general if $\Gamma \neq \Delta$ then $H(\Gamma, 0) = \underline{a}_{Q\mathbb{C}}$ and for $s \in H(\Gamma, -1)$

$$E(g, s : \Gamma) = \sum_{Q(\mathbb{Q})\backslash G(\mathbb{Q})} \exp \langle u + \rho_Q, H_Q(\gamma g) \rangle$$

where

$$s = u + X(\Gamma), \quad u \in \underline{a}_{Q\mathbb{C}},$$

provided the series converges. This is so if $\operatorname{Re} u$ belongs to $\mathcal{C}(\underline{a}_Q) + \rho_Q$.

We denote by $S(\Gamma)$ the set of $s \in H(\Gamma, -1)$ of the form $s = u + X(\Gamma)$. We need to investigate the divisor of poles and zeroes of $E(g, s : \Gamma)$. To that end, following Bernstein and Zelevinsky, we will introduce the concept of segment: a segment is an r-tuple of real numbers $(s_1, s_2, \cdots, s_r)$ with

$$s_1 - s_2 = -1, \ s_2 - s_3 = -1, \cdots, s_{r-1} - s_r = -1.$$

Note that the set $\{s_1, s_2, \cdots, s_r\}$ completely determines the r-tuple $(s_1, s_2, \cdots, s_r)$ so we may also think of a segment as a set of real numbers. This gives a meaning to the following definition:

Definition. Let S *and* T *be segments. Then we say that* S *and* T *are not linked if* $S \cup T$ *is not a segment or* $S \subseteq T$ *or* $T \subseteq S$. *We say that* S *dominates* T *if the following condition is satisfied: write* S *and* T *in the form*

$$S = (a, a, \cdots, a) + \left(-\frac{u-1}{2}, -\frac{u-1}{2}+1, \cdots, \frac{u-1}{2}\right)$$

$$T = (b, b, \cdots, b) + \left(-\frac{v-1}{2}, -\frac{v-1}{2}+1, \cdots, \frac{v-1}{2}\right)$$

then $a \geq b$.

Pictorically S dominates T if the "axis of symmetry" of S coincides with, or is on the right of the axis of symmetry of T:

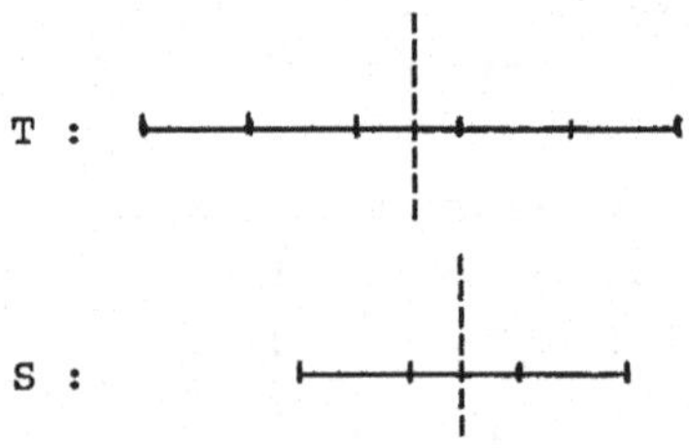

If Γ has connected components $\Gamma_1, \Gamma_2, \cdots, \Gamma_r$ then any point s in $H(\Gamma, -1)$ has the form

$$s = (S_1, S_2, \cdots, S_r)$$

where $S_1, S_2, \cdots, S_r$ are segments. Furthermore s belongs to $S(\Gamma)$ if and only if for each $i < j \leq r$ the segment S_i dominates S_j.

(4.3) Proposition. *Suppose that* h *is a point of* $S(\Gamma)$:

$$h = (S_1, S_2, \cdots, S_r).$$

Suppose that $\alpha^\vee(h) \in \mathbb{Z}$ *for all* α *and that for any pair* (i, j) S_i *and* S_j *are not linked. Then* $E(g, s : \Gamma)$ *is holomorphic at* h. *Furthermore let* (S_i, S_j) *be a pair of identical segments with* $i < j$. *Let* α *be a root connecting* Γ_i *and* Γ_j. *Then* $E(g, s : \Gamma)$ *vanishes on* $H(\Gamma, -1) \cap H(\alpha, 0)$.

Proof. If α and β are two positive roots then $\alpha^\vee - \beta^\vee$ is constant on $H(\Gamma, -1)$ if and only if α and β (viewed as pairs of integers) connect the two same components of Γ. This being so fix a pair (i, j) with $i < j$ and consider the two segments S_i and S_j. Let X_{ij} be the set of roots α which connect Γ_i and Γ_j and are such that $\alpha^\vee(h) = 1$. Define similarly Y_{ij} to be the set of roots β which connect Γ_i and Γ_j and are such that $\beta^\vee(h) = 0$.

Thus $\alpha^\vee - 1 = \beta^\vee$ on $H(\Gamma, -1)$ for $\alpha \in X_{ij}$ and $\beta \in Y_{ij}$. Suppose first that $S_i = S_j$. Then $\#Y_{ij} = \#X_{ij} + 1$. Choose in any way a root α_{ij} in Y_{ij}. Then, for s in $H(\Gamma, -1)$:

$$\alpha_{ij}^\vee(s) \prod_{X_{ij}} (\alpha^\vee(s) - 1) = \prod_{Y_{ij}} \beta^\vee(s).$$

Suppose now that $S_i \neq S_j$. Then we claim that $\#X_{ij} = \#Y_{ij}$ and thus, for $s \in H(\Gamma, -1)$:

$$\prod_{X_{ij}} (\alpha^\vee(s) - 1) = \prod_{Y_{ij}} \beta^\vee(s).$$

First if X_{ij} or Y_{ij} is not empty then $S_i \cup S_j$ is a segment. Then $S_i \subset S_j$ or $S_j \subset S_i$. We treat the case $S_i \subset S_j$ and leave the case $S_i \supset S_j$ to the reader. First $\#Y_{ij} = \#S_i$. Secondly S_j has at least one element on the left of S_i:

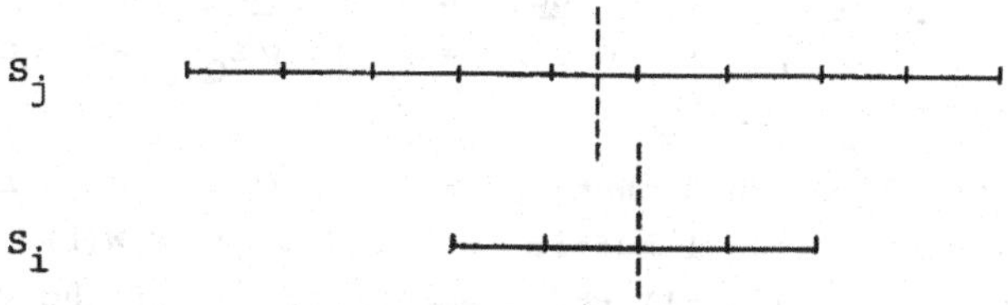

that shows that $\#X_{ij} = \#S_i$.

Now any singular hyperplane passing through h must have the form $H(\alpha, 1)$ where $\alpha \in X_{ij}$ for a suitable (i, j). Similarly any hyperplane $H(\alpha, 0)$ is a zero for $E(g, s)$; it passes through h if and only if $\alpha \in Y_{ij}$ for an appropriate (i, j). Our contention follows from these remarks.

We also remark that we have proved the following corollary:

<u>Corollary</u>. <u>Under the assumptions of Proposition</u> (4.3), <u>the quotient</u> $E(g, s : \Gamma)/\prod \alpha_{ij}^\vee(s)$, <u>the product over all pairs</u> (i, j) <u>with</u> $S_i = S_j$, <u>is holomorphic at</u> h. <u>Its value there is</u> $H_h(g, h)$.

(4.4) We come at last to the proof of Proposition (3.5). Suppose Ω is an orbit of W in the group of weights. Then there is an element h of Ω satisfying the following conditions:

(1) h belongs to $S(\Gamma) \subset H(\Gamma, -1)$, where Γ is a set of simple roots. We set $H(\Gamma, 0) = \underline{a}_\Omega$.

(2) If Γ has connected components $\Gamma_1, \Gamma_2, \cdots, \Gamma_r$ and h is written accordingly

$$h = (S_1, S_2, \cdots, S_r)$$

then for each $i < j$ S_i and S_j are not linked (and S_i dominates

S_j).

(3) If $S_i = S_j$ then for each k, $i < k < j$, $S_k = S_i$.

The first conditions are the ones of the previous subsection. The third one is merely for convenience. We will prove Propostion (3.5) by showing that, if Γ is not connected, then

$$(4) \quad E(g, s : \Gamma)/\Pi\, \alpha^{\vee}_{ij}(s)$$

is not square integrable at $s = h$. To see that let us examine the constant term of $E_N(g, s : \Gamma)$. It has the form:

$$(5) \quad \Sigma\, m(w, s : \Gamma) \exp \langle ws + \rho, H(g) \rangle,$$

the sum over all $w \in W(\underline{a}_P)$ such that $w\gamma > 0$ for $\gamma \in \Gamma$. Among the w's we have those which fix h. They are described in the following lemma:

Lemma. Suppose $w\gamma > 0$ for all $\gamma \in \Gamma$ and $wh = h$. Then $w\Gamma = \Gamma$. In particular $w\,H(\Gamma, -1) = H(\Gamma, -1)$, $w\,X(\Gamma) = X(\Gamma)$, $w\,\underline{a}_Q = \underline{a}_Q$ and w induces an element of $W(\underline{a}_Q)$ on $\underline{a}_Q$.

Proof. Let $W(\Gamma)$ be set of w such that $w\Gamma = \Gamma$. Suppose $wh = h$ and $w\gamma > 0$ for all $\gamma \in \Gamma$. It will suffice to find a $w_0 \in W(\Gamma)$ such that $w_0 h = h$ and $w w_0$ fixes all the elements of one of the Γ_i's (recall Γ_i is a subset of $\{1, 2, \cdots, n\}$ and $W \simeq \mathfrak{S}_n$). For then the lemma will follow by induction on r.

Choose a segment of maximal length among the S_i's. Consider the set of segments identical to that one:

$$(6) \quad S_i, S_{i+1}, \cdots, S_\ell.$$

Let Γ_ℓ be $\{p, p+1, \cdots, q\}$. Suppose first $wp = p$. Then $wq > p$. If $wq = q$ then w fixes every index in Γ_ℓ and we are done.

Suppose $wq \neq q$. We have $s_{wq} = s_q$. Thus s_{wq} must be an element of some segment S_k with $k > \ell$. In particular $S_k \cup S_\ell$ is a segment. Since S_ℓ is maximal $S_k \subset S_\ell$. Since (6) is the set of all segments identical to S_i, $S_k \neq S_\ell$:

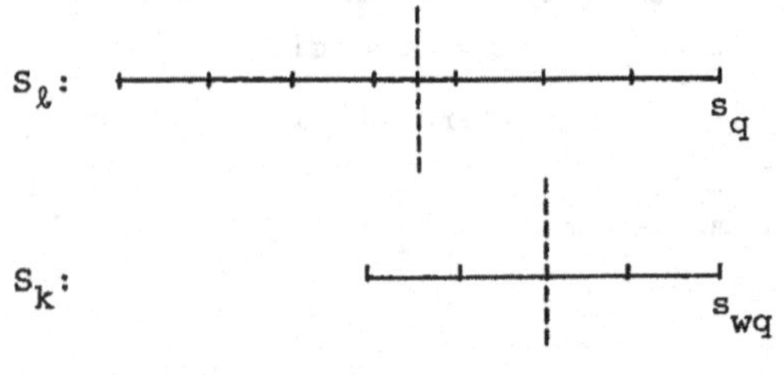

This contradicts the fact that S_ℓ dominates S_k. Hence $wq = q$.

Let now X be the set of the first indices in $\Gamma_i, \Gamma_{i+1}, \cdots, \Gamma_\ell$. If $wX = X$ then there is an element w_0 of $W(\Gamma)$ which permutes $\Gamma_i, \Gamma_{i+1}, \cdots, \Gamma_\ell$ (hence fixes h) and is such that $w w_0$ is the identity on X. In particular $w w_0 p = p$ and we are reduced to the previous case.

Suppose that $wX \neq X$. There is at least one p' in X such that $w p' \notin X$. Replacing again w by an appropriate $w w_0$ we may assume $w p \notin X$. Then s_{wp} belongs to a segment S_k not equal to S_ℓ and $S_k \cup S_\ell$ is a segment. Again $S_k \subset S_\ell$, $S_k \neq S_\ell$:

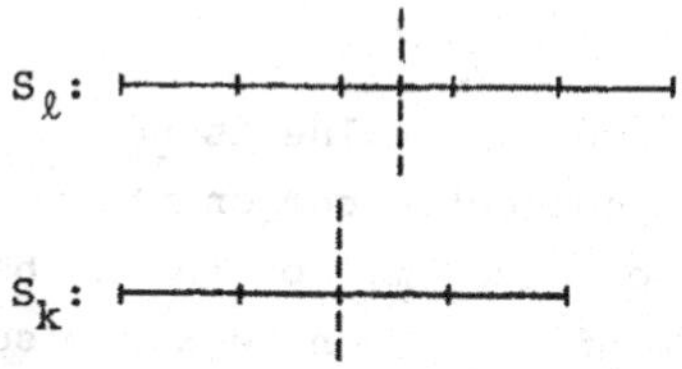

We cannot have $k < \ell$ because S_k would dominate S_ℓ. Thus $k > \ell$. Since $q > p$ we have $w q > w p$ and $s_q = s_{wq}$ belongs to a segment S_j with $j > \ell$:

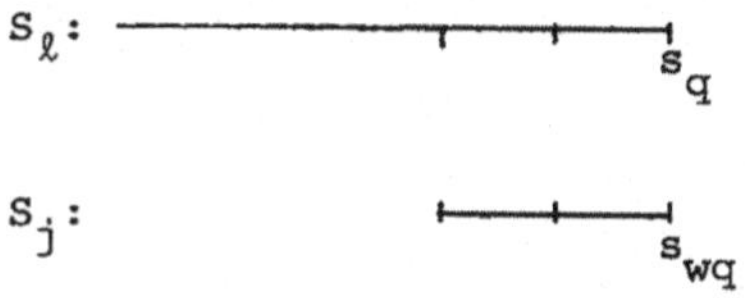

Again this contradicts the fact S_ℓ dominates S_j. Hence $w X = X$ and we are done.

In the expression (5) for the constant term of $E(g, s : \Gamma)$ we therefore distinguish the subsums

$$(7) \quad \sum_{w\Gamma = \Gamma,\ wh = h} m(w, s : \Gamma) \exp \langle ws + \rho, H(g) \rangle,$$

and

$$(8) \quad \sum_{w\gamma > 0,\ \gamma \in \Gamma,\ wh \neq h} m(w, s : \Gamma) \exp \langle ws + \rho, H(g) \rangle$$

We shall prove the following Proposition:

Proposition. _The quotient of_ (7) _by the product_ $\Pi\, \alpha^\vee_{ij}(s)$ _is holomorphic and non zero at_ h.

This will imply Proposition (3.5). For it will first follow that the quotient of (8) by the same product is holomorphic at h and its value there of the form

$$\Sigma \exp < k + \rho, H(g) > P_k(H(g))$$

where the P_k's are polynomial and the k's not equal to h. On the other hand the quotient of (7) by the product will have at h a value of the form

$$\exp < h + \rho, H(g) > P(H(g))$$

where P is a non zero polynomial. Hence h is an exponent of the value at h of the quotient

$$E(g, s : \Gamma)/ \Pi \alpha_{ij}^{\vee}(s).$$

However, h is not in $-\mathcal{O}(a_P)$, and therefore this value is not square integrable, if Γ has more than one connected component: indeed, write again h in the form (2). Let also m_i be the number of elements of the i-th connected component of Γ. Then we can also write

$$h = (\underbrace{a_1, a_1, \cdots, a_1}_{m_1}, \underbrace{a_1, a_2, \cdots, a_2}_{m_2}, \cdots) + X(\Gamma) .$$

Since h is in $S(\Gamma)$ we have

$$a_1 \geq a_2 \geq a_3 \geq \cdots .$$

On the other hand:

$$m_1 a_1 + m_2 a_2 + \cdots m_r a_r = 0.$$

In particular $a_1 \geq 0$. Since $X(\Gamma)$ has the form

$$(- \frac{m_1-1}{2} , - \frac{m_1-1}{2} + 1, \cdots \frac{m_1-1}{2} , \cdots)$$

we have, if $h = (s_1, s_2, \cdots, s_n)$,

$$s_1 + s_2 + \cdots + s_{m_1} = m_1 a_1 \geq 0$$

and h is not in $-\mathcal{O}(a_P{}^G)$ for this would require

$$s_1 + s_2 + \cdots + s_{m_1} < 0.$$

(4.5) It remains to prove Proposition (4.4). To begin with if s is in $H(\Gamma, -1)$ and $w\Gamma = \Gamma$ then

$$s = u + X(\Gamma), \quad u \in H(\Gamma, 0) = \underline{a}_{Q\mathbb{C}}$$

(1)

$$ws = wu + + X(\Gamma)$$

and $u \mapsto wu$ is an element of $W(\underline{a}_Q)$.

Writing

(2) $h = v + X(\Gamma)$

we see that in (7) the sum is over the fixator of v in $W(\underline{a}_Q)$. Furthermore the product $\Pi\,\alpha^{v}{}_{ij}(s)$ can be written as a product

(3) $\Pi\,\alpha^{v}(u)$, $\alpha \in \Sigma'$,

where Σ' is a certain subset of $\Sigma(\underline{a}_Q, Q)$. Recall that if $h = (S_1, S_2, \cdots, S_r)$ then the set of segments identical to a given segment form a sequence $S_i, S_{i+1}, \cdots, S_\ell$. So Σ' is the set of combinations of a subset Δ' of $\Sigma(\underline{a}_Q, Q)$. Furthermore:

(4) $\underline{a}_Q = \underline{b} \oplus \underline{a}'$,

where $\underline{b}$ is the set of u such that $\alpha^{v}(u) = 0$ for all $\alpha \in \Sigma'$ and $\underline{a}'$ the linear span of Σ'. Then Σ' is the set of positive roots of a root system in $\underline{a}'$ (a product of systems of type A_i) and the fixator of v in $W(\underline{a}_Q)$ fixes $\underline{b}$ pointwise and induces on $\underline{a}'$ the Weyl group W' of that root system:

(5) if $u = b + u'$, $w \in W(\underline{a}_Q)$, $w\,h = h$, then $w\,u = b + w'u'$ and $W' \in W'$.

We claim now that if w is in $W(\underline{a}_P)$, $w\Gamma = \Gamma$ and $wh = h$, then for $s \in H(\Gamma, -1)$

$$m(w, s : \Gamma) = c(w', u')$$

where we have used the notations of (1) and (5) and $c(w', \circ)$ is a certain meromorphic function on $\underline{a}'_{\mathbb{C}}$. It suffices to verify this when w' is a simple reflection in W'; then w interchanges two identical segments S_i and S_{i+1} in h.

Let Γ_i and Γ_{i+1} be

$$\Gamma_i = \{p, p+1, \cdots, p+m-1\},\ \Gamma_{i+1} = \{p+m, p+m-1, \cdots, p+2m-1\}.$$

Let α be the simple root of $\Sigma(\underline{a}_Q, Q)$ such that

$$\alpha^{v}(u) = u_i - u_{i+m},\ p < i \leq p+m-1.$$

Then α is a simple root in Σ'. Recall that

$$m(w, s) = \Sigma\, m(s_i - s_j) = \Pi\, L(s_i - s_j)/L(s_i - s_j + 1),$$

$$p \leq i \leq p+m-1,\ p+m \leq j \leq p+2m-1.$$

If now s is in $H(\Gamma, -1)$ and we use notations (1) and (5) then this reduces to

$$m(w, s : \Gamma) = \Pi\, L(\alpha^{v}(u) + 1 - i)/L(\alpha^{v}(u) + i),\ 1 \leq i \leq m.$$

Since $\alpha^{\vee}(u) = \alpha^{\vee}(u')$, our assertion follows. Note that this can also be written as:

$$c(w', u') = \Pi\, L(i - \alpha^{\vee}(u'))/L(\alpha^{\vee}(u') + i),\ 1 \leq i \leq m.$$

In particular $c(w', u') = -1$ if $\alpha^{\vee}(u') = 0$.

Now to prove Proposition (4.4) we have to show that the quotient

$$\Sigma\, c(w', u')\, \exp < w'u', H > / \Pi \alpha^{\vee}(y'),$$

where w' ranges over W' and α over Σ', is holomorphic at 0, with a non zero value for a suitable H.

(4.6) Finally what we have to prove is the following lemma on a reduced root system:

Lemma. Consider a reduced root system on a vector space a. Let Σ be a set of positive roots, W the Weyl group of the system. Suppose $c(w, s)$ is a meromorphic function on $W \times a_{\mathbb{C}}$ satisfying the cocycle condition

$$c(w_1 w_2, s) = c(w_1, w_2 s)\, c(w_2, s),\quad c(e, s) = 1,$$

and such that there is a meromorphic function F_α on $\mathbb{C}$ with

$$c(w, s) = F(\alpha^{\vee}(s)),$$

if w is the reflection with respect to the simple root α. Finally suppose $F_\alpha(0) = -1$. Then the quotient of

$$N(s, Y) = \sum_{W} c(w, s)\, \exp < ws, Y >$$

by

$$\Pi \alpha^{\vee}(s),\ \alpha \in \Sigma$$

is holomorphic at 0. Furthermore there is a Y such the value of the quotient at 0 is non zero.

Proof. To begin with, $N(s, Y) = 0$ if $\alpha^{\vee}(s) = 0$ for some $\alpha > 0$. Thus, if we write

$$N(s, Y) = H(s, Y) \prod_{\alpha>0} \alpha^{\vee}(s)$$

then $H(s, Y)$ is holomorphic at 0. We have to see that $H(0, Y) \neq 0$ for some Y. To see that let us apply to the previous equality the differential operator

$$(\partial_\rho)^m$$

where $m = \#\Sigma$ and

$$\rho = \frac{1}{2} \Sigma \alpha,\ \alpha > 0$$

and then set $s = 0$.

We get

$$\partial_\rho^{\,m} N(s, Y)\big|_{s=0}$$

$$= cH(0, Y) \prod_{\alpha>0} \alpha^{\vee}(\rho)$$

with $c \neq 0$. Since

$$\prod_{\alpha>0} \alpha^{\vee}(\rho) \neq 0$$

it suffices to show that

$$\partial_\rho^{\,m} N(s, Y)\big|_{s=0} \neq 0$$

for some Y. Now we have

$$\partial_\rho \exp < ws, Y >$$

$$= \exp < ws, Y > < w\ , Y >.$$

Hence

$$\partial_\rho^{\,m} N(s, Y)\big|_{s=0}$$

$$= \sum_{w} c(w, 0) < w\rho, Y >^{m} + Q(Y)$$

where Q is a polynomial of degree $< m$. Thus it will suffice to show the first sum is not identically zero, that is

$$U = \sum_{w} (-1)^{\ell(w)} (w\rho)^{m}$$

is a non zero element of $\mathrm{sym}^{m}(\underline{a})$. The pairing on $\underline{a} \times \underline{a}^{\vee}$ extends to a W-invariant pairing, noted $< \circ, \circ >$, on

$$\mathrm{sym}^{m}(a) \times \mathrm{sym}^{m}(a^{\vee}).$$

It will be enough to show that

$$< U, \prod_{\alpha>0} \alpha^{\vee} > \neq 0.$$

But this is

$$\sum_{w} (-1)^{\ell(w)} < (w\rho)^{m}, \prod_{\alpha>0} \alpha^{\vee} >.$$

There is a non zero constant c such that

$$< X^{m}, \prod_{\alpha>0} \alpha^{\vee} > = c \prod_{\alpha>0} < X, \alpha^{\vee} >.$$

Hence the above expression is:

$$c \sum_{w} (-1)^{\ell(w)} \prod_{\alpha>0} < w\rho, \alpha^{\vee} >$$

$$= c \sum_{w} \prod_{\alpha} < \rho, \alpha^{\vee} >$$

$$= c\, \#W \prod_{\alpha} < \rho, \alpha^{\vee} > \neq 0.$$

So we are done.

REFERENCES.

[B-Z] I.N. Bernstein and A.V. Zelevinsky: Induced representations on reductive p-adic groups. I. Ann. scient. Éc. Norm. Sup., 4e série, t. 10, 1977, p. 441 to 472.

[J-S] H. Jacquet and J. Shalika: Sur le spectre résiduel du groupe linéaire, C.R. Acad. Sc. Paris, t. 293, 1981, Série I-40, p. 541 to 543.

[L] R. Langlands: On the Functional Equations Satisfied by Eisenstein Series, Springer-Verlag, Lec. Notes Math., Vol. 544, 1976.

[S] B. Speh: Unitary representations of GL(n, $\mathbb{R}$) with non-trivial ($\underline{g}$, K) cohomology, preprint, 1980.

[Z] A.V. Zelevinsky: Induced representations of reductive p-adic groups II. On irreducible representations of GL(n), Ann. scient. Éc. Norm. Sup., 4e série, t. 13, 1980, p. 165 to 210.

ON LIFTING

David Kazhdan
Harvard University

Introduction

Let F be a local nonarchimedian field, $\varepsilon\colon F^* \to \mathbb{C}^*$ be a character of order n, $G = GL_n(F)$. We denote by the same letter ε the character of G given by $\varepsilon(g) \overset{\text{def}}{=} \varepsilon(\det g)$, $g \in G$.

Let $\mathfrak{E}(G)$ be the set of equivalent classes of smooth irreducible representations of G and $\mathfrak{E}^{\varepsilon}(G) = \{\pi \in \mathfrak{E}(G) \mid \pi \otimes \varepsilon \simeq \pi\}$. One of the main results of this paper is a description of the set $\mathfrak{E}^{\varepsilon}(G)$. Let $L \supset F$ be the cyclic extension which corresponds to ε by the local class field theory and $\mathfrak{G} = \mathrm{Gal}(L:F)$, $(\simeq \mathbb{Z}/n\mathbb{Z})$.

<u>Theorem A.</u> There exists a "natural" one-to-one correspondence between $\mathfrak{E}^{\varepsilon}(G)$ and $\mathfrak{G}$-orbits on $\hat{L}^*$ where $\hat{L}^* \overset{\text{def}}{=} \mathrm{Hom}(L^*, \mathbb{C}^*)$.

We will formulate now a global variant of this theorem. So let F be a global field, C_F be the adele class group of F, $\varepsilon\colon C_F \to \mathbb{C}^*$ be a character of order n and $L \supset F$ be the corresponding cyclic extension, $\mathfrak{G} = \mathrm{Gal}(L:F)$. Let $\mathbb{A}$ be the adele ring of F, $G_{\mathbb{A}} = Gl_n(\mathbb{A})$, $\mathfrak{E}_a(G)$ be the set of equivalent classes of irreducible automorphic representations of $G_{\mathbb{A}}$ and $\mathfrak{E}_a^{\varepsilon}(G) = \{\pi \in \mathfrak{E}_a(G) \mid \pi \otimes \varepsilon \simeq \pi\}$ where we consider ε as a character of $G_{\mathbb{A}}$.

"<u>Theorem B.</u>" There exists a "natural" one-to-one correspondence between $\mathfrak{E}_a^{\varepsilon}(G)$ and $\mathfrak{G}$-orbits on $\hat{C}_L$ where $\hat{C}_L \overset{\text{def}}{=} \mathrm{Hom}(C_L, \mathbb{C}^*)$.

We put the word "theorem" in quotation marks because the proof of this statement is based on Arthur's results ([Ar]) and we want to restrict ourselves to more elementary tools. So we will actually prove a special case of this theorem which we will now formulate. For any place p of F we denote by F_p the completion of F at p and by $\varepsilon_p: F_p^* \to \mathbb{C}^*$ the restriction of ε to F_p^*. Let p_1, p_2 be two non-archimedean places such that $\varepsilon_{p_1}, \varepsilon_{p_2}$ are primitive characters of order n, $\tilde{\mathfrak{X}}_a^\varepsilon(G) = \{\pi \in \mathfrak{X}_a^\varepsilon(G) \mid$ the local components of π at p_1 and p_2 are cuspidal$\}$ and $\hat{C}_L^0 = \{\chi \in \hat{C}_L \mid \chi_{p_1}^\sigma \neq \chi_{p_1}, \ \chi_{p_2}^\sigma \neq \chi_{p_2}$ for any $\sigma \in \mathfrak{G} - \{e\}\}$ where χ_{p_1}, χ_{p_2} are local components of χ at p_1 and p_2.

<u>Theorem C</u>. There exists a natural one-to-one correspondence between $\tilde{\mathfrak{X}}_a^\varepsilon(G)$ and $\mathfrak{G}$-orbits on $\hat{C}_L^0$.

We will actually prove first Theorem C and deduce from it Theorem A. For more detailed formulations of those theorems see §1.

I want to express my gratitude to J. Bernstein and S. Kudla for helpful remarks.

Section 1

Let F be a local nonarchimedean field, $\mathcal{O} \subset F$ the ring of integers, $p = (t)$ the maximal ideal in $\mathcal{O}$, $q = \#(\mathcal{O}/p)$, $\nu: F^* \to \mathbb{Z}$ be the valuation such that $\nu(t) = 1$ and $|\ |: F \to \mathbb{R}$ be the norm such that $|t| = q^{-1}$. Let L be a cyclic extension of F with Galois group $\mathfrak{G} \simeq \mathbb{Z}/n\mathbb{Z}$. We fix an embedding $\varepsilon: \mathfrak{G} \hookrightarrow \mathbb{C}^*$. Let $N: L^* \to F^*$ be the norm map. By the local class field theory we may identify $F^*/N(L^*)$ with $\mathfrak{G}$ and therefore we consider ε as a character of F^*. Let $G = GL_n(F)$, $K = GL_n(\mathcal{O})$. We denote by the same letter ε the character of G given by $\varepsilon(g) \overset{\text{def}}{=} \varepsilon(\det g)$. We denote by $\mathfrak{X}(G)$ the set of equivalence classes of smooth irreducible representations of G and by $\mathfrak{X}^\varepsilon(G) \subset \mathfrak{X}(G)$ the subset of equivalence classes of representations (π, V) such that $\pi \otimes \varepsilon \simeq \pi$. Let $\hat{L}^*$ be the group of quasicharacters of L^*. $\mathfrak{G}$ acts naturally on L^* and $\hat{L}^*$. We denote by $\hat{L}^*/\mathfrak{G}$ the space of $\mathfrak{G}$

orbits on L^*.

One of the main results of this paper is the one-to-one correspondence between $\mathfrak{E}^{\varepsilon}(G)$ and $L^*/\mathfrak{G}$. To formulate this correspondence we have to introduce some notations.

We fix a generator σ_0 of $\mathfrak{G}$ and define $L_0 = \{\ell \in L \mid \ell^{\sigma_0}(-1)^{\frac{n(n-1)}{2}} = \ell\}$. Let ℓ^0 be a nonzero element in L_0. It is clear that L_0 does not depend on a choice of σ_0 and for all $\ell \in L_0$ we have $\ell\ell^0 \in F$. For any $\ell \in L$ we define $\tilde{\Delta}(\ell) \overset{\text{def}}{=} \prod_{1 \le i < j \le n} (\ell^{\sigma_0^i} - \ell^{\sigma_0^j})$. It is clear that $\tilde{\Delta}(\ell) \in L_0$. We define a function Δ_{ℓ^0} on L by $\Delta_{\ell^0}(\ell) = |N(\ell)|^{\frac{-n(n-1)}{2}} |\tilde{\Delta}(\ell)| \cdot \varepsilon(\tilde{\Delta}(\ell)\ell^0)^{\frac{n(n-1)}{2}}$.

We denote by $L' \subset L$ the set of elements $\ell \in L$ such that $\Delta(\ell) \neq 0$ and call them regular elements.

For any $(\pi,V) \in \mathfrak{E}^{\varepsilon}(G)$ there exists a nonzero operator $A_\pi: V \to V$ such that $A_{\pi_0}\pi(g) = \varepsilon(g)\pi(g)A_\pi$ and such A_π is uniquely determined up to multiplication by a scalar. If $V^K \neq \langle 0 \rangle$ and $\varepsilon|_K \equiv 1$, we normalize A_π by the condition $A_\pi|_{V^K} \equiv \mathrm{Id}$.

Let $S(G)$ be the space of locally constant functions f on G with compact support and let dg be the Haar measure of G. For any $\pi \in \mathfrak{E}^{\varepsilon}(G)$ we denote by χ_π^ε the linear function on $S(G)$ (i.e., the distribution) given by $\chi_\pi^\varepsilon(f) = \mathrm{Tr}(\pi(f)\cdot A_\pi)$ where $\pi(f) \overset{\text{def}}{=} \int_G f(g)\pi(g)\,dg$, dg is a Haar measure on G. Using the technique of [H-Ch 1] it is not difficult to show that the restriction of χ_π^ε on the set G' of regular elements is a locally constant measure which we will write as $\chi_\pi^\varepsilon(g)dg$ and $\chi_\pi^\varepsilon(g)$ is a locally constant function on G' which does not depend on a choice of a Haar measure dg. It is clear that we have $\chi_\pi^\varepsilon(x^{-1}gx) = \varepsilon(x)\chi_\pi^\varepsilon(g)$ for all $g \in G'$, $x \in G$.

We now fix an embedding $L^* \subset G$.

Lemma 1. Let $g \in G'$ be an element such that $\chi_\pi^\varepsilon(g) \neq 0$ for some $\pi \in \mathfrak{E}^{\varepsilon}(G)$. Then g is conjugate to a regular element in L^*.

<u>Proof</u>: Let $Z_G(g)$ be the centralizer of g in G. Since $\chi_\pi^\varepsilon(g) \neq 0$ we have $\varepsilon|_{Z_G(g)} \equiv 1$. As g is regular, Z_G is isomorphic to a product of the multiplicative group of fields $Z_G(g) = \prod_{i=1}^r L_i^*$ where $\sum_{i=1}^r |L_i:F| = n$ and $\varepsilon(\prod_{i=1}^r \ell_i) = \varepsilon(\prod_{i=1}^r N_i(\ell_i))$ where $N_i: L_i^* \to F^*$ are the norm maps. Since $\varepsilon|_{Z_G(g)} \equiv 1$ we have $N_i(L_i^*) \subset \operatorname{Ker} \varepsilon = N(L^*)$ for all i, $1 \leq i \leq r$. But as follows from the local class field theory it is possible only if $L_i \supset L$ for all i, $1 \leq i \leq r$. Since $\sum_{i=1}^r [L_i:F] = [L:F]$ we have $Z_G(g) \simeq L^*$. But in this case g is conjugate to an element in L^*.

We now formulate our main results.

<u>Theorem A.</u>

(a) For any $\chi \in \hat{L}^*$ there exists a nonzero constant c_χ and a representation $\pi_\chi \in \mathcal{E}^\varepsilon(G)$ such that

$$\chi_{\pi_\chi}^\varepsilon(\ell) = c_\chi \sum_{\sigma \in \mathfrak{G}} \Delta(\ell^\sigma)^{-1} \chi(\ell^\sigma).$$

(b) The equivalence class of π_χ is uniquely defined by this condition.

(c) $\pi_\chi \simeq \pi_{\chi'}$ if and only if there exists $\sigma \in \mathfrak{G}$ such that $\chi' = \chi^\sigma$.

(d) Any representation $\pi \in \mathcal{E}^\varepsilon(G)$ is equivalent to π_χ for some $\chi \in \hat{L}^*$.

(e) π_χ is cuspidal if and only if $\chi^\tau \neq \chi$ for all $\tau \in \mathfrak{G}$, $\tau \neq e$.

For any $\chi \in \hat{L}^*$ we define $\mathfrak{G}_\chi = \{\tau \in \mathfrak{G} \mid \chi^\tau = \chi\}$ and denote by $L_1 \subset L$ the subfield of $\mathfrak{G}_\chi$-invariant elements. Let $N_1: L^* \to L_1^*$ be the norm map. As is well known, there exists $\chi_1 \in \hat{L}_1^*$ such that $\chi = \chi_1 \circ N_1$. Let $n' = [L_1:F]$, $m = [L:L_1] = n/n'$, $G_1 = GL_{n'}(F)$, $M = G_1^m$ and let ε_1 be the character of G_1 given by $\varepsilon_1(g_1) \overset{\text{def}}{=} \varepsilon(\det g)$. For any representation $\tau \in \mathcal{E}(G_1)$ we denote by $\rho_\tau \in \mathcal{E}(M)$ the tensor product $\rho_\tau = \tau \otimes (\tau \otimes \varepsilon_1) \otimes \ldots \otimes (\tau \otimes \varepsilon_1^{m-1})$. Let $P \subset G$ be a parabolic subgroup with M as a Levi component. For any $\tau \in \mathcal{E}(M)$ we define $\pi_\tau = \mathrm{Ind}_P^G \rho_\tau$ where Ind_P^G is the unitary induction ([B-Z]). Let $\tau_{\chi_1} \in \mathcal{E}^{\varepsilon_1}(G_1)$ be the cuspidal representation which corresponds to χ_1, as in Theorem 1.

"Proposition 1." $\pi_\chi = \pi_{\tau_{\chi_1}}$.

We put it in quotation marks because the proof of this proposition is based on "Theorem B" which we do not prove in this paper.

If $L \supset F$ is not a field but a product of cyclic extensions L_i, $1 \le i \le r$, of F we define for any $\chi \in \hat{L}^*$ a representation $\pi_\chi \in \mathcal{E}(G)$ in the following way.

Let $n_i = [L_i:F]$, and let $M \subset G$ be the Levi subgroup of the form $M = \prod_{i=1}^r GL_{n_i}(F)$. We write χ as $\chi = (\chi_1,\ldots,\chi_r)$ for $\chi_i \in \hat{L}_i^*$, $1 \le i \le r$. For any i, $1 \le i \le r$ we define $\pi_{\chi_i} \in \mathcal{E}(GL_{n_i}(F))$, as in Theorem A and then define $\pi_\chi \in \mathcal{E}(G)$ as the unitary representation induced from $\otimes_{i=1}^r \pi_{\chi_i}$ on M.

We will now formulate a global analogue of Theorem A. We will actually first prove this global result and then deduce our Theorem A.

Let F be a global field $L \supset F$, a cyclic extension with the Galois group $\mathfrak{G}$. We choose a generator σ of $\mathfrak{G}$. This gives us an identification $\mathfrak{G} \simeq \mathbb{Z}/n\mathbb{Z}$. We fix an embedding $\varepsilon\colon \mathfrak{G} \hookrightarrow \mathbb{C}^*$. Let $\mathbb{A}$ be the ring of F-adeles, C_F, C_L be the idele class groups for F and L and $N\colon C_L \to C_F$ be the norm map. By the class field theory we may identify $C_F/N(C_L)$ with $\mathfrak{G}$ and therefore consider ε as a character of C_F or the group $F_{\mathbb{A}}^*$ of ideles.

Let $\underline{G} \simeq \underline{GL_n}$ be the general linear group of order n and $\underline{Z} \subset \underline{G}$ the center of $\underline{G}$. We denote with the same letter ε the character on $G_{\mathbb{A}}$ given by $\varepsilon(g) \overset{\det}{=} \varepsilon(\det g)$.

Let $\Sigma(F)$ be the set of valuations on F. For any $p \in \Sigma(F)$ we denote by F_p the completion of F at p and we will write $G_p \overset{\text{def}}{=} GL_n(F_p)$, $L_p = L \otimes_F F_p$ and so on. Any character $\chi \in \hat{L}^*_{\mathbb{A}}$ can be written as a product $\chi = \underset{p\in\Sigma(F)}{\otimes} \chi_p$, $\chi_p \in \hat{L}^*_p$. For example, we will denote by $\varepsilon_p \in \hat{L}^*_p$ the local components of ε. Analogously any irreducible representation $\pi \in \mathfrak{E}(G_{\mathbb{A}})$ of $G_{\mathbb{A}}$ can be written as a restricted tensor product $\pi = \underset{p\in\Sigma(F)}{\otimes} \pi_p$, $\pi_p \in \mathfrak{E}(G_p)$. ([GGP])

We denote by $\mathfrak{E}_a(G_{\mathbb{A}})$ the set of equivalent classes of automorphic representations and by $\mathfrak{E}^{\varepsilon}_a(G_{\mathbb{A}}) \subset \mathfrak{E}_a(G_{\mathbb{A}})$ the subset of irreducible representations π of $G_{\mathbb{A}}$ such that $\pi \otimes \varepsilon \simeq \pi$.

We can formulate now the global analogues of theorem A. Let $\chi \in \hat{C}_L$ be an idele class character. For any $p \in \Sigma(F)$ we have (by Theorem A) the representation $\pi_{\chi_p} \in \mathfrak{E}(G_p)$. We denote by π_χ the restricted tensor product $\pi_\chi \overset{\text{def}}{=} \underset{p\in\Sigma(F)}{\otimes} \pi_{\chi_p}$.

"<u>Theorem B</u>."

(a) For any $\chi \in \hat{C}_L$ we have $\pi_\chi \in \mathfrak{E}^{\varepsilon}_a(G_{\mathbb{A}})$.

(b) Given $\chi, \chi' \in \hat{C}_L$ we have $\pi_\chi \simeq \pi_{\chi'}$ if and only if $\chi' = \chi^\tau$ for some $\tau \in \mathfrak{G}$.

(c) For any $\pi \in \mathfrak{E}^{\varepsilon}_a(G_{\mathbb{A}})$ there exists $\chi \in \hat{C}_L$ such that $\pi \simeq \pi_\chi$.

We put the word "Theorem" in quotation marks because the proof of this statement is based on Arthur's results (see [Ar]) and we want to restrict ourselves here to more elementary tools. So we will be dealing with a special case.

Let $p_1, p_2 \in \Sigma(F)$ be two places such that L_{p_1}, L_{p_2} are fields, and $D \supset F$ be a central division algebra of degree n^2 which is ramified only at p_1 and p_2. As is well known, there exist embeddings $L \hookrightarrow D$. We fix one. Let $\tilde{G}$ be the multiplicative group of D and $\varepsilon: \tilde{G}_{\mathbb{A}} \to \mathbb{C}^*$ be the character given by $\varepsilon(\tilde{g}) = \varepsilon(N(g))$ where $N: \tilde{G} \to F$

s the reduced norm.

Let Z be the center of $\widetilde{G}$. We define $P\widetilde{G} \overset{def}{=} Z\setminus\widetilde{G}$. For any $\in \Sigma(F)$, $p \neq p_1, p_2$ D_p is isomorphic to $M_n(F_p)$ and we will identify $\widetilde{G}_p$ with G_p for $p \neq p_1, p_2$. In particular, for any $\widetilde{\pi} \in \mathcal{E}(\widetilde{G}_{\mathbb{A}})$ we will write $\widetilde{\pi} = \otimes_p \widetilde{\pi}_p$ where $\widetilde{\pi}_p \in \mathcal{E}(G_p)$ for $p \neq p_1, p_2$. For any $\in \hat{C}_F$ we denote by $\widetilde{W}_\theta$ the space $L_2^\theta(\widetilde{G}_F \setminus \widetilde{G}_{\mathbb{A}})$ *), by $\widetilde{\rho}_\theta$ the natural representation of $\widetilde{G}_{\mathbb{A}}$ on $\widetilde{W}_\theta$, and by $\mathcal{E}_a^\theta(\widetilde{G}_{\mathbb{A}}) \subset \mathcal{E}(\widetilde{G}_{\mathbb{A}})$ the set of irreducible components of $\widetilde{\rho}_\theta$. As follows from [DKV] every component $\in \mathcal{E}_a^\theta(G_{\mathbb{A}})$ occurs in $\widetilde{\rho}_\theta$ with the multiplicity one. We denote by $_a^{\varepsilon,\theta} \subset \mathcal{E}_a^\theta(\widetilde{G}_{\mathbb{A}})$ the subset of $\widetilde{\pi}$ such that $\widetilde{\pi} \otimes \varepsilon \simeq \widetilde{\pi}$.

Let $\hat{C}_L^0 = \{\chi \in \hat{C}_L \mid$ for any $\sigma \in \mathfrak{G} - \{e\}$, $\chi_{p_i}^\sigma \neq \chi_{p_i}$ for $i = 1,2\}$.

<u>Theorem C'</u>.

a) For any $\chi \in \hat{C}_L^0$ there exists a unique $\widetilde{\pi}_\chi \in \mathcal{E}_a^{\theta_\chi,\varepsilon}$ such that $\widetilde{\pi}_\chi = \otimes(\widetilde{\pi}_\chi)_p$, where $(\widetilde{\pi}_\chi)_p$ is equivalent to π_{χ_p} for $p \neq p_1, p_2$ and θ_χ is the restriction of χ to $C_F \subset C_L$.

b) Any $\widetilde{\pi} \in \mathcal{E}_a^{\theta,\varepsilon}$ has the form $\widetilde{\pi}_\chi$ for some $\chi \in \hat{C}_L^0$.

<u>Remark</u>. The uniqueness in (a) follows from [DKV]. It also follows rom [DKV] that Theorem C' is equivalent to Theorem C in the Introduction.

ection 2

We will use the language of [BZ]. Let G be an ℓ-group, $G_0 \subset G$ e a normal subgroup such that $G/G_0 \simeq \mathbb{Z}/n\mathbb{Z}$. We fix an embedding $/G_0 \overset{\varepsilon}{\hookrightarrow} \mathbb{C}^*$ and also denote by ε the composition $G \to G/G_0 \overset{\varepsilon}{\hookrightarrow} \mathbb{C}^*$. et $\mathcal{E}(G)$ (resp. $\mathcal{E}(G_0)$) be the sets of equivalent classes of irreduible admissible representations of G (resp. G_0). The group $\mathfrak{G} \overset{def}{=}$ $/G_0$ acts naturally on $\mathcal{E}(G_0)$. $(\sigma,\rho) \to \rho^\sigma$, $\rho \in \mathcal{E}(G_0)$, $\sigma \in \mathfrak{G}$, and e denote by $\mathcal{E}^0(G_0) \subset \mathcal{E}(G_0)$ the subset of $\rho \in \mathcal{E}(G_0)$ such that

---

*) $L_2^\theta(\widetilde{G}_F \setminus \widetilde{G}_{\mathbb{A}})$ is the space of measurable complex valued functions $f\colon \widetilde{G}_F \setminus \widetilde{G}_{\mathbb{A}} \to \mathbb{C}$ such that $f(zx) = \theta(z)f(x)$, $x \in \widetilde{G}_F \setminus \widetilde{G}_{\mathbb{A}}$, $z \in Z_{\mathbb{A}}$ and $\int_{Z_{\mathbb{A}}\widetilde{G}_F\setminus\widetilde{G}_{\mathbb{A}}} |f(x)|^2\, dx < \infty$ where dx is a Haar measure on $\widetilde{G}_{\mathbb{A}}$.

$\rho^\sigma \neq \rho$ for all $\sigma \in \mathfrak{G}$, $\sigma \neq e$. We denote by $\mathcal{E}^\varepsilon(G) \subset \mathcal{E}(G)$ the subset of $\pi \in \mathcal{E}(G)$ such that $\pi \otimes \varepsilon \simeq \pi$. For any $(\pi,V) \in \mathcal{E}(G)$ we denote by $A_\pi : V \to V$ a nonzero operator such that $A_\pi \circ \pi(g) = \varepsilon(g)\pi(g) \circ A_\pi$. It is clear that A_π is well defined up to a multiplication by a scalar. Let $S(G)$ (resp. $S(G_0)$) be the set of locally constant functions with compact support on G (resp. on G_0), $S(G)'$, $S(G_0)'$ the dual spaces and dg, dg_0 be Haar measures on G and G_0. For any $(\pi,V) \in \mathcal{E}^\varepsilon(G)$ we denote by t_π^ε the functional on $S(G)$ given by $t_\pi^\varepsilon(f) \overset{\text{def}}{=} \mathrm{Tr}(\pi(f) \circ A_\pi)$ where $\pi(f) \overset{\text{def}}{=} \int_G f(g)\pi(g)\, dg$. For any $\rho \in \mathcal{E}(G_0)$ we define $t_\rho \in S(G_0)'$ by $t_\rho(f) \overset{\text{def}}{=} \mathrm{Tr}\, \rho(f)$, $f \in S(G_0)$ and $t_\rho^0 \overset{\text{def}}{=} \sum_{\tau \in \mathfrak{G}} \varepsilon(\tau) t_{\rho^\tau}$.

<u>Lemma 1</u>. There exists a one-to-one correspondence

$$\alpha: \mathcal{E}^0(G_0)/\mathfrak{G} \to \mathcal{E}^\varepsilon(G)$$

such that for any $\pi \in \mathcal{E}^\varepsilon(G)$, and $\bar\rho \in \mathcal{E}^0(G_0)/\mathfrak{G}$,

(a) $\pi = \alpha(\bar\rho)$ if and only if the restriction of π to G_0 is the direct sum $\bigoplus_{\rho \in \bar\rho} \rho$.

(b) $\pi = \alpha(\bar\rho)$ if and only if there exists $c \in \mathbb{C}^*$ such that $t_\pi^\varepsilon = c t_\rho^0$, for some $\rho \in \bar\rho$.

<u>Proof</u>: For any $\bar\rho \in \mathcal{E}^0(G_0)/\mathfrak{G}$ we choose a representative $\rho \in \mathcal{E}^0(G_0)$ of $\bar\rho$ and define $\alpha(\bar\rho) \overset{\text{def}}{=} \mathrm{Ind}_{G_0}^G \rho$. It is clear that $\alpha(\bar\rho)$ does not depend on the choice of ρ and that $\alpha(\bar\rho)_{G_0} = \bigoplus_{\rho \in \bar\rho} \rho$. We have $\alpha(\bar\rho) \otimes \varepsilon = (\mathrm{Ind}_{G_0}^G \rho) \otimes \varepsilon = \mathrm{Ind}_{G_0}^G (\rho \otimes \varepsilon) = \alpha(\bar\rho)$ since $\varepsilon|_{G_0} \equiv 1$. Now take $(\pi,V) \in \mathcal{E}^\varepsilon(G)$. The restriction of (π,V) to G_0 is a direct sum $(\pi,V) = \oplus_{i=1}^N (\rho_i, W_i)$ of nonequivalent representations, $N \le n$ (see [BZ]). Since $\varepsilon_\pi : V \to V$ commutes with the action of G_0 we have $\varepsilon_\pi|_{W_i} = \lambda_i \mathrm{Id}$, $1 \le i \le N$. Let $x \in G$ be an element such that $\varepsilon(x) = \eta \overset{\text{def}}{=} e^{2\pi i/n}$.

It is clear then that $\pi(x)W_i = W_{i'}$, where $\lambda_{i'} = \eta\lambda_i$. So we see that $m = n$ and we may assume that $\pi(x^k)W_1 = W_k$, $1 \le k \le n$. It is clear that the set $\{\rho_i\}$, $1 \le i \le n$ is an orbit of $\mathfrak{G}$ in $\mathcal{E}^0(G_0)$ and $\pi = \mathrm{Ind}_{G_0}^{G} \rho_i$ for any $1 \le i \le n$. So we see that $\alpha\colon \mathcal{E}(G_0)^0/\mathfrak{G} \to \mathcal{E}(G)$ is one-to-one. Part (a) is obvious. To prove (b) take $f \in S(G_0)$. Then

$$t_\pi^\varepsilon(f) \overset{\text{def}}{=} \mathrm{Tr}\,\pi(f)\cdot\varepsilon_\pi = \sum_{i=1}^{n} \lambda_i \,\mathrm{Tr}\,\rho_i(f) = \lambda_1 t_{\rho_1}^0(f).$$

The lemma is proved.

Let F be a local nonarchimedean field and $L \supset F$ a cyclic extension of degree n. $\mathfrak{G} = \mathrm{Gal}(L:F)$. Let $N_L\colon L^* \to F^*$ be the norm map. By the local class field theory we may identify $F^*/\mathrm{Im}\,N_L$ with $\mathfrak{G}$. Let $\mathfrak{G} \hookrightarrow \mathbb{C}^*$ be an embedding. We denote by ε_L the composition $\varepsilon_L\colon F^* \to F^*/\mathrm{Im}\,N_L \hookrightarrow \mathbb{C}^*$.

Let $G \overset{\text{def}}{=} GL_n(F)$, $G' \subset G$ the subset of regular elements, $Z \overset{\text{def}}{=}$ the center of G $(\simeq F^*)$, $B \subset G$ the upper triangular subgroup, $U \subset B$ the unipotent radical of B, $PG = G/Z$, $K = GL_n(\mathcal{O}_F) \subset G$ and PK the image of K in PG. We denote by dg, dx the Haar measures on G and PG such that $\int_K dg = \int_{PK} dx = 1$. We denote by ε the character of G given by $\varepsilon(g) \overset{\text{def}}{=} \varepsilon_L(\det g)$, $g \in G$ and define $G_0 \overset{\text{def}}{=} \ker \varepsilon \subset G$. It is clear that $Z \subset G_0$ and therefore we may consider ε as a character of PG. We also will identify G/G_0 with $F^*/\mathrm{Im}\,N_L \simeq \mathfrak{G}$. It is clear that our triple (G,G_0,ε) satisfies the condition of Lemma 1. So we will use freely the notations like $\mathcal{E}^\varepsilon(G)$ or $\mathcal{E}^0(G_0)$. For any character $\theta \in \hat{Z}$ we define $\mathcal{E}^\theta(G) = \{\pi \in \mathcal{E}(G) \mid \pi(z) = \theta(z)\mathrm{Id}\ \forall z \in Z\}$.

In this paragraph we will restrict ourselves to the case when L is the unramified extension.

For any $s = (s_1,\ldots,s_n) \in \mathbb{C}^{*n}$ we denote by $\chi_s\colon B \to \mathbb{C}^*$ the character given by $\chi_s(b) \overset{\text{def}}{=} \prod_{i=1}^n s_i^{\nu(b_{ii})}$ where $\{b_{ii}\}$ are the diagonal entries of $b \in B$. We denote by $\tilde{\pi}_s$ the representation of G obtained by the unitary induction from (B,χ_s) to G.

The following result is well known [Car]:

Lemma 2.

(a) $\tilde{\pi}_s$ has a unique subquotient (π_s, V_s^K) such that $V_s^K \neq \{0\}$.

(b) π_s is equivalent to $\pi_{s'}$ if and only if there exists a permutation $\sigma \in S_n$ such that $s' = s^\sigma$.

(c) For any irreducible representation (π,V) of G such that $V^K \neq \{0\}$ there exists $s \in \mathbb{C}^{*n}$ such that π is equivalent to π_s.

(d) $\pi_s(z) = N(s)^{\nu(z)}\mathrm{Id}$ for $\forall\ z \in Z$ where $N(s) \overset{\text{def}}{=} s_1 \ldots s_n$.

For any $s \in \mathbb{C}^{*n}$, $s = (s_1,\ldots,s_n)$ we define $s^* = (s_1^*,\ldots,s_n^*)$ by $s_i^* = \bar{s}_i^{-1}$. For any $s \in \mathbb{C}^{*n}$ there exists (unique up to a scalar) Hermitian form $<\ ,\ >$ on $V_s \times V_{s^*}$ such that $<\pi_s(g)v, \pi_{s^*}(g)v^*> = <v,v^*>$ for all $g \in G$, $v \in V_s$, $v^* \in V_{s^*}$. Fix elements $v_0 \in V_s^K$, $v_0^* \in V_{s^*}^K$ such that $<v_0,v_0^*> = 1$. We denote by $\Gamma_s\colon G \to \mathbb{C}$ the spherical function $\Gamma_s \overset{\text{def}}{=} <\pi_s(g)v_0, v_0^*>$.

For any character $\theta \in \hat{Z}$ $(\simeq \hat{F}^*)$ we define

$$\mathfrak{E}^{\varepsilon,\theta}(G) \overset{\text{def}}{=} \mathfrak{E}^{\varepsilon}(G) \cap \mathfrak{E}^{\theta}(G) \quad \text{and} \quad \mathfrak{E}_K^{\varepsilon,\theta}(G) = \{(\pi,V) \in \mathfrak{E}^{\varepsilon,\theta}(G) \mid V^k \neq \{0\}\}$$

It is clear that for any ramified character θ we have $\mathfrak{E}_K^{\varepsilon,\theta}(G) = \emptyset$.

Proposition 1. For any unramified character $\theta \in \hat{Z}$, $\#(\mathfrak{E}_K^{\varepsilon,\theta}(G)) = 1$.

Proof. Let $\eta = e^{\frac{2\pi i}{n}} \in \mathbb{C}^*$. For any $s = (s_1,\ldots,s_n)$ we define $\eta s = (\eta s_1,\ldots,\eta s_n)$. Since θ is unramified we have $\theta(z) = \lambda^{\nu(z)}$ for some $\lambda \in \mathbb{C}^*$. Let $\mu \in \mathbb{C}^*$ be such that $\mu^n\eta^{n(n-1)/2} = \lambda$ and $s_\mu = (\mu,\eta\mu,\eta^2\mu,\ldots,\eta^{n-1}\mu)$. Of course, μ is not uniquely defined but the representation π_{s_μ} does not depend on a choice of μ. We denote it by π_θ. As follows from Lemma 2 any $\pi \in \mathfrak{E}_K^{\varepsilon,\theta}(G)$ has the form $\pi \simeq \pi_s$ where s is such that (a) $N(s) = \lambda$ and (b) there exists $\sigma \in S_n$ such that $s^\sigma = \eta s$. It is clear now that $\mathfrak{E}_K^{\varepsilon,\theta} = \{\pi_\theta\}$. The proposition is proved.

We denote the spherical function Γ_{s_μ} by Γ_θ.

Let $\mathfrak{E}_K^{0,\theta}(G_0)$ be the set of equivalent classes of irreducible representations (ρ,W) of G_0 such that

(a) $\rho(z) = \theta(z)\mathrm{Id}\ \ \forall\ z \in Z \subset G_0$

(b) $W^K \neq \{0\}$

(c) For any $\sigma \in \mathfrak{G} - \{e\}$ we have $\rho^\sigma \neq \rho$.

Corollary. For any unramified character θ we have $\#(\mathfrak{E}_K^{0,\theta}(G_0)) = 1$.

Proof. Let (ρ,W) be an element in $\mathfrak{E}_K^{0,\theta}(G_0)$ and (π,V) the induced representation of G. As follows from Lemma 1 π is irreducible. It is clear then that $\pi \in \mathfrak{E}_K^{\varepsilon,\theta}(G)$. Therefore $(\pi,V) \simeq (\pi_\theta,V_\theta)$ and we may realize W as a G_0-invariant subspace of V_θ. Since $\dim W^K = \dim V_\theta^K = 1$ we have $W = \pi_\theta(G_0)V_\theta^K$. The corollary is proved.

We denote the representation of G_0 on $\pi_\theta(G_0)V_\theta^K$ by (ρ_θ,W_θ).

We now fix $\lambda \in \mathbb{C}^*$ and denote by $\theta \in \hat{Z}$ the character given by $\theta(z) \overset{\text{def}}{=} \lambda^{\nu(z)}$. Let $S_\theta(G)$ be the space of locally constant functions $f\colon G \to \mathbb{C}$ such that $f(zg) = \theta(z)^{-1}f(g)$, $z \in Z$, $g \in G$ and $|f|$ has compact support mod Z. We define the product $\circledast$ in $S_\theta(G)$ by

$$(f_1 \circledast f_2)(g) \overset{\text{def}}{=} \int_{x\in PG} f_1(x)f_2(x^{-1}g)\,dx, \qquad \text{for } f_1,f_2 \in S_\theta(G),$$

and we denote by $\mathfrak{H}_\theta \subset S_\theta(G)$ the subalgebra of two-sided K-invariant functions.

We define $S_\theta(G_0) = \{f \in S_\theta(G) \mid \operatorname{supp} f \in G_0\}$, $\mathfrak{H}_\theta^0 = \mathfrak{H}_\theta \cap S_\theta(G_0)$. For any $(\rho,W) \in \mathfrak{E}^\theta(G_0)$, $f \in S_\theta(G_0)$ we define

$$\rho(f) = \int_{x\in PG} f(x)\rho(x)\,dx \in \operatorname{End} W \quad \text{and} \quad \langle\rho,f\rangle \overset{\text{def}}{=} \operatorname{Tr}\rho(f).$$

Lemma 3. Supp $\Gamma_\theta \subset G_0$.

Proof. Since $\pi_\theta \circ \varepsilon = \pi_\theta$ we have $\Gamma_\theta(g)\varepsilon(g) = \Gamma_\theta(g)$. The lemma is proved.

For any matrix $A \in M_n(F)$ we denote by $|A|$ the maximum of norms of matrix coefficients of A. For any number $r \in \mathbb{Z}^+$ define $Y_r = \{A \in M_n(F) \mid |A| \le q^r\}$, denote by $\tilde{\varphi}_r$ the characteristic function of Y_r and by $\varphi_r \in \mathcal{H}_\theta(G)$ the function on G given by

$$\varphi_r(g) = \begin{cases} \lambda^{-\frac{\nu(\det g)}{n}} \tilde{\varphi}_r\left(t^{-\frac{\nu(\det g)}{n}} g\right) & \text{if } \nu(\det g) \in n\mathbb{Z} \\ 0 & \text{otherwise.} \end{cases}$$

We denote by σ_r the representation of $PGL(n,\mathbb{C})$ on the symmetric power $\mathrm{Sym}^{rn}\mathbb{C}^n$ of the standard space $\mathbb{C}^n$, which acts there by the formula

$$\sigma_r(g)(e_1 \otimes \ldots \otimes e_{rn}) = \det(g)^{-r}(ge_1 \otimes \ldots \otimes ge_{rn}), \qquad e_1,\ldots,e_{rn} \in \mathbb{C}^n.$$

Proposition 2. Assume that $z_1 \ldots z_n = \lambda$. Then $\varphi_r\Gamma_z$ is a function on PG and $\int_{PG} \varphi_r\Gamma_z(x)\,dx = q^{\frac{r(n-1)n}{2}}\,\mathrm{Tr}\,\sigma_r(\delta^z)$ where $\delta^z \in GL(n,\mathbb{C})$ is a diagonal matrix with $\delta^z_{i,i} = z_i$.

Proof. Let $\psi_0 \colon F^+ \to \mathbb{C}^*$ be an additive character such that $\psi_0|_{\mathcal{O}} \equiv 1$ and $\psi_0(t^{-1}) \ne 1$. We denote by $\psi \colon U \to \mathbb{C}^*$ the character given by $\psi(u) = \psi_0(u_{12} + \ldots + u_{n-1,n})$ where u_{ij} are matrix elements of u. As is well known, any element $g \in G$ can be written as a product

$g = u\delta^a k$ with $u \in U$, $a = (a_1,\ldots,a_n) \in \mathbb{Z}^n$, $k \in K$, where δ^a is the diagonal matrix with $\delta^a_{ii} = t^{a_i}$. The element a is uniquely determined by g and we denote it $a(g)$.

Any $a \in \mathbb{Z}^n$ defines a character λ_a of $\mathbb{C}^{*n}$ by $\lambda_a(z) \overset{\text{def}}{=} \prod_{i=1}^{n} z_i^{a_i}$ for $a = (a_1,\ldots,a_n)$, $z = (z_1,\ldots,z_n)$. We will identify $\mathbb{C}^{*n}$ with the diagonal $\Delta_{\mathbb{C}}$ subgroup in $GL_n(\mathbb{C})$. Let $\mathbb{Z}^{n+} \subset \mathbb{Z}^n$ be the subset of $a = (a_1,\ldots,a_n)$ such that $a_1 \geq a_2 \geq \ldots \geq a_n$. It is clear that for any $a \in \mathbb{Z}^{n+}$, λ_a is a dominant character of $\Delta_{\mathbb{C}}$ (in respect to the upper triangle subgroup $B_{\mathbb{C}} \subset GL_n(\mathbb{C})$), and we denote by σ_a the irreducible representation of $GL(n,\mathbb{C})$ with the highest weight λ_a. We denote by $\mathbb{Z}^n_0 \subset \mathbb{Z}^n$ the subgroup of $a = (a_1,\ldots,a_n)$ such that $a_1 + \ldots + a_n = 0$, and define $\mathbb{Z}^+_0 = \mathbb{Z}^{n+} \cap \mathbb{Z}^n_0$.

For any $z \in \mathbb{C}^{*n}$ we denote by W_z the function on G given by

$$W_z(g) = \begin{cases} \psi(u)\underset{\sim}{\Delta}(a)^{1/2}\, \mathrm{Tr}\, \sigma_a(\delta^z) & \text{if } g = u\delta^a k \text{ with } a \in \mathbb{Z}^{n+} \\ 0 & \text{otherwise} \end{cases}$$

where $\underset{\sim}{\Delta}$ is the modulus character of $B\left(\underset{\sim}{\Delta}(a) = q^{\sum_{j=1}^{n} (n+1-2j)a_j}\right)$. As is well known (see [Sat])

$$\Gamma_z(g) = \int_K W_z(kg)\, dk.$$

We now assume that $z = (z_1,\ldots,z_n)$ is such that $z_1 \ldots z_n = \lambda$. Since φ_r is a two-sided K-invariant function on G we have

$$\int_{PG} (\varphi_r\Gamma_z)(x)\, dx = \int_{PG} (\varphi_r W_z)(x)\, dx$$

$$= \sum_{a\in\mathbb{Z}^+_0} \underset{\sim}{\Delta}(z)\, \mathrm{Tr}\, \sigma_a(\delta^z) \int_U \psi(u)\tilde{\varphi}_r(u\delta^a)\, du,$$

where du is the Haar measure on U such that $\int_{U\cap K} du = 1$, and where the last equality follows from the definition of φ_r.

To finish our computations we will use the following result.

Lemma 4.

$$\int_U \psi(u)\tilde{\varphi}_r(u\delta^a)\,du = \begin{cases} q^{\frac{-n(n-1)}{2}r} & \text{if } a = (a_1,-r,-r,\dots,-r),\ a_1 \geq - \\ 0 & \text{otherwise.} \end{cases}$$

Proof. We will use the induction in n. We identify $G_{n-1} \overset{\text{def}}{=} GL_{n-1}(F)$ with the subgroup of G of matrices $g = (g_{ij})$ such that $g_{nj} = g_{jn} = 0$ for $j < n$ and $g_{nn} = 1$ and define $U_{n-1} = U_n \cap G_{n-1}$, $V = \{u = (u_{ij}) \in U \mid u_{ij} = 0 \text{ for } i < j < n\}$. $V_K = V \cap K$. It is clear that any element $u \in U$ can be uniquely written as a product $u = u_{n-1}v$, $u_{n-1} \in U_{n-1}$, $v \in V$.

We denote by $W \subset V$ the set of $v = (v_{ij}) \in V$ such that $v_{in} = 0$ for $i < n-1$ and $v_{n-1,n} \in t^{-1}\mathcal{O}$.

For any $a \in \mathbb{Z}^n$ we denote by $X_r(a) \subset U$ the subset of all $u \in U$ such that $u\delta^a \in Y_r$ (see the definition of $\tilde{\varphi}_r$). It is easy to check the following properties of $X_r(a)$.

Claim. (1) $X_r(a)$ is left V_k-invariant

(2) If $a_n < -r$ then $X_r(a) = \emptyset$.

(3) If $a_n = -r$ then $X_r(a) = V_K X_r^{(n-1)}(a)$ where $X_r^{(n-1)}(a) \overset{\text{def}}{=} X_r(a) \cap U_{n-1}$.

(4) If $a_n > -r$ then $X_r(a)$ is left W-invariant.

It is clear now that $\int_U \psi(u)\tilde{\varphi}_r(u\delta^a)\,du = 0$ if $a_n \neq -r$ and in the case when $a_n = -r$

$$\int_U \psi(u)\tilde{\varphi}_r(u\delta^a)\,du = \int_{u\in X_r(a)} \psi(u)\,du$$

$$= \int_{u_{n-1}\in X_r^{(n-1)}(a)} \psi(u_{n-1})\,du.$$

The statement of Lemma 4 follows now from the inductive assumptions.

Let $a_0 = ((n-1)r,-r,\dots,-r)$. As follows from Lemma 4,

$$\int_U \psi(u)\tilde{\varphi}_r(u\delta^a)\,du = 0 \quad \text{for} \quad a \in \mathbb{Z}_0^+ - \{a_0\}$$

nd $\int_U \psi(u)\tilde{\varphi}_r(u\delta^{a_0})\,du = q^{\frac{rn(n-1)}{2}}$. Then Proposition 2 is proved.

<u>Corollary</u>. For any integer $r \geq 0$ we have

$$\langle \pi_\theta, \varphi_r \rangle = (\eta q)^{\frac{rn(n-1)}{2}}.$$

<u>Proof</u>. By the definition $\langle \pi_\theta,\varphi_r\rangle = \operatorname{Tr} \pi_\theta(\varphi_r)$. Since $\varphi_r \in \mathcal{H}_\theta$ e have $\operatorname{Im} \pi_\theta(\varphi_r) \subset V^K$. Therefore

$$\operatorname{Tr} \pi_\theta(\varphi_r) = \int_{x\in PG} \varphi_r(x) \langle \pi_\theta(g)V_\theta, V_\theta\rangle\,dx = \int_{x\in PG} \varphi_r(x)\Gamma_\theta(x)\,dx.$$

et $\mu \in \mathbb{C}^*$ be such that $\mu^n\eta^{\frac{n(n-1)}{2}} = \lambda$ and $s = s_\mu = (\mu,\mu\eta,\dots,\mu\eta^{n-1})$. e consider s as a diagonal matrix in $GL(n,\mathbb{C})$. Let $\tilde{s} \in GL(n,\mathbb{C})$ be he matrix $\tilde{s} = (\tilde{s}_{ij})$, $1 \leq i,j \leq n$ such that $\tilde{s}_{ij} = 1$ if $i - j \equiv$ (mod n-1) and zero otherwise. It is clear that s and $\mu\tilde{s}$ are conugate in $GL(n,\mathbb{C})$.

As follows from the proposition $\langle \pi_\theta,\varphi_r\rangle = q^{\frac{rn(n-1)}{2}} \operatorname{Tr} \sigma_r(s)$. ut it is easy to see that $\operatorname{Tr} \sigma_r(s) = (-1)^{\frac{r(n-1)}{2}} = \eta^{\frac{rn(n-1)}{2}}$. The orollary is proved.

ection 3

We will use here the notations from §2. Now let $L \supset F$ be the nramified extension of degree n. We extend the valuation $\nu: F^* \to \mathbb{Z}$ o $\nu: L^* \to \mathbb{Z}$. Let $\mathfrak{G}$ be the Galois group $\mathfrak{G} = \operatorname{Gal}(L:F)$. We will identify with $\mathbb{Z}/n\mathbb{Z}$ in such a way that Fr corresponds to $1 \in \mathbb{Z}/n\mathbb{Z}$, where

$Fr \in \mathfrak{G}$ is the element which acts on the residue field by $x \to x^q$. We denote by $\varepsilon\colon \mathfrak{G} \hookrightarrow \mathbb{C}^*$ the embedding such that $\varepsilon(Fr) = \eta (= e^{2\pi i/n})$.

As in §1 we denote by $L_0 \subset L$ the set of all $\ell \in L$ such that $\ell^{Fr} = (-1)^{n(n-1)/2}\ell$. We fix $\ell_0 \in L_0$ such that $|\ell_0| = 1$ (it is easy to show the existence of such ℓ_0). For any $\ell \in L$ we define

$$\widetilde{\Delta}(\ell) \overset{\text{def}}{=} \prod_{1\le i<j\le n} (\ell^{Fr^i} \cdot (\ell^{Fr^j})^{-1} - 1) \in L_0$$

and

$$\Delta(\ell) = |\widetilde{\Delta}(\ell)| \cdot \varepsilon(\widetilde{\Delta}(\ell))^{\frac{n(n-1)}{2}}$$

where $\varepsilon\colon L^* \to \mathbb{C}^*$ is the character given by $\varepsilon(x) = \eta^{\nu(x)}$, and $|x| \overset{\text{def}}{=} q^{-\nu(x)}$ for $x \in L^*$. It is easy to see that the definition of Δ coincides with one from §1.

Let $L' \subset L$ be the subset of $\ell \in L$ such that $\Delta(\ell) \neq 0$. We call such elements regular. We denote by the same letter the character $\varepsilon\colon G \to \mathbb{C}^*$ given by $\varepsilon(g) = \varepsilon(\det g)$. It is clear that we may consider ε as a character of PG.

We will fix an embedding $L^* \hookrightarrow G$ such that $\mathcal{O}_L^* = L^* \cap K$ and for any $\ell \in L'$ we denote by I_ℓ the linear functional on $S(G)$ given by

$$I_\ell(g) \overset{\text{def}}{=} \int_{x\in PG} \varepsilon(x) f(x^{-1}\ell x)\, dx.$$

<u>Theorem 1</u>. For any $f \in \mathcal{H}_\theta$, $\ell \in L'$ we have

$$I_\ell(f) = \Delta(\ell)^{-1} \langle \pi_\theta, f\rangle,$$

where $\pi_\theta(f) \overset{\text{def}}{=} \operatorname{Tr} \pi_\theta(f)$.

We start with the following result. Let ${}^0\mathcal{H}_\theta = \{f \in \mathcal{H}_\theta \mid \langle \pi_\theta, f\rangle\} = 0$.

<u>Proposition 1</u>. For any $f \in {}^0\mathcal{H}_\theta$, $\ell \in L'$ we have $I_\ell(f) = 0$.

Proof. Since $\operatorname{supp}\Gamma_\theta \subset G_0$ (Lemma 2.3), we may assume that $\operatorname{supp} f \subset G_0$. We choose $\gamma \in G$ such that $\varepsilon(\gamma) = \eta$ and define $\tilde{f} \in \mathbb{C}_\theta(G_0)$ by $\tilde{f}(g) = \sum_{i=0}^{n-1} \eta^{-i} f(\gamma^{-i} g \gamma^{i})$, $g \in G_0$.

Lemma 1. For any $\rho \in \mathfrak{E}_\theta(G_0)$ we have $\langle\rho,\tilde{f}\rangle = 0$.

Proof of the Lemma. It is clear that $\langle\rho,\tilde{f}\rangle = \sum_{\sigma\in\mathfrak{G}} \varepsilon(\sigma)^{-1}\langle\rho^\sigma,f\rangle$. It is clear now that if $\langle\rho,\tilde{f}\rangle \neq 0$ then there exists $\sigma_0 \in \mathfrak{G}$ such that $\rho^{\sigma_0} \in \mathfrak{E}_K^{0,\theta}(G_0)$ and $\langle\rho,\tilde{f}\rangle = \varepsilon(\sigma_0)^{-1}\langle\rho^{\sigma_0},f\rangle$. But by the corollary to Proposition 2.2, we have $\mathfrak{E}_K^{\varepsilon,\theta}(G_0) = \rho_\theta$. Lemma 1 is proved.

It follows now from [DKV] that all regular orbital integrals of $\tilde{f}$ are zero for all $f \in {}^0\mathcal{H}_\theta$. But it is easy to see that

$$I_\ell(f) = \int_{x\in PG_0} \tilde{f}(x^{-1}\ell x)\, dx.$$

The proposition is proved.

Corollary (to the proposition). There exists a function φ on L' such that $I_\ell(f) = \varphi(\ell)\langle\pi_\theta,f\rangle$ for all $f \in S_\theta(G)$, $\ell \in L'$.

$$\text{Define } \varphi_\theta \in \mathcal{H}_\theta \text{ by } \varphi_\theta(g) = \begin{cases} \theta^{-1}(z) & \text{for } g \in zk,\ z \in Z,\ k \in K \\ 0 & \text{otherwise.} \end{cases}$$

It is clear that $\langle\pi_\theta,\varphi_\theta\rangle = 1$ and therefore

$$\varphi(\ell) = \int_{\substack{x\in PG \\ \ell^x\in K}} \varepsilon(x)\, dx$$

for $\ell \in L' \cap K$.

Theorem 1 is equivalent to the equality

$$\varphi(\ell) = \Delta(\ell)^{-1}, \qquad \ell \in L' \cap K.$$

Let $\mathfrak{f} = \mathcal{O}/p$ be the residue field.

Definition. We say that an element $k \in K$ is $\mathfrak{f}$-semisimple if $k^a = 1$ for some $a \in \mathbb{Z}$, $(a,q) = 1$. We say that k is $\mathfrak{f}$-unipotent if $k^{q^N} \to 1$ for $N \to \infty$.

Lemma 2. Any element $k \in K$ has unique decomposition $k = su$ where s is $\mathfrak{f}$-simple, u is $\mathfrak{f}$-unipotent and s, u commute.

Proof. Uniqueness is clear. To construct such a decomposition we choose an increasing sequence of integers m_i, $1 \le i < \infty$ such that $q^{m_i} \equiv 1 (\mathrm{mod}\ C)$ where $C = \prod_{a=1}^{n} (q^a-1)$ (= prime to q part of the order of $GL(n,\mathfrak{f})$). It is clear that the sequence $k^{q^{m_i}}$ is convergent to an element $s \in K$ of order dividing C and that $(ks^{-1})^{q^{m_i}} \to 1$ for $i \to \infty$. Therefore $u \overset{\mathrm{def}}{=} ks^{-1}$ is $\mathfrak{f}$-unipotent. The lemma is proved.

We call s (resp. u) the $\mathfrak{f}$-ss (resp. $\mathfrak{f}$-un) part of k. For any $k \in K$ we denote by $X_k \subset G$ the set of all $x \in G$ such that $k^x \in K$.

Lemma 3. $X_k \subset Z(s)K$ where $Z(s)$ is the centralizer of s in G.

Proof. Take $x \in X_k$ and consider $k' \overset{\mathrm{def}}{=} k^x$ and its decomposition $k' = s'u'$ to $\mathfrak{f}$-ss and $\mathfrak{f}$-un parts. Let $\bar{s}, \bar{s}'$ be the images of s, s' in $Gl(n,\mathfrak{f})$ under the natural map: $GL(n,\mathcal{O}) \to GL(n,\mathfrak{f})$ and let $P_s(t), P_{s'}(t) \in \mathcal{O}[t]$, $P_{\bar{s}}(t), P_{\bar{s}'}(t) \in \mathfrak{f}[t]$ be the characteristic polynomials of $s, s', \bar{s}, \bar{s}'$. It is clear that $P_{\bar{s}}(t), P_{\bar{s}'}(t)$ are reductions mod p of $P_s(t), P_{s'}(t)$. On the other hand, $s' = \lim_{i\to\infty}(k')^{q^{m_i}} = \lim_{i\to\infty}(k^{q^{m_i}})^x = s^x$ and $P_s(t) = P_{s'}(t)$. Therefore, $P_{\bar{s}}(t) = P_{\bar{s}'}(t)$. $\bar{s}, \bar{s}' \in GL_n(\mathfrak{f})$ are elements of order dividing C, $(C,q) = 1$. There exists $\bar{y} \in GL(n,\mathfrak{f})$ such that $\bar{s} = \bar{s}'^{\bar{y}}$. Let $y \in K$ be a representative of $\bar{y}$. As follows from the Hensel lemma there exist $k_1 \in \widetilde{K}$ such that $s'^{y} = s^{k}$. Therefore $s' = s^{k_1 y^{-1}}$ and $x \in Z(s)k_1 y^{-1}$. The lemma is proved.

We will now finish the proof of Theorem 1. We assume that the analogous theorem is known for the groups $GL_m(R)$ where R are local fields and $m < n$.

Fix $\ell \in L' \cap K$ and write it as a product $\ell = su$ of $\mathfrak{f}$-ss and $\mathfrak{f}$-un parts. It is clear that the centralizer $Z(s)$ is isomorphic to $GL_m(R)$ where m is a divisor of n and $R \subset L_{un}$ is the unramified extension of F of degree n/m. We write G' for $Z(s)$, let Z' be the center of G', and define $PG' = G'/Z'$, and $K' = G' \cap K$. Also let PK' be the image of K' in PG'.

Lemma 4. If $s \notin Z$, then $\varphi(\ell) = \Delta(\ell)^{-1}$.

Proof. By Lemma 3 we have

$$\int_{\substack{x \in PG \\ \ell^x \in K}} \varepsilon(x)\, dx = \int_{\substack{x' \in PG' \\ \ell^{x'} \in K'}} \varepsilon(x')\, dx'$$

where dx' is the Haar measure of PG' such that $\int_{PK'} dx' = 1$. It follows from the inductive assumptions that the last integral is equal to $\Delta(\ell)^{-1}$ and Theorem 1 is proved for $\ell \in L' \cap K$ with $\mathfrak{f}$-ss part $s \notin Z$. The lemma is proved.

Assume now that $s \in Z$. Choose $z \in Z$ such that $|\ell - z| \leq |\ell - z'|$ for any $z' \in Z$. Since $\varphi(\ell z) = \varphi(\ell)$ for $z \in Z \cap K$ we may replace ℓ by ℓz^{-1} and may therefore assume that $|\ell - Id| \leq |\ell - z|$ for $z \in Z$. Define $r \in \mathbf{Z}^+$ by $|\ell - Id| = q^{-r}$ and define $\ell_1 \in M_n(\mathcal{O})$ by $\ell_1 = Id + t^{-r}(\ell - Id)$.

Lemma 5. $\ell_1 \in K$ and $s_1 \notin Z$ where s_1 is the f-ss part of ℓ_1.

Proof. We have $\ell \in \mathcal{O}_{L'}$, $\ell - 1 = t^r x$, $x \in \mathcal{O}_L^*$. It is clear that $\bar{x} \in m$ where $\bar{x} \in \mathcal{O}_L/p\mathcal{O}_L$ is the reduction of x mod p. The lemma is proved.

Lemma 6. $\ell^x \in K$ if and only if $\ell_1^x \in Y_r$, $\ell \in L' \cap K$, $x \in PG$.

Proof. $\ell^x \in K \iff \ell^x - \mathrm{Id} \in M_n(\mathcal{O}) \iff (\ell - \mathrm{Id})^x \in M_n(\mathcal{O}) \iff (\ell_1 - \mathrm{Id})^x \in Y_r \iff \ell_1^x \in Y_r$. The lemma is proved.

Now we can finish the proof of the theorem. We have

$$\varphi(\ell) = \int_{\substack{x \in PG \\ \ell^x \in K}} \varepsilon(x)\, dx = \int_{\substack{x \in PG \\ \ell_1^x \in K}} \varepsilon(x)\, dx = I_{\ell_1}(\varphi_r)$$

$$\overset{(1)}{=} \langle \pi_\theta, \varphi_r \rangle \varphi(\ell_1) \overset{(2)}{=} (\eta q)^{\frac{rn(n-1)}{2}} \Delta(\ell_1)^{-1} = \Delta(\ell)^{-1}$$

where (1) follows from the corollary to Lemma 4 and (2) follows from Proposition 2.2. Theorem 1 is proved.

We will now reformulate Theorem 1 in terms of Langland's correspondence.

Let ${}^L G \simeq GL_n(\mathbb{C})$ and ${}^L T$ be the semidirect product ${}^L T_0 \times \mathfrak{G}$ where ${}^L T_0 = \mathbb{C}^{*n}$ and $\mathfrak{G}$ $(\simeq \mathbb{Z}/n\mathbb{Z})$ acts on ${}^L T_0$ by cyclic permutations. Then ${}^L G$ and ${}^L T$ are groups dual to G and L^* (see [Bor]). There is a canonical isomorphism between the semisimple conjugacy classes $s \subset {}^L G$ and irreducible unramified representations π_s of G. For any $f \in \mathcal{H}$ $(\simeq \mathcal{H}(G,K))$ we denote by $S(f)$ the function on ${}^L G$ given by $S(f)(s) \overset{\text{def}}{=} \langle \pi_s, f \rangle$. The following result is due to Satake (see ([Car]).

Lemma 7. S induces an isomorphism between $\mathcal{H}$ and the algebra of regular class functions on ${}^{L}G$. Analogously we may identify the Hecke algebra $\mathcal{H}_L = \mathcal{H}(L^*,\mathcal{O}_L^*)$ with the regular functions on ${}^{L}T_0 \times \{Fr\} \subset {}^{L}T$ invariant under conjugation by ${}^{L}T_0$.

Let $i: {}^{L}T \hookrightarrow {}^{L}G$ be the imbedding which identifies T_0 with diagonal matrices and such that $i(Fr) = \tilde{s}$ (see the end of §2) and $i^*: \mathcal{H} \to \mathcal{H}_L$ be the adjoint homomorphism.

The following statement is evidently equivalent to Theorem 1.

Theorem 1'. For any $f \in \mathcal{H}$, $\ell \in L'$ we have $\Delta(\ell)I_\ell(f) = i^*(f)(\ell)$.

Let m be a divisor of $n, n' = n/m$, $L_1 \supset F$ be the unramified extension of F of degree m, $\mathfrak{G}_1 = Gal(L_1,F)$ and $H = GL_{n'}(L_1)$. The dual group ${}^{L}H$ is the semidirect product ${}^{L}H = {}^{L}H_0 \times \mathfrak{G}_1$ where ${}^{L}H_0$ is the product of m-copies of $GL_{n'}(\mathbb{C})$ and $\mathfrak{G}_1$ acts on ${}^{L}H_0$ by cyclic permutation. We denote by $i_H: {}^{L}H \hookrightarrow {}^{L}G$ the natural embedding. Let $K_H = GL_{n'}(\mathcal{O}_{L_1})$. $\mathcal{H}^1 = \mathcal{H}(H,K_H)$. As before, we identify $\mathcal{H}^1$ as regular H_0-invariant functions on ${}^{L}H_0 \times \{Fr\} \subset {}^{L}H$ and denote by $i_H^*: \mathcal{H} \to \mathcal{H}^1$ the restriction map.

We fix an embedding $L_1 \hookrightarrow M_m(F)$ such that $\mathcal{O}_{L_1} \subset M_m(\mathcal{O})$. Then we have embeddings $M_{n'}(L_1) \hookrightarrow M_n(F)$ and $H \subset G$. For any $h \in H$ we denote by $\overline{Ad(h)}$ the adjoint action of h on the quotient space $M_n(F)/M_{n'}(L_1)$ and define $D(h) \overset{def}{=} \det(\overline{Ad}(h) - Id)$ and $\Delta(h) = |Dh|^{1/2}\eta_1^{\frac{m(m-1)}{4}\nu(D(h))}$ where $\eta_1 \overset{def}{=} \ell^{2\pi i/m}$. It is easy to see that $\nu(D(h)) \in 2\mathbb{Z}$ and therefore $\Delta(h)$ is well-defined and $\Delta(h) \in \mathbb{R}$.

Let G' be the set of regular elements in G and $H' \overset{def}{=} H \cap G'$. It is clear that $\Delta(h) \neq 0$ for $h \in H'$. For any $h \in H'$ we denote by Z_h the centralizer of h in G. It is clear that $Z_h \subset H$. We denote by $I_h \in \mathcal{H}'$ and $J_h \in (\mathcal{H}^1)'$ the linear functionals given by

$$I_h(f) \overset{\text{def}}{=} \int_{x \in Z_h \backslash G} \varepsilon_1(x) f(x^{-1}hx)\, dx, \qquad f \in \mathcal{H}$$

and

$$J_h(\varphi) = \int_{y \in Z_h \backslash H} \varphi(y^{-1}hy)\, dy, \qquad \varphi \in \mathcal{H}^1,$$

where dx, dy are invariant measures and $\varepsilon_1 \cdot G \to \mathbb{C}^*$ is the character given by $\varepsilon_1(x) = \eta_1^{\nu(\det x)}$.

<u>Conjecture</u>. For any $h \in H'$ there exists a constant $c \in \mathbb{R}$ (which depends on a choice of dx, dy) such that for all $f \in \mathcal{H}$ we have

$$I_h(f) = cJ_h(i_h^* f)\Delta(h)^{-1}$$

<u>Theorem 2</u>. The conjecture is true for split $h \in H'$.

<u>Proof</u>. Fix a minimal parabolic subgroup $P = M \cdot U \subset G$ such that $h \in P$. We may assume that $h \in M$. Let $P_H = P \cap H$, $M_H = M \cap H$, $U_H = U \cap H$. It is clear that $P_H = M_H U_H$ is a Borel subgroup in H and $U_H \subset B_H$ is its unipotent radical, M is isomorphic to the product of n' copies of $GL_m(F)$ and $M_H = (L_1^*)^{n'}$. Let $\tilde{\mathcal{H}} = \mathcal{H}(M, M \cap K)$, $\tilde{\mathcal{H}}^1 = \mathcal{H}(M_H, M_H \cap K)$ be the corresponding Hecke algebras.

Let ${}^L M \simeq GL_m(\mathbb{C})^{n'}$, ${}^L M_H \simeq (\mathbb{C}^{*m})^{n'} \times \mathfrak{G}_1$ be the dual groups and $j: {}^L M \hookrightarrow {}^L G$, $j_H: {}^L M_H \hookrightarrow {}^L H$, be the natural embedding. Then there exists an embedding $i_H: {}^L M_H \hookrightarrow {}^L M$ such that the diagram

$$\begin{array}{ccc} {}^L M_H & \overset{j_H}{\hookrightarrow} & {}^L H \\ \Big\downarrow i_H & & \Big\downarrow i \\ {}^L M & \overset{j}{\hookrightarrow} & {}^L G \end{array}$$

commutes. It is easy to see that i_H is just a product of embeddings $\mathbb{C}^{*m} \times \mathfrak{G}_1 \hookrightarrow GL_m(\mathbb{C})$ of the type considered earlier.

The following result is well known (see [Car]).

Lemma 7. For any $f \in \mathcal{H}$ and regular $m \in M$ we have

$$\int_U f(mu)\, du = (j^*f)(m).$$

Theorem 2 now follows from Theorem 1.

For any $f \in \mathbb{C}_\theta(G)$ we denote by $\delta(f)$ the function on the set M_H' of regular elements in $M_H \subset G$ defined by

$$\delta(f)(m) \stackrel{\text{def}}{=} \Delta(m) \int_{Z_m \setminus G} \varepsilon_1(x) f(x^{-1}mx)\, dx \qquad \text{for } m \in M_H'.$$

Proposition 2. For any $f \in \mathbb{C}_\theta(G)$, $\delta(f)$ is a restriction to M_H' of a locally constant function $\bar{f}$ on M_H.

Proof. We assume that an analogous statement is known for the group $GL_a(R)$, where R is a finite extension of F and $a < n$. Fix $f \in \mathbb{C}_\theta(G)$. We have to prove that for any $m_0 \in M_H$ there exists an open neighborhood V of m_0 in M_H and a constant $c \in \mathbb{C}$ such that $\bar{f}(m) = c$ for all $m \in V \cap M_H'$.

Assume first that $m_0 \notin Z$. Let $G_1 \subset G$ be the centralizer of m_0 in G. It is clear that $G_1 \simeq GL_a(R)$ for $a < n$. We denote by δ_1 the map $\delta_1 \colon \mathbb{C}(G_1) \to \mathbb{C}(M_H')$ analogous to δ.

Let $C \subset G$ be an open compact set such that $m_0^x \notin \operatorname{supp} f$ for all $x \neq G_1C$ (it is easy to see the existence of such C). Then there exists a neighborhood V_1 of m_0 in M_H such that $m^x \notin \operatorname{supp} f$ for all $m \in V_1$, $x \notin G_1C$.

Let $\pi \colon G_1 \times C \to G$ be the product map and $\pi^*(dg)$ be the preimage of a Haar measure dg on G. Since $\pi^*(dg)$ is G_1-invariant we can write it as a product $\pi^*(dg) = dg_1 \times dc$ where dg_1 is a Haar measure on G_1 and dc is a measure on C. For any $f \in \mathbb{C}_\theta(G)$ we now define

$\tilde{f} \in \mathbb{C}(G_1)$ by $\tilde{f}(g_1) \overset{\text{def}}{=} \int_C \varepsilon(c) f(c^{-1}g_1c)\,dc$. It is clear that $\delta(f)(m) = \delta_1(\tilde{f})(m)$ for $m \in V_1 \cap M_H'$. The proposition follows now from the inductive assumptions.

Now assume that $m_0 \in Z$. Then we may assume that $m_0 = e$. The same arguments as in the proof of Theorem 2 reduce the proposition to the case where $L_1 = L$ and $H = L^*$.

Let $u_0 \in G$ be a regular unipotent element and let $Z(u_0) \subset G$ be its centralizer in G. We denote by $\alpha: \mathbb{C}_\theta(G) \to \mathbb{C}$ the linear functiona given by

$$\alpha(f) = \int_{Z(u_0)\backslash G} \varepsilon(x) f(x^{-1}u_0x)\,dx.$$

As is well known (see [H]) this integral is absolutely convergent.

<u>Lemma 8</u>. If $f \in \mathbb{C}_\theta(G)$ is such that $\alpha(f) = 0$ then there exists a neighborhood V of e in G such that $\bar{f}(\ell) = 0$ for $\ell \in L' \cap V$.

<u>Proof</u>. Let N be the set of unipotent elements, S the space of locally constant functions on N with compact support and $\pi: S_\theta(G) \to S$ the restriction. Let S' be the dual space to S and $\Lambda = \{\lambda \in S' \mid Ad'(g)\lambda = \varepsilon(g)\lambda \;\; \forall g \in G\}$, where Ad is the natural adjoint action of G on S. Since G has only a finite number of orbits on N it is easy to see that $\Lambda = \mathbb{C}\lambda_0$, $\lambda_0 \in S'$ where $\alpha = \pi^*(\lambda_0)$ (see [Sh]).

Let $S_\varepsilon = \{\tilde{f} \in S \mid \lambda(\tilde{f}) = 0$ for all $\lambda \in \Lambda\}$. Then S_ε is spanned by functions of the form $Ad(g)\tilde{F} - \varepsilon(g)\tilde{F}$, $\tilde{F} \in S$, $g \in G$ (see [Sh]).

Let $f \in \mathbb{C}_\theta^\infty(G)$ be such that $\pi^*\lambda(f) = 0$. We denote by $\tilde{f} \in S$ the restriction of f to N. Since $\lambda_0(\tilde{f}) = \alpha(f) = 0$ we have $\tilde{f} \in S_\varepsilon$ and therefore we can write

$$\tilde{f} = \sum_{i=1}^{r} (Ad(g_i)\tilde{F}_i - \varepsilon(g_i)\tilde{F}_i), \qquad g_i \in G, \;\; \tilde{F}_i \in S, \;\; 1 \le i \le r.$$

For any i, $1 \le i \le r$ we choose $F_i \in \mathbb{C}_\theta^\infty(G)$ such that $\tilde{F}_i = \pi(F_i)$

and consider

$$f_1 \overset{\text{def}}{=} f - \sum_{i=1}^{r} (\mathrm{Ad}\, g_i(F_i) - \varepsilon(g_i)F_i) \quad C_\theta^\infty(G).$$

Then $\pi(f_1) = 0$. Since f_1 is a locally constant function with compact (mod Z) support there exists a neighborhood V of e in G such that for any $g \in V$, $x \in G$, $f_1(x^{-1}gx) = 0$. Therefore $\overline{f}_1(\ell) = 0$ for all $\ell \in L' \cap V$. On the other hand, it is clear that $\overline{f} = \overline{f}_1$. The lemma is proved.

Now we can finish the proof of the proposition. Let $\varphi_0 \in \mathfrak{H} \subset \mathbb{C}_\theta(G)$ be the unit of the Hecke algebra. Then by Theorem 1 $\overline{f}\big|_{L'\cap\mathfrak{O}_L^*} = 1$. Therefore $\alpha_0 \overset{\text{def}}{=} \alpha(\varphi_0) \neq 0$. For any $f \in S_\theta(G)$ we can write $f = \alpha_0^{-1}\alpha(f)\cdot\varphi_0 + f'$ where $\alpha(f') = 0$. Therefore there exists a neighborhood V of e in G such that $\overline{f}(\ell) = \alpha_0^{-1}\alpha(f)$ for $\ell \in V \cap L'$.

The proposition is proved.

Section 4

Let F be a global field and $L \supset F$ be a cyclic extension of degree n, $\mathfrak{G} = \mathrm{Gal}(L:F)$. We fix a generator $\sigma \in \mathfrak{G}$. This gives us an identification of $\mathfrak{G} \longrightarrow \mathbb{Z}/n\mathbb{Z}$. We will be using the notation of §1. So $\tilde{G}$ is the multiplicative group of a central division algebra over F of degree n^2 which is ramified in two places $p_1, p_2 \in \Sigma(F)$, $\theta \in \hat{C}_F$ is a character of the idele class group C_F, $\tilde{W}_\theta = L_2^\theta(\tilde{G}_F \setminus \tilde{G}_{\mathbb{A}})$, etc. For any $p \in \Sigma(F)$ we denote by $\tilde{V}_p$ the space of functions f_p on $\tilde{G}_p$ such that

(a) $f_p(z_p g_p) = \theta_p(z_p)^{-1} f_p(g_p)$, $z_p \in Z_p$, $g_p \in \tilde{G}_p$ where θ_p is the p-component of θ,

(b) $|f|$ has compact support on $Z_p \setminus \tilde{G}_p$ and

(c) f_p is smooth (that is, $f_p \in C^\infty(G_p)$ if p is archimedean and is locally constant otherwise).

Let $S \subset \Sigma(F)$ be a finite set such that S contains all the archimedean places, all places where L or θ or D are ramified, and such that the ring $\mathcal{O}_S \subset F$ of S-integers is a principal ideal ring. For any $p \neq p_1, p_2$ we may identify $\tilde{G}_p$ with $G_p = GL_n(F_p)$. Let $K_p^0 \overset{\text{def}}{=} GL_n(\mathcal{O}_p) \subset \tilde{G}_p$. For any $p \notin S$ we denote by $f_p^0 \in \tilde{V}_p$ the unique function such that $f_p^0\big|_{K_p^0} \equiv 1$ and $\sup f_p^0 \subset Z_p K_p^0$. We denote by $\tilde{V}$ the space of functions on $\tilde{G}_{\mathbb{A}}$ which are linear combinations of products $\bigotimes_{p\in\Sigma(F)} f_p$ where $f_p \in \tilde{V}_p$ for all $p \in \Sigma(F)$ and $f_p = f_p^0$ for almost all p.

Let $\underline{\varepsilon}: \tilde{W}_\theta \to \tilde{W}_\theta$ be the operator of multiplication by the function ε (see the end of §1). For any $f \in \tilde{V}$ we denote by $\tilde{\rho}_\theta(f) \in \operatorname{End} \tilde{W}_\theta$ the operator $\tilde{\rho}_\theta(f) \overset{\text{def}}{=} \int_{x\in Z_{\mathbb{A}}\backslash G_{\mathbb{A}}} f(x)\tilde{\rho}_\theta(x)\, dx$, where dx is the Tamagawa measure on $P\tilde{G}_{\mathbb{A}}$.

<u>Proposition 1</u>. For any $f \in \tilde{V}$, $\tilde{\rho}_\theta(f)$ is of trace class and

$$\operatorname{Tr}(\tilde{\rho}_\theta(f)\cdot\underline{\varepsilon}) = \sum_{\ell\in F^*\backslash L'} \int_{L^*_{\mathbb{A}}\backslash\tilde{G}_{\mathbb{A}}} \varepsilon(x) f(x^{-1}\ell x)\, dx.$$

<u>Proof</u>. Let $K_f(x,y)$ be the function on $Z_{\mathbb{A}}\backslash\tilde{G}_{\mathbb{A}} \times Z_{\mathbb{A}}\backslash\tilde{G}_{\mathbb{A}}$ given by $K_f(x,y) = \sum_{\gamma\in P\tilde{G}_F} f(x^{-1}\gamma y)$. It is clear that for any $F \in \tilde{W}_\theta$

$$((\tilde{\rho}_\theta(f)\cdot\underline{\varepsilon})F)(x) = \int_{PG_F\backslash P\tilde{G}_{\mathbb{A}}} K_f(x,y)\varepsilon(y)F(y)\, dy.$$

It is now clear that $\tilde{\rho}_\theta(f)\cdot\underline{\varepsilon}$ is of trace class and

$$\operatorname{Tr}(\tilde{\rho}_\theta(f)\cdot\underline{\varepsilon}) = \int_{P\tilde{G}_F\backslash P\tilde{G}_{\mathbb{A}}} K_f(x,x)\varepsilon(x)\, dx = \sum_{\gamma\in(P\tilde{G}_F)} I_\gamma(f),$$

where γ runs through the set of representatives for conjugacy classes in $P\tilde{G}_F$, $Z(\gamma)$ is the centralizer of γ and

$$I_\gamma(f) = \int_{Z_F(\gamma)\backslash P\widetilde{G}_{\mathbb{A}}} \varepsilon(x) f(x^{-1}\gamma x)\, dx.$$

<u>Lemma 1</u>. If $I_\gamma \not\equiv 0$ then γ is conjugate to a regular element in L.

<u>Proof</u>. Let $N_{\mathbb{A}}: \widetilde{G}_{\mathbb{A}} \to F^*_{\mathbb{A}}$ be the reduced norm, let $N_L: L^*_{\mathbb{A}} \to F^*_{\mathbb{A}}$ be the norm, and let $pr: F^*_{\mathbb{A}} \to C_F$ be the natural projection. Define

$$\overline{N} \overset{\text{def}}{=} pr \circ N: \widetilde{G}_{\mathbb{A}} \to C_F, \quad \overline{N}_L \overset{\text{def}}{=} pr \circ N_L: L^*_{\mathbb{A}} \to C_F \quad \text{and} \quad C_F^L \overset{\text{def}}{=} \operatorname{Im} \overline{N}_L.$$

As follows from the definition of $\varepsilon: \widetilde{G}_{\mathbb{A}} \to \mathbb{C}^*$, $I_\gamma \equiv 0$ if $\overline{N}(Z_{\mathbb{A}}(\gamma)) \not\subset C_F^L$. It is easy to see that $Z(\gamma)$ is isomorphic to the multiplicative group D_1^* of a central division algebra D_1 degree n'^2 over an extension L_1 of F of degree m, $n = mn'$. Moreover, the restriction of N to $Z_{\mathbb{A}}(\gamma)$ is the composition $N = N_{L_1} \circ N_1$, where $N_1: D_{1_{\mathbb{A}}} \to L^*_{1_{\mathbb{A}}}$ is the reduced norm and $N_{L_1}: L^*_{1_{\mathbb{A}}} \to F^*_{\mathbb{A}}$ is the norm. Since N_1 is surjective we see that $I_\gamma \not\equiv 0$ implies $\operatorname{Im}(pr \circ N_1) \subset C_F^0$. Since $[L_1:F] \leq [L:F]$ it follows from class field theory that we have the inclusion $\operatorname{Im}(pr \circ N_1) \subset C_F^0$ only if $L_1 = L$, $n' = 1$. The lemma is proved.

We can now prove the proposition.

Let $\underline{PL}^*$ be the quotient group $\underline{PL}^* = \underline{F}^* \backslash L^*$. As is well known (see [W]) the Tamagawa measure of $\underline{PL}^*$ is equal to n. As follows from Lemma 1 we have

$\operatorname{Tr}(\widetilde{\rho}_\theta(f) \circ \underline{\varepsilon})$

$$= \sum_{\ell' \in L'/\mathfrak{G}} \nu_L \cdot \int_{L^*_{\mathbb{A}}\backslash \widetilde{G}_{\mathbb{A}}} \varepsilon(x) f(x^{-1}\ell x)\, dx = \sum_{\ell' \in F^*\backslash L'} \int_{L^*_{\mathbb{A}}\backslash \widetilde{G}_{\mathbb{A}}} \varepsilon(x) f(x^{-1}\ell x)\, dx.$$

The proposition is proved.

Let $L_0 = \{\ell \in L \mid \ell^\sigma = (-1)^{\frac{n(n-1)}{2}} \ell\}$ where (as before) σ is a generator of $\mathfrak{G}$. Since $\mathcal{O}_S$ is a principle ideal ring there exists an element $\ell^0 \in L_0$ such that $\ell_p^0 \in \mathcal{O}_p^*$ for $p \in S$. For any $p \in \Sigma(F)$ we denote by Δ_p the function $\Delta_{\ell_p^0}$ on L_p defined in §1. (Actually, we have defined $\Delta_{\ell_p^0}$ only for nonarchimedean places but it is clear how to extend the definition to the archimedean case). For any $\ell \in L$, $p \in \Sigma(F)$ we may (and will) consider ℓ as an element in F_p.

Lemma 2. For any regular $\ell \in L'$ we have $\Delta_p(\ell) = 1$ for almost all $p \in \Sigma(F)$ and $\prod_{p\in\Sigma(F)} \Delta_p(\ell) = 1$.

Proof. Clear.

Let $V \subset \tilde{V}$ be the linear subspace spanned by the functions of the form $f = \bigotimes_{p\in\Sigma(F)} f_p$ where for $p \in S$ the support of f_p is in the set $\tilde{G}_p'$ of regular elements in $\tilde{G}_p$.

For any $p \in \Sigma(F)$ we denote by $\tilde{\tilde{V}}_p$ the space of smooth functions $\overline{f}_p$ on L_p such that $\overline{f}_p(z_p\ell_p) = \theta_p^{-1}(z_p)\overline{f}(\ell_p)$, $z_p \in F_p^*$, $\ell_p \in L_p$ and $|\overline{f}_p|$ has compact support on $F_p^* \setminus L_p^*$. For $p \notin S$ we denote by $\overline{f}_p^0 \in \tilde{\tilde{V}}_0$ the function such that $\operatorname{supp} \overline{f}_p^0 \subset F_p^* \cdot \mathcal{O}_{L_p}^*$ and $\overline{f}_p^0\big|_{\mathcal{O}_{L_p}^*} \equiv 1$. We denote by $\tilde{\tilde{V}}$ the space of functions on $L_{\mathbb{A}}^*$ spanned by products $\overline{f} = \bigotimes_{p\in\Sigma(F)} \overline{f}_p$, where $\overline{f}_p \in \tilde{\tilde{V}}_p$ for all $p \in \Sigma(F)$ and $\overline{f}_p = \overline{f}_p^0$ for almost all p.

For any $p \in \Sigma(F)$, $f_p \in \tilde{V}_p$ we denote by $\overline{f}_p$ the function on L_p' given by

$$\overline{f}_p(\ell_p) = \Delta_p(\ell_p) \cdot \int_{x_p \in L_p^* \setminus G_p} \varepsilon_p(x_p) f_p(x_p^{-1}\ell_p x_p)\, dx_p$$

where ε_p is the restriction of the character $\varepsilon\colon \tilde{G}_{\mathbb{A}} \to C^*$ to $\tilde{G}_p$.

If $\operatorname{supp} f_p \subset \tilde{G}_p'$ then $\overline{f}_p$ has compact support in L_p' and therefore can be extended (by zero) to a smooth function on L_p. We denote this function on L_p also by $\overline{f}_p$. If $p \notin S$, then for any $f_p \in \tilde{V}_p$ we may extend $\overline{f}_p$ to a locally constant function on L_p (see Propositi

.2) and $\overline{(f_p^0)} = \overline{f}_p^0$ (see Theorem 3.1). Therefore for any

$= \bigotimes_{p\in\Sigma(F)} f_p \in V$ we may define $\overline{f} \in \tilde{\tilde{V}}$ as a product $\overline{f} = \bigotimes_{p\in\Sigma(F)} \overline{f}_p$.

t is clear that this map extends to a linear operator $A: V \to \tilde{\tilde{V}}$. Let

$_\theta = L_2^\theta(L^* \setminus L_{\mathbb{A}}^*)$ and $\overline{\rho}$ be the natural representation of $I_{\mathbb{A}}^*$ on $\overline{W}_\theta$.

or any $\overline{f} \in \tilde{\tilde{V}}$ we define

$$\rho(\overline{f}) \overset{\text{def}}{=} \int_{PL_{\mathbb{A}}^*} \overline{f}(\ell)\overline{\rho}(\ell)\, d\ell,$$

where $d\ell$ is the Haar measure on $PL_{\mathbb{A}}^*$ such that $\int_{PL_F^*\setminus PL_{\mathbb{A}}^*} d\ell = 1$.

Proposition 2. For any $f \in V$ we have

$$\mathrm{Tr}(\rho_\theta(f)\cdot\varepsilon) = \mathrm{Tr}\,\overline{\rho}(\overline{f}).$$

Proof. Assume that $f = \bigotimes_{p\in\Sigma(G)} f_p$. By Proposition 1 we have

$$\mathrm{Tr}(\rho(f)\cdot\underline{\varepsilon}) = \sum_{\ell\in L'} I_\ell(f), \quad \text{where for any } \ell \in L'$$

$$I_\ell(f) = \int_{x\in L_{\mathbb{A}}^*\setminus\tilde{G}_{\mathbb{A}}} \varepsilon(x) f(x^{-1}\ell x)\, dx$$

$$= \prod_{p\in\Sigma(F)} \int_{x_p\in L_p^*\setminus G_p} \varepsilon_p(x_p) f_p(x_p^{-1}\ell x_p)\, dx_p$$

$$\overset{(1)}{=} \prod_{p\in\Sigma(F)} \Delta_p(\ell) \int_{x_p\in L_p^*\setminus\tilde{G}_p} \varepsilon_p(x_p) f_p(x_p^{-1}\ell x_p)\, dx_p$$

$$\overset{\text{def}}{=} \overline{f}(\ell),$$

here (1) follows from Lemma 2.

Since $S \neq \emptyset$ we see that $\overline{f}(\ell) = 0$ for all $\ell \in L^* - L'$. There-ore we have

$$\mathrm{Tr}(\rho(f)\cdot\underline{\varepsilon}) = \sum_{\ell\in F^*\setminus L^*} \overline{f}(\ell) = \sum_{\ell\in F^*\setminus L^*} \overline{f}(\ell) = \mathrm{Tr}\,\overline{\rho}(\overline{f}).$$

he proposition is proved.

We will now describe the image of A. For any $p \in \Sigma(F)$ we have $L_p = (L_{1p})^{n'_p}$ where L_{1p} is an extension of F_p of degree m_p, $m_p \cdot n'_p = n$. Let $\mathfrak{G}_p$ be the group of automorphisms of L_p. It's clear that $\mathfrak{G}_p$ is a semidirect product of a symmetric group $S_{n'}$ and n'-copies of $\mathbb{Z}/m\mathbb{Z}$ and $\mathfrak{G}$ can be considered as a subgroup in $\mathfrak{G}_p$. For $p \in \Sigma(F) - S$ we denote by $\overline{V}_p \subset \tilde{\tilde{V}}_p$ the subspace of $\mathfrak{G}_p$-invariant functions. If $p \in S$, $p \neq p_1, p_2$ we define $\overline{V}_p$ to be the subspace of $\mathfrak{G}_p$-invariant functions $\overline{f}_p$ such that $\operatorname{supp} \overline{f}_p \subset L'_p$, where $L'_p \overset{\text{def}}{=} L^*_p \cap \tilde{G}'_p$. Now let $p = p_1$ or p_2.

Let $x_p \in \tilde{G}_p$ be an element such that $x_p^{-1} \ell_p x_p = \ell_p^{\sigma}$ for all $\ell_p \in L^*_p \subset \tilde{G}_p$ where σ is a generator of $\mathfrak{G}$. Let $N_p\colon \tilde{G}_p \to F^*_p$ be the reduced norm. We define

$$\overline{V}_p = \{\overline{f}_p \in \tilde{V}_p \mid \operatorname{supp} \overline{f}_p \subset L'_p \text{ and } \overline{f}_p(\ell_p^{\sigma}) = \varepsilon_p(N_p(x_p)) \cdot (-1)^{n-1}\}.$$

We denote by $\overline{V} \subset \tilde{\tilde{V}}$ the subspace spanned by functions $\overline{f}$ of the form

$$\overline{f} = \bigotimes_{p \in \Sigma(F)} \overline{f}_p, \qquad \overline{f}_p \in \overline{V}_p.$$

<u>Lemma 3</u>. $\operatorname{Im} A = \overline{V}$.

<u>Proof</u>. Let $A_p\colon V_p \to \tilde{\tilde{V}}_p$ be the corresponding local map $f_p \to \overline{f}_p$. It is sufficient to prove that $\operatorname{Im} A_p = \overline{V}_p$ for all $p \in \Sigma(F)$. It is clear that $\operatorname{Im} A_p \subset \overline{V}_p$ and that $\operatorname{Im} A_p = \overline{V}_p$ for $p \in S$. So we assume that $p \in S$. Using the argument from the proof of Theorem 3.2 we easily reduce to the case when L_p is a field. Using the inductive assumptions (as in the proof of Proposition 3.2) we see that any $\overline{f}_p \in \overline{V}_p$ such that $\overline{f}_p(1) = 0$ is in the image of A_p. Also, $\overline{f}^0_p = A(f^0_p)$. Since any $\overline{f}_p \in \overline{V}_p$ can be written as a sum $\overline{f}_p = c\overline{f}^0_p + \overline{f}'_p$ where $\overline{f}'_p(1) = 0$, the lemma is proved.

<u>Proposition 3</u>. Let $\chi', \chi'' \in \hat{C}_L$ be two idele class characaters

such that for almost all $p \in \Sigma(F)$ the corresponding local characters $\chi'_p, \chi''_p \in \hat{L}^*_p$ are conjugate by an element in $\mathfrak{G}_p$. Then χ' and χ'' are conjugate by an element in $\mathfrak{G}$.

Proof. Let $W_{L/F}$ be the relative Weil group (see [T]). Then C_L is an open subgroup in $W_{L/F}$ and the quotient group is isomorphic to $\mathfrak{G}$. Let ρ', ρ'' be representations of $W_{L/F}$ induced from χ' and χ''.

Lemma 4. The representations ρ' and ρ'' are equivalent.

Proof. Let $S' \subset \Sigma(F)$ be a finite subset of $p \in \Sigma(F)$ such that for any $p \notin S'$, L_p, χ'_p, χ''_p are unramified and χ'_p is conjugate to χ''_p by an element in $\mathfrak{G}_p$. For any $p \in \Sigma(F)$ we have an embedding $\mathrm{Gal}^0(\bar{F}_p:F_p) \hookrightarrow W_{L/F}$ (see [T]) defined up to an inner automorphism where $\mathrm{Gal}^0(\bar{F}_p:F_p) \subset \mathrm{Gal}(\bar{F}_p:F_p)$ is the subgroup of elements which act on the residue field as a power of Frobenius. Let Fr_p be an element in $\mathrm{Gal}^0(\bar{F}_p:F_p)$ which acts as Frobenius on the residue field and w_p be its image in $W_{L/F}$. It is clear that for $p \in \Sigma(F) - S'$, $\mathrm{Tr}\,\rho'(w_p) = \mathrm{Tr}\,\rho''(w_p) = 0$ if p does not split completely in L. Assume now that $p \in \Sigma(F) - S'$ splits completely $p = p_1 \dots p_n$ where $p_1, \dots, p_n$ are distinct prime ideals in L.

For any i, $1 \le i \le n$ let L_i be the completion of L at p_i, $\mathcal{O}_i$ be the ring of integers of L_i and t_i be a generator of the maximal ideal in $\mathcal{O}_i$. It is clear that L_p is canonically isomorphic to the direct sum $L_p = \sum_{i=1}^n L_i$, $\mathrm{Tr}\,\rho'(w_p) = \sum_{i=1}^n \chi'(t_i)$ and $\mathrm{Tr}\,\rho''(w_p) = \sum_{i=1}^n \chi''(t_i)$. Therefore $\mathrm{Tr}\,\rho'(w_p) = \mathrm{Tr}\,\rho''(w_p)$ for all $p \in \Sigma(F) - S'$. It now follows from the Chebotarev density theorem that ρ' and ρ'' are equivalent. The lemma is proved.

We can now finish the proof of the proposition. Really it is clear that the restriction of ρ' on C_L is equivalent to the direct sum of

χ'^{σ}, $\sigma \in \mathfrak{G}$ and the restriction of ρ'' to C_L is equivalent to the sum of χ''^{σ}, $\sigma \in \mathfrak{G}$. Therefore $\chi'' = \chi'^{\sigma}$ for some $\sigma \in \mathfrak{G}$. The proposition is proved.

Define $\hat{C}_L^{\theta} = \{\chi \in \hat{C}_L \mid \chi\big|_{C_F} = \theta\}$. For any $\chi \in \hat{C}_L^{\theta}$ we denote by $\underline{\chi} \in \overline{V}'$ the linear functional on $\overline{V}$ given by

$$\underline{\chi}(f) \overset{\text{def}}{=} \int_{PL_{\mathbb{A}}^*} \chi(\ell) f(\ell)\, d\ell .$$

The group $\mathfrak{G}$ acts naturally on $\hat{C}_L^{\theta}$. Let

$$\hat{C}_L^{\theta 0} \overset{\text{def}}{=} \{\chi \in \hat{C}_L^{\theta} \mid \chi_{p_i}^{\sigma} \neq \chi_{p_i},\ \ i = 1,2 \ \text{ for all } \ \sigma \in \mathfrak{G} - \{e\}\}.$$

<u>Lemma 5</u>.

(a) $\underline{\chi}^{\sigma} = \underline{\chi}$ for all $\chi \in \hat{C}_L^{\theta}$, $\sigma \in \mathfrak{G}$.

(b) $\underline{\chi} \equiv 0$ for all $\chi \in \hat{C}_L^{\theta} - \hat{C}_L^{\theta 0}$.

(c) Let $\chi_1, \ldots, \chi_N \in \hat{C}_L^{\theta 0}$ be characters such that $\chi_i^{\sigma} \neq \chi_j$ for all $1 \leq i \neq j \leq N$, $\sigma \in \mathfrak{G}$. Then $\underline{\chi}_i \in \overline{V}'$ are linearly independent.

<u>Proof</u>. (a) is clear.

(b) For any $p \in \Sigma(F)$ we denote by $\underline{\chi}_p \in \overline{V}_p'$ the functional on $\overline{V}_p$ given by

$$\underline{\chi}_p(f_p) \overset{\text{def}}{=} \int_{x_p \in F_p^* \backslash L_p^*} \chi_p(x_p) f_p(x_p)\, dx_p .$$

Since $\chi \in \hat{C}_L^{\theta} - \hat{C}_L^{\theta 0}$ there exist $\sigma \in \mathfrak{G} - \{e\}$ and $i = 1$ or 2 such that $\chi_{p_i}^{\sigma} = \chi_{p_i}$. Let $x_{\sigma} \in \tilde{G}_{p_i}$ be an element such that $x_{\sigma}^{-1} \ell x_{\sigma} = \ell^{\sigma}$ for all $\ell \in L_{p_i}^* \subset \tilde{G}_{p_i}$. By the definition we have $\underline{\chi}_{p_i}^{\sigma} = \eta_{\sigma} \cdot \underline{\chi}_{p_i}$ where $\eta_{\sigma} = \varepsilon_{p_i}(N_{p_i}(x_{\sigma}))$. Since D_{p_i} is a division algebra we have $\eta_{\sigma} \neq 1$ and therefore $\underline{\chi}_{p_i} = 0$.

(c) Follows immediately from Proposition 3.

Let $\mathfrak{X}_a^{\varepsilon,\theta} \subset \mathfrak{X}(\tilde{G}_{\mathbb{A}})$ be the set of equivalent classes irreducible representation (π, M) of $\tilde{G}_{\mathbb{A}}$ which occur in the decomposition of $\tilde{W}_\theta$ and such that $\pi \otimes \varepsilon \simeq \pi$ (see §1). As follows from [DKV] for any (π, M) $\in \mathfrak{X}_a^{\varepsilon,\theta}$ there exists unique (up to a scalar) $\tilde{G}_{\mathbb{A}}$-covariant embedding $M \hookrightarrow \tilde{W}_\theta$ and the image of M is an $\underline{\varepsilon}$-invariant subspace. We denote by $\underline{\varepsilon}_\pi : M \to M$ the induced automorphism.

For any $\pi \in \mathfrak{X}_a^{\varepsilon,\theta}$ we denote by t_π^ε the functional on V given by $t_\pi^\varepsilon(f) \overset{\text{def}}{=} \operatorname{Tr}(\pi(f) \cdot \underline{\varepsilon}_\pi)$, $f \in V$. Analogously, for any $\chi \in \hat{C}_L^\theta$ we denote by t_χ the functional on V given by $t_\chi(f) \overset{\text{def}}{=} \underline{\chi}(f)$, $f \in V$.

<u>Lemma 6</u>. We have

$$\sum_{\pi \in \mathfrak{X}_a^{\varepsilon,\theta}} t_\pi^\varepsilon = \sum_{\chi \in \hat{C}_L^{\theta 0}/\mathfrak{G}} n t_\chi .$$

<u>Proof</u>. It is clear that $\operatorname{Tr}(\tilde{\rho}_\theta(f)\underline{\varepsilon}) = \sum_{\pi \in \mathfrak{X}_a^{\varepsilon,\theta}} t_\pi^\varepsilon(f)$ for any $f \in V$. On the other hand, we have $\operatorname{Tr} \bar{\rho}(\bar{f}) = \sum_{\chi \in \hat{C}_L^\theta} \underline{\chi}(\bar{f})$ for all $f \in \bar{V}$. So by Lemma 5 we have $\operatorname{Tr} \bar{\rho}(\bar{f}) = n \sum_{\chi \in \hat{C}_L^{\theta 0}/\mathfrak{G}} \underline{\chi}(\bar{f})$. Lemma 6 now follows from Proposition 2.

For any $p \in \Sigma(F) - S$ we have a canonical isomorphism $\tilde{G}_p \xrightarrow{\sim} GL_n(F_p)$, and we denote by $\mathcal{H}_p$ the Hecke algebra $\mathcal{H}_p \overset{\text{def}}{=} \mathcal{H}_{\theta_p}(\tilde{G}_p, K_p)$ where $K_p \simeq GL_n(\mathcal{O}_p)$ (see §3). For any finite set $\Lambda \subset \Sigma(F)$, $\Lambda \supset S$ we denote by $\mathcal{H}^\Lambda$ the restricted tensor product $\mathcal{H}^\Lambda \overset{\text{def}}{=} \bigotimes_{p \in \Sigma(F)-\Lambda} \mathcal{H}_p$ and by $V^\Lambda \subset V$ the subspace of two-sided K^Λ-invariant functions where $K^\Lambda \overset{\text{def}}{=} \prod_{p \in \Sigma(F)-\Lambda} K_p$. The algebra $\mathcal{H}^\Lambda$ acts by convolution on V^Λ. If $\Lambda \subset \Lambda'$, then we have natural embeddings $\mathcal{H}^{\Lambda'} \xrightarrow{i_{\Lambda,\Lambda'}} \mathcal{H}^\Lambda$ and $V^\Lambda \xrightarrow{j_{\Lambda',\Lambda}} V^{\Lambda'}$ and for any $\varphi \in \mathcal{H}^{\Lambda'}$, $f \in V^\Lambda$ we have $j_{\Lambda',\Lambda}(i_{\Lambda,\Lambda'}(\varphi) \cdot f) = \varphi \cdot j_{\Lambda',\Lambda}(f)$. It is clear that $V = \cup_\Lambda V^\Lambda$.

For any Λ we denote by R_Λ the set of one-dimensional representations of $\mathcal{H}^\Lambda$. Given Λ', Λ'' and $r' \in R_{\Lambda'}$, $r'' \in R_{\Lambda''}$ we say that r' and r'' are equivalent if there exists a finite set $\Lambda \supset \Lambda', \Lambda''$ such that $r' \circ i_{\Lambda',\Lambda} = r'' \circ i_{\Lambda'',\Lambda}$. The set of these equivalence classes we denote by R and will call "characters of the Hecke algebra".

Given a functional μ in V' and a character $r \in R$ of the Hecke algebra we say that μ is an r-eigenfunctional if for all sufficiently large Λ we have

$$\mu(\varphi \circ f) = r(\varphi)\mu(f), \qquad f \in V^\Lambda, \quad \varphi \in \mathcal{H}^\Lambda.$$

We say that μ is an $\mathcal{H}$-eigenfunctional if there exists $r \in R$ such that μ is an r-eigenfunctional. We will denote this r by $r(\mu) \in R$. It is clear from the definition that $r(\mu)$ is well-defined if $\mu \neq 0$.

Given a sequence of functionals $\mu_k \in V'$, $1 \le k \le \infty$ such that the series $\sum_{k=1}^\infty \mu_k(v)$ is absolutely convergent for any $v \in V$ we define $\sum_{k=1} \mu_k \in V'$ by

$$\left(\sum_{k=1}^\infty \mu_k\right)(v) \overset{\text{def}}{=} \sum_{k=1}^\infty \mu_k(v).$$

<u>Lemma 7</u>. Let $\mu_k', \mu_k'' \in V'$ be two sequences of $\mathcal{H}$-eigenfunctionals. Let $r_k' \overset{\text{def}}{=} r(\mu_k')$ and $r_k'' \overset{\text{def}}{=} r(\mu_k'')$, $1 \le k < \infty$. Assume that $r_{k_1}' \neq r_{k_2}'$, $r_{k_1}'' \neq r_{k_2}''$ for $1 \le k_1 \neq k_2 < \infty$ and that $\sum_{k=1}^\infty \mu_k' = \sum_{k=1}^\infty \mu_k''$. Then there exists a unique permutation $s\colon \mathbb{Z}^+ \to \mathbb{Z}^+$ such that $\mu_k'' = \mu_{s(k)}'$.

<u>Proof</u>. Clear.

<u>Lemma 8</u>.

(a) For any $\pi \in \mathcal{E}_a^{\varepsilon,\theta}$, t_π^ε is an $\mathcal{H}$-eigenfunctional.

Let $r_\pi \overset{\text{def}}{=} r(t_\pi^\varepsilon)$.

(b) For any $\pi, \pi' \in \mathcal{E}_a^{\varepsilon,\theta}$, $\pi \neq \pi'$ we have $r_\pi \neq r_{\pi'}$.

c) For any $\chi \in \hat{C}_L^0$, t_χ is an $\mathfrak{H}$-eigenfunctional.

Let $r_\chi \overset{\text{def}}{=} r(t_\chi)$.

d) Given $\chi', \chi'' \in \hat{C}_L^0$ we have $r_{\chi'} = r_{\chi''}$ if and only if there exists $\sigma \in \mathfrak{G}$ such that $\chi'' = \chi'^{\sigma}$.

Proof.

(a) Clear.

(b) Follows immediately from the strong multiplicity one theorem for $\widetilde{W}_\theta$ (see [DKV]).

(c) Follows from Theorem 3.2.

(d) Follows from Theorem 3.2 and Proposition 3.

Theorem 1. There exists one-to-one correspondence $\alpha : \hat{C}_L^{\theta 0} \xrightarrow{\sim} \mathfrak{E}_a^{\varepsilon,\theta}$ such that for all $\chi \in \hat{C}_L^{\theta 0}$ we have

$$t^{\varepsilon}_{\alpha(\chi)} = nt_\chi .$$

Proof. By Lemma 6 we know that

$$\sum_{\pi \in \mathfrak{E}_a^{\varepsilon,\theta}} t_\pi^\varepsilon = \sum_{\chi \in \hat{C}_L^0/\mathfrak{G}} nt_\chi .$$

Theorem 1 now follows from Lemmas 6 and 7.

Let $p \in \Sigma(F)$, $p \neq p_1, p_2$ be a place such that L_p is a field.

Corollary. For any $\chi_p \in \hat{L}_p^*$ there exists $\pi_p \in \mathfrak{E}^{\varepsilon_p}(G_p)$ and $c \in \mathbb{C}^*$ such that $t_{\pi_p}^{\varepsilon_p} = ct_{\chi_p}$, where ε_p is the p-component of ε, and the functional t_{π_p} is defined in the beginning of §2.

Proof. There exists a $\chi \in \hat{C}_L^0$ such that our χ_p is the

p - component of χ. By Theorem 1 there exists $\pi \otimes \pi_p \in \mathfrak{E}_a^{\varepsilon,\theta}$ such that $t_\pi^\varepsilon = nt_\chi^0$. It is clear that the p - component of π has the required property.

Section 5.

In this paragraph F will be a nonarchimedian local field and L/F a cyclic extension of degree n. We will use notation from §2. In particular, $\varepsilon\colon G \to \mathbb{C}^*$ is equal to $\varepsilon_L \circ \det$. Let V be the space of locally constant functions on G' with compact support, $\overline{V}$ the space of $\mathfrak{G}$-invariant locally constant functions on L^* with compact support and $A\colon V \to \overline{V}$ be the map defined in the previous paragraph $A(f) = \overline{f}$.

For any character $\chi \in L^*$ we denote by $t_\chi \in V'$ the linear functional given by $t_\chi(f) = \int_{L^*} \chi(x)\overline{f}(x)\,dx$. In this paragraph we will prove Theorem A. As follows from the technique of [DKV] it is suffici ent to prove it in the case where char F = 0. So we restrict ourselves to this case.

Proposition 1. For any $\chi \in \hat{L}^*$ there exists a unique $\pi \in \mathfrak{E}^\varepsilon(G)$ and $c \in \mathbb{C}^*$ such that $t_\pi^\varepsilon(f) = ct_\chi(f)$ for all $f \in V$.

Proof. The existence follows from the corollary to Theorem 4.1. Now assume that $\pi' \in \mathfrak{E}^\varepsilon(G)$ is another representation and $\gamma \in \mathbb{C}^*$ is a constant such that $t_\pi^\varepsilon(f) = \gamma t_{\pi'}^\varepsilon(f)$ for all $f \in V$.

Let ρ, ρ' be irreducible components of the restriction of π, π' to G_0 ($\simeq \operatorname{Ker} \varepsilon$). As follows from Lemma 2.1 there exist $\gamma, \gamma' \in \mathbb{C}^*$ such that

$$t_\pi^\varepsilon = \gamma \sum_{\sigma \in \mathfrak{G}} \varepsilon(\sigma) t_{\rho^\sigma}, \qquad t_{\pi'}^\varepsilon = \gamma' \sum_{\sigma \in \mathfrak{G}} \varepsilon(\sigma) t_{\rho'^\sigma}.$$

Therefore

$$\gamma \sum_{\sigma \in \mathfrak{G}} \varepsilon(\sigma) t_{\rho^\sigma}(f) = \gamma' \sum_{\sigma \in \mathfrak{G}} \varepsilon(\sigma) t_{\rho'^\sigma}(f) \qquad \text{for all } f \in V, \ \operatorname{supp} f \subset G_0.$$

Since $\operatorname{char} F = 0$ we have $t_{\rho^\sigma}, t_{\rho'^\sigma} \in L^1_{loc}(G_0)$ ([H - Ch.]). Therefore we have $\sum_{\sigma\in\mathfrak{G}} \varepsilon(\sigma) t_{\rho^\sigma} = \gamma' \sum_{\sigma\in\mathfrak{G}} \varepsilon(\sigma) t_{\rho'^\sigma}$. By Lemma 2.1 all characters ρ^σ, $\sigma \in \mathfrak{G}$ (resp. ρ'^σ, $\sigma \in \mathfrak{G}$) are distinct. It now follows from the linear independence of distinct characters that $\rho' = \rho^\sigma$ for some $\sigma \in \mathfrak{G}$. But then $\pi' \simeq \pi$. The proposition is proved.

For any $\chi \in \hat{L}^*$ we denote by $\pi_\chi \in \mathfrak{E}^\varepsilon(G)$ the representation defined in Proposition 1.

<u>Proposition 2</u>. For any $\pi \in \mathfrak{E}^\varepsilon(G)$ there exists $\chi \in \hat{L}^*$ such that $\pi = \pi_\chi$.

<u>Proof</u>. As we have seen in the proof of Proposition 1 for any $\pi \in \mathfrak{E}^\varepsilon(G)$ there exists $\varphi_\pi \in L^1_{loc}(G_0)$ such that $t^\varepsilon_\pi(f) = \int_{G_0} \varphi_\pi(g) f(g)\, dg$ for any $f \in S(G_0)$. Since $\pi \otimes \varepsilon \simeq \pi$ we see that

$$t^\varepsilon_\pi(f) = \int_{G_0} \varphi_\pi(g) f(g)\, dg \qquad \text{for all } f \in S(G).$$

We fix an embedding $L^* \hookrightarrow G$ and define as before $L' = L^* \cap G'$. Let φ'_π be the restriction of φ_π to L' and let $\mu_\pi = \Delta \cdot \varphi'_\pi$ where $\Delta: L \to \mathbb{C}$ is defined in §1.

<u>Lemma 1</u>.

(a) If $g \in G'$ is not conjugate to an element in L' then $\varphi_\pi(g) = 0$.

(b) μ_π is a restriction of a locally constant function on L^*.

<u>Proof</u>. (a) Since $g \in G'$ is not conjugate to an element in L' there exists $x_0 \in G$ such that $\varepsilon(x_0) \neq 1$ and $x_0^{-1} g x_0 = g$ (see the proof of Lemma 1.1). But it is clear that $\varphi_\pi(x^{-1}gx) = \varepsilon(x)\varphi_\pi(g)$ for

all $x \in G$. Therefore $\varphi_\pi(g) = 0$.

(b) Take $\ell_0 \in L^*$. Let M be the centralizer of ℓ_0 in $M_n(F)$ and H be the centralizer of ℓ_0 in G_0. Let $N \subset M$ be the subset of nilpotent elements, $X \subset S(M)'$ be the subspace of H-invariant distributions with support on N. It is clear that $\dim X < \infty$. We denote by $Y \subset S(M)'$ the image of X under the Fourier transform. As is well known $Y \subset L^1_{loc}(M)$. We fix a neighborhood U of ℓ_0 in G_0 such that for any $g \in U$ the logarithm $\ln(g\ell_0^{-1})$ is well defined and denote by J the space of functions j on U of the form $j(u) = y(\ln(u\ell_0^{-1}))$ for some $y \in Y$. Harish-Chandra ([H-Ch.]) has shown that for any $\rho \in \mathfrak{E}(G_0)$ there exists a neighborhood $U_\rho \subset U$ of ℓ_0 in G and a function $j_\rho \in J$ such that $\theta_\rho(g) = j_\rho(g)$ for all $g \in U_\rho \cap G'$ where θ_ρ is the character of ρ. Let $J^\varepsilon \subset J$ be the subspace of functions $j \in J$ such that $j(g^x) = \varepsilon(x)j(g)$ for $x \in Z_G(\ell)$ and $g \in U_1 \cap G'$ where $U_1 \subset G$ is a sufficiently small neighborhood of ℓ. It is easy to see that $\dim J^\varepsilon = 1$. As follows from Lemma 2.1 the germ of φ_π at ℓ lies in J^ε, for any $\pi \in \mathfrak{E}^\varepsilon(G)$. Therefore for any $\pi, \pi' \in \mathfrak{E}^\varepsilon(G)$ there exist $\gamma, \gamma' \in \mathbb{C}$, $(\gamma,\gamma') \neq (0,0)$ and a neighborhood $\tilde{U}$ of ℓ_0 in L such that $\gamma\varphi_\pi(\ell) = \gamma'\varphi_{\pi'}(\ell)$ for all $\ell \in \tilde{U} \cap L'$. Fix any $\chi \in \hat{L}^*$ such that $\sum_{\sigma\in\mathfrak{G}} \chi^\sigma(\ell) \neq 0$ and take $\pi' = \pi_\chi$. By the definition we have $\varphi_{\pi_\chi}(\ell) = c \sum_{\sigma\in\mathfrak{G}} \chi^\sigma(\ell)$, $c \in \mathbb{C}^*$. We may assume that $\chi^\sigma(\ell) = \chi^\sigma(\ell_0)$ for all $\ell \in \tilde{U}$. Therefore $\varphi_\pi(\ell) =$ constant for $\ell \in \tilde{U} \cap G'$.

The lemma is proved.

Now we can prove the proposition. For any $\pi \in \mathfrak{E}^\varepsilon(G)$ we will consider μ_π as a locally constant function on L^*. It is clear that μ_π is a $\mathfrak{G}$-invariant function and $\mu_\pi(zx) = \theta_\pi(z)\mu_\pi(x)$, $z \in F^*$, $x \in L^*$ where θ_π is the central character of π. Since $F^* \setminus L^*$ is compact we can write μ_π as a finite sum of characters $\mu_\pi = \sum_{i=1}^N c_i \sum_{\sigma\in\mathfrak{G}} \chi_i^\sigma$,

$_i \in \hat{L}^*$. Therefore $\varphi_\pi = \sum_{i=1}^{N} c_i \varphi_{\pi_{\chi_i}}$. Since different characters of are linearly independent we have $\pi = \pi_\chi$ for some $\chi \in L^*$. The proposition is proved.

Proof of Theorem A. The only statement which is left for us to prove is part (e).

We start with the following result.

Let $\mathcal{E}_2(G) \subset \mathcal{E}(G)$ be the subset of square integrable (mod Z) representations and $\mathcal{E}_2^\varepsilon(G) \overset{\text{def}}{=} \mathcal{E}^\varepsilon(G) \cap \mathcal{E}_2(G)$.

Lemma 2. Any $\pi \in \mathcal{E}_2^\varepsilon(G)$ is cuspidal.

Proof. If it is not cuspidal then there exists a proper parabolic subgroup $P = MU$, $M = GL_{n'}(F)^{m'}$, $mn' = n$ and a cuspidal representation $\tau \in \mathcal{E}(GL_{n'}(F))$ such that π is the unique square integrable subquotient of $\mathrm{Ind}^G(\tau, \tau\otimes\delta, \tau \ldots \tau\otimes\delta^{m-1})$ where δ is a modulus character. Let $\Gamma_{M,G}$ be the Jacquet functor ([BZ]). Then $\Gamma_{M,G}(\pi) = \otimes \tau\circ\delta \otimes \ldots \otimes \tau\circ\delta^{m-1}$ and it is clear that $\Gamma_{M,G}(\pi\otimes\varepsilon) = \Gamma_{M,G}(\pi) \otimes \varepsilon$ where we consider ε as a character of M. Since the restriction of to the center of M is not Id we have $\Gamma_{M,G}(\pi\otimes\varepsilon) \neq \Gamma_{M,G}(\pi)$. Therefore $\pi \otimes \varepsilon \neq \pi$. This contradiction proves the lemma.

To finish the proof we have to use the global technique. So we change our notation. Now F is a global field, $p_0 \in \Sigma(F)$ is a place such that F_{p_0} is "our" local field, $L \supset F$ is a cyclic extension such that L_{p_0} is "our" extension. We fix three nonarchimedean places p_1, p_2, p' such that the completions of L at those places are fields. We also fix two central division algebras D, 'D over F, $\dim_F D = \dim_F D' = n^2$ such that $D_p = M_n(F_p)$ for $p \neq p_1, p_2$ and $'D_p = M_n(F_p)$, $p \neq p_0, p', p_1, p_2$. $'D_{p_i} = D_{p_i}$, $i = 1, 2$ and $'D_{p_0}$, $'D_{p'}$, are division algebras. Let $\hat{C}^0_L \subset \hat{C}_L$ be the subset of characters $\chi \in \hat{C}_L$ such that for any $\sigma \in \mathfrak{G} - \{e\}$, $\chi^\sigma_p \neq \chi_p$ for $p = p_1, p_2, p_0, p'$, $'\underline{G}$ be the multiplicative group of $'D$ and $'\mathfrak{X}^\varepsilon_a \subset \mathfrak{X}('\tilde{G}_{\mathbb{A}})$ be the set of automorphic representations π' of $'\tilde{G}_{\mathbb{A}}$ such that $\pi' \otimes \varepsilon \simeq \pi'$ where the character $\varepsilon\colon {}'\tilde{G}_{\mathbb{A}} \to \mathbb{C}^*$ is defined in the same way as in §4. We define the subspace $'V \subset S(\tilde{G}_{\mathbb{A}})$, and linear functionals $'t_\chi \in (V)'$ as in §4. Given a pair $\pi \in \mathfrak{X}(\tilde{G}_{\mathbb{A}})$, $'\pi \in \mathfrak{X}('\tilde{G}_{\mathbb{A}})$, $\pi = \otimes_p \pi_p$, $'\pi = \otimes_p {}'\pi_p$ we say that $\pi \sim \pi'$ if $\pi_p = {}'\pi_p$ for $p \neq p_0, p'$.

Lemma 3.

(a) There exists an embedding $j\colon \mathfrak{X}_a('\tilde{G}_{\mathbb{A}}) \hookrightarrow \mathfrak{X}(\tilde{G}_{\mathbb{A}})$ such that $j('\pi) \sim {}'\pi$ for all $'\pi \in \mathfrak{X}_a('\tilde{G}_{\mathbb{A}})$ and

$$\mathrm{Im}\, j = \{\pi \in \mathfrak{X}_a(G_{\mathbb{A}}) \mid \pi = \otimes_p \pi_p,\ \pi_{p_0},\ \pi_{p'} \text{ are square integrable}\},$$

(b) There exists a one-to-one correspondence

$$\alpha' : \hat{C}^0_L/\mathfrak{G} \to {}'\mathfrak{X}^\varepsilon_a$$

such that

$$'t^\varepsilon_{'\alpha(\chi)} = n' t_\chi$$

where $'\mathfrak{X}^\varepsilon_a = \{\pi \in {}'\mathfrak{X}_a('G'_{\mathbb{A}}) \mid \pi \otimes \varepsilon \simeq \pi\}$ and the linear functionals are $'t^\varepsilon_{'\alpha(\chi)}$ defined as in §4.

Proof. (a) is contained in [DKV] and (b) is completely parallel to Theorem 4.1.

Corollary. For $\chi_{p_0} \in \hat{L}^*_{p_0}$ the representation $\pi_{\chi_{p_0}}$ is square integrable if and only if $\chi^\sigma_{p_0} \neq \chi_{p_0}$ for all $\sigma \neq e$.

Theorem A now follows from this corollary and Lemma 2. Theorem C now follows from Theorem 4.1.

We can deduce "Proposition 1.1" from "Theorem B" in precisely the same way.

References

[Ar] J. Arthur. On a trace formula. To appear.

[Bor] A. Borel. Automorphic L-functions in "Automorphic forms, representations and L-functions II". Amer. Math. Soc., Providence (1979), 27-63.

[BZ] J. Bernstein and A. Zelvinsky. Induced representations of reductive p-adic groups. Ann. Sci. Ecole Norm. Sup. (4) 10 (1977), 441-472.

[Car] P. Cartier. Representations of p-adic groups in "Automorphic forms, representations and L-functions I". Amer. Math. Soc. Providence (1979), 111-157.

[DKV] P. Deligne, D. Kazhdan, M.-F. Vigneras. GL(n) and simple algebras. To appear.

[H] R. Howe. The Fourier transform and germs of characters. Math. Ann. 208(1974), 305-322.

[H-Ch] Harish-Chandra. Admissible invariant distributions on reductive p-adic groups, Queen's Papers in Pure and Applied Math. 48(1978), 281-346.

[Sat] I. Satake. Theory of spherial functions on reductive algebraic groups over p-adic fields. Inst. Hautes Etudes Sci., Publ. Math. 18(1963), 1-69.

[Sh] J. Shalika. A theorem on semisimple p-adic groups, Ann. of Math. 95(1972), 226-242.

[T] J. Tate. Number theoretic background in "Automorphic forms, representations and L-functions II". Amer. Math. Soc., Providence (1979), 3-27.

[W] A. Weil. Adeles and algebraic groups. Inst. for Adv. Study, Princeton, 1961.

ON PRINCIPAL VALUES ON P-ADIC MANIFOLDS

R. Langlands and D. Shelstad(*)

In the paper [L] a project for proving the existence of transfer factors for forms of SL(3), especially for the unitary groups studied by Rogawski, was begun, and it was promised that it would be completed by the present authors. Their paper is still in the course of being written, but the present essay can serve as an introduction to it. It deals with SL(2) which has, of course, already been dealt with systematically [L-L], the existence of the transfer factors being easily verified. Thus it offers no new results, but develops, in a simple context, some useful methods for computing the principal value integrals introduced in [L].

We describe explicitly the Igusa fibering, form and integrand associated to orbital integrals on forms of SL(2), taking the occasion to clarify the relation of this fibering to the Springer-Grothendieck resolution (cf. §3). The Igusa data established, there are two problems: (i) to show that certain principal values are zero, (ii) to compare principal values on two twisted forms of the same variety. To deal with the first we have, in §1, computed directly some very simple principal values on $\mathbf{P}^1$, and shown that principal values behave like ordinary integrals under standard geometric operations such as fibering and blowing-up. The second problem is dealt with in a similar way, by using Igusa's methods to establish, in a simple case, a kind of comparison principle (Lemma 4.B).

The endoscopic groups for a form of SL(2) are either tori or SL(2). For tori the solution of the first problem (Lemma 4.A) leads immediately to the existence of transfer factors, and the hypotheses of [L_1, pp. 102, 149] are trivially satisfied. If G is anisotropic over F and the endoscopic group is SL(2) the solution of the second problem (Lemma 4.B with $\kappa \equiv 1$) and the characterization of stable orbital integrals (cf. [V]) yields the existence of transfer factors as well as the local hypothesis of [L_1, p. 102]. The analogous results at archimedean places are known in general (cf. [L_1, Lemma 6.17]). The global hypothesis [L_1, p. 149] follows from [L_1, Lemma 7.22].

The principal values which arise for forms of SL(2) are computed without difficulty, but we expressly avoid such calculations. The aim of the project begun in [L], and continued here, is to develop methods for proving the existence of transfer factors which appeal only to geometric techniques of some generality and thus have some prospect of applying to all groups. One encouraging sign is the smoothness with which they mesh with the notion of κ-orbital integral. They can be easily applied to the study of the germ at regular unipotent elements. A further

(*)Partially supported by NSF Grants MCS 81-02392-01, MCS 81-08814-01.

test, perhaps not easy to carry out, would be the semi-regular elements, already studied for GL(n) by Repka [R].

Throughout this paper F will be a nonarchimedean local field of characteristic zero, with residue field of q elements; $|\cdot|_F = |\cdot|$ will denote the valuation on F and ϖ a prime element; $\bar{F}$ will be an algebraic closure of F.

§1. REMARKS.

The following lemmas concern the simplest of the principal value integrals which arise in §1 of [L].

Let $N = N(m_1, \ldots, m_n)$ be the box

$$|u_j| \leq q^{-m_j} \qquad (1 \leq j \leq n) \tag{1.1}$$

in F^n. Consider the (multi-valued) differential form

$$\nu_{(c_1,\ldots,c_n)} = \prod_{j=1}^{n} u_j^{c_j} \frac{du_1}{u_1} \wedge \ldots \wedge \frac{du_n}{u_n} , \tag{1.2}$$

where $c_1, \ldots, c_n$ are rational numbers. Let $\Theta_1, \ldots, \Theta_n$ be quasicharacters on $F^\times$. Writing

$$\Theta_j = \theta_j |\cdot|^{t_j} \tag{1.3}$$

with θ_j unitary and t_j a real number, we assume

$$t_j + c_j \neq 0 \text{ if } \theta_j \equiv 1 \qquad (1 \leq j \leq n) . \tag{1.4}$$

Set

$$h_{(\Theta_1,\ldots,\Theta_n)}(u_1,\ldots,u_n) = \prod_{j=1}^{n} \Theta_j(u_j) . \tag{1.5}$$

Then (1.4) allows us to define the principal value integral

$$\oint_N h_{(\Theta_1,\ldots,\Theta_n)} |\nu_{(c_1,\ldots,c_n)}| \tag{1.6}$$

following [L, Lemma 1.3]. Thus consider for $\mathrm{Re}(s_j) \gg 0$ $(1 \leq j \leq n)$

(1.7) $$\int_N \prod_{j=1}^{n} |u_j|^{s_j} h_{(\Theta_1,\ldots,\Theta_n)} |\nu_{(c_1,\ldots,c_n)}|$$

$$= \prod_{j=1}^{n} \int_{|u_j|\le q^{-m_j}} \theta_j(u_j)|u_j|^{s_j+t_j+c_j} \frac{du_j}{|u_j|}$$

$$= \prod_{j=1}^{n} \sum_{n=m_j}^{\infty} \left(\int_{|u_j|=q^{-n}} \theta_j(u_j) \frac{du_j}{|u_j|} \right) q^{-(s_j+t_j+c_j)n}$$

$$= \varepsilon(1-\frac{1}{q})^n \prod_{j=1}^{n} \frac{(\theta_j(\varpi)q^{-(s_j+t_j+c_j)})^{m_j}}{1-\theta_j(\varpi)q^{-(s_j+t_j+c_j)}} ,$$

where

$$\varepsilon = \begin{cases} 1 & \text{if each } \theta_j \text{ is unramified} \\ 0 & \text{otherwise.} \end{cases}$$

The analytic continuation of this function is, thanks to (1.4), analytic at $s_1 = \ldots = s_n = 0$; (1.6) is the value at $s_1 = \ldots = s_n = 0$. Thus:

LEMMA 1.A.

$$\oint_{N(m_1,\ldots,m_n)} h_{(\Theta_1,\ldots,\Theta_n)} |\nu_{(c_1,\ldots,c_n)}| = \varepsilon(1-\frac{1}{q})^n \prod_{j=1}^{n} \frac{(\theta_j(\varpi)q^{-(t_j+c_j)})^{m_j}}{1-\theta_j(\varpi)q^{-(t_j+c_j)}} .$$

To define now $\oint_X h|\nu|$ we assume:

(1.8) X is an F-manifold, h is a $\mathbf{C}$-valued function supported on a compact open subset of X, ν is a differential form on X, and

(1.9) the support of h is the disjoint union of neighborhoods $\mathbf{U}$ with the following properties:

(1.10) there are local coordinates $u_1, \ldots, u_n$ on X such that $\mathbf{U}$ is given by (1.1) for some $m_1, \ldots, m_n$,

(1.11) on $\mathbf{U}$, $\nu = \alpha\nu_{(c_1,\ldots,c_n)}$ with $|\alpha|$ constant, and $h = \gamma h_{(\Theta_1,\ldots,\Theta_n)}$ with γ constant, where $c_1, \ldots, c_n$ and $\Theta_1, \ldots, \Theta_n$ satisfy (1.4). Then

$$\oint_X h|\nu| \overset{\text{def.}}{=} \sum_{\mathbf{U}} \gamma|\alpha| \oint_{N(m_1,\ldots,m_n)} h_{(\theta_1,\ldots,\theta_n)} |\nu_{(c_1,\ldots,c_n)}| .$$

Our definition is that for the case $r = s = 1$ in the proof of Proposition 1.2 n [L]. The integral is independent of the choice for $\{\mathbf{U}, u_1,\ldots,u_n\}$ ([L, Proposition 1.2]). Note especially that the conditions (1.9) - (1.11) are local, i.e., hey are satisfied if we can find around each point in the support of h a neighborhood $\mathbf{U}$ satisfying (1.10) and (1.11). We will allow α to take values in a finite Galois extension L of F; in that case, $|\alpha| = |Nm_F^L \alpha|^{1/[L:F]}$.

The following remark will simplify a later argument. Given X, h and ν as n (1.8), a patch $\mathbf{U}$ as in (1.9), and integers M_j $(1 \le j \le r)$ such that $M_j \ge m_j$, et $\overline{\mathbf{U}} = \overline{\mathbf{U}}(M_1,\ldots,M_r)$ be the subset

$$|u_j| = q^{-M_j} \quad (1 \le j \le r), \quad |u_j| \le q^{-m_j} \qquad (r+1 \le j \le n)$$

of $\mathbf{U}$. Then:

LEMMA 1.B.

$\oint h|\nu|$ exists and equals the value at $s_1 = \ldots = s_n = 0$ of $\overline{\mathbf{U}}$

$$\gamma|\alpha| \int_{\overline{\mathbf{U}}} \prod_{j=1}^{n} |u_j|^{s_j} h_{(\theta_1,\ldots,\theta_n)} |\nu_{(c_1,\ldots,c_n)}| .$$

Moreover, if the support of h is the disjoint union of a collection $\mathbf{S}$ of such neighborhoods then

$$\oint_X h|\nu| = \sum_{\overline{\mathbf{U}} \in \mathbf{S}} \oint_{\overline{\mathbf{U}}} h|\nu| .$$

Proof: The first assertion follows from the definitions, and the second from the independence of $\oint_X h|\nu|$ from the choice of decomposition for the support of h.

We consider an example. Let $U_0, \ldots, U_n$ be homogeneous coordinates on $\mathbf{P}^n$. Suppose that $\theta_0, \ldots, \theta_n$ are quasicharacters on $F^\times$ such that $\prod_{j=0}^{n} \theta_j \equiv 1$, and that $c_0, \ldots, c_n$ are rational numbers such that $\sum_{j=0}^{n} c_j = 0$. Assume

(1.12) $t_j + c_j \ne 0$ if $\theta_j \equiv 1$ $(0 \le j \le n)$, where $\theta_j = \theta_j|.|^{t_j}$. Let ν be the form on $\mathbf{P}^n$ given on $U_k \ne 0$ by

$$(1.13) \qquad (-1)^k \prod_{j=0}^{n} U_j^{c_j} \frac{dU_0}{U_0} \wedge \dots \wedge \widehat{\frac{dU_k}{U_k}} \wedge \dots \wedge \frac{dU_n}{U_n} ,$$

where $\widehat{\ }$ indicates deletion. Let h be the function

$$(1.14) \qquad h(U_0, \dots, U_n) = \prod_{j=0}^{n} \Theta_j(U_j) \quad .$$

Then (1.12) ensures that $\oint_{\mathbf{P}^n(F)} h|\nu|$ is well-defined.

LEMMA 1.C.

$$\oint_{\mathbf{P}^n(F)} h|\nu| = 0 \quad .$$

<u>Proof for $n = 1$:</u> Set $\Theta = \Theta_0$, $\theta = \theta_0$, $t = t_0$, $c = c_0$, $u = U_0$ on $U_1 = 1$ and $u = U_1$ on $U_0 = 1$. Then

$$(1.15) \qquad \oint_{\mathbf{P}^1(F)} h|\nu|$$

$$= \oint_{|u| \le 1} \theta(u)|u|^{t+c} \frac{du}{|u|} + \oint_{|u| \le q^{-1}} \theta^{-1}(u)|u|^{-(t+c)} \frac{du}{|u|}$$

$$= \varepsilon(1 - \frac{1}{q}) \left[\frac{1}{1 - \theta(\varpi) q^{-(t+c)}} + \frac{\theta(\varpi)^{-1} q^{t+c}}{1 - \theta(\varpi)^{-1} q^{t+c}} \right]$$

$$= 0 \quad .$$

The proof for $n > 1$ will be by reduction to the case $n = 1$; it follows Lemma 1.F.

Consider X, h and ν as in (1.8) - (1.11) and a neighborhood $\mathbf{U}$ as in (1.9). To compute $\oint_{\mathbf{U}} h|\nu|$ we may change coordinates and assume that $m_1 = \dots = m_n = 1$. This will be done for the next lemma.

Suppose that we blow up X at $u_1 = \dots = u_n = 0$ to obtain the F-manifold $\overline{X}$ and projection $\pi : \overline{X} \longrightarrow X$. Let $\overline{\mathbf{U}} = \pi^{-1}(\mathbf{U})$, $\overline{h} = h \circ \pi$ and $\overline{\nu} = \pi^*(\nu)$.

LEMMA 1.D.

Assume that $\sum_{j=1}^{n}(t_j+c_j)\neq 0$ if $\prod_{j=1}^{n}\theta_j\equiv 1$. Then $\oint_{\bar{\mathbf{U}}}\bar{h}|\bar{\nu}|$ exists and equals $\int_{\mathbf{U}} h|\nu|$.

Proof: Near $u_1=\ldots=u_n=0$, $\bar{X}$ is given by $u_iU_j=U_iu_j$ $(i,j=1,\ldots,n)$, where $U_1,\ldots,U_n$ are homogeneous coordinates on $\mathbf{P}^{n-1}(F)$.

On $U_i=1$ we have the coordinates $U_1,\ldots,U_{i-1}$, $z=u_i$, $U_{i+1},\ldots,U_n$. Then $u_j=zU_j$ $(j\neq i)$. Thus

$$\bar{\nu}=\alpha z^{(c_1+\ldots+c_n)}\prod_{j\neq i}U_j^{c_j}\frac{dU_1}{U_1}\wedge\ldots\wedge\frac{dz}{z}\wedge\ldots\frac{dU_n}{U_n}$$

and

$$\bar{h}=\gamma\prod_{j=1}^{n}\theta_j(z)\prod_{j\neq i}\theta_j(U_j)$$

$$=\gamma|z|^{(t_1+\ldots+t_n)}\prod_{j\neq i}|U_j|^{t_j}\prod_{j=1}^{n}\theta_j(z)\prod_{j\neq i}\theta_j(U_j)$$

on this patch.

Let $x\in\mathbf{U}$ have coordinates $u_1,\ldots,u_n$. Set $S(x)=\{i : |u_i|=\max_{1\leq j\leq n}|u_j|\}$. For $S\subseteq\{1,2,\ldots,n\}$, let $\mathbf{U}_S=\{x\in\mathbf{U} : S(x)=S\}$. Then $\mathbf{U}$ is the disjoint union of the $\mathbf{U}_S$. Let $\bar{\mathbf{U}}_S=\pi^{-1}(\mathbf{U}_S)$. Fix $i\in S$. Then $\bar{\mathbf{U}}_S$ is contained in $U_i=1$; it consists of the points in $U_i=1$ with $|z|\leq 1$, $|U_j|=1$ $(j\in S, j\neq i)$ and $|U_k|<1$ $(k\notin S)$. The assumption in the statement of the lemma ensures that $\oint_{\bar{\mathbf{U}}_S}\bar{h}|\bar{\nu}|$ and $\oint_{\bar{\mathbf{U}}}\bar{h}|\bar{\nu}|$ are well-defined.

On the other hand, $\oint_{\mathbf{U}}h|\nu|$ is the value at $s_1=\ldots=s_n=0$ of

$$\int_{\mathbf{U}}\prod_{j=1}^{n}|u_j|^{s_j}h|\nu|=\sum_S\int_{\mathbf{U}_S}\prod_{j=1}^{n}|u_j|^{s_j}h|\nu|$$

$$=\sum_S\int_{\bar{\mathbf{U}}_S}|z|^{s_1+\ldots+s_n}\prod_{j\neq i}|U_j|^{s_j}\bar{h}|\bar{\nu}|\;,$$

where for each $S\subseteq\{1,2,\ldots,n\}$ we have fixed $i\in S$. Then, by Lemma 1.B,

$$\oint_{\mathbf{U}} h|\nu| = \sum_S \oint_{\overline{\mathbf{U}}_S} \bar{h}|\bar{\nu}| = \oint_{\overline{\mathbf{U}}} \bar{h}|\bar{\nu}| ,$$

and we are done.

COROLLARY 1.E.

Under the assumption of Lemma 1.D, $\oint_{\overline{X}} \bar{h}|\bar{\nu}|$ is well-defined and equals $\oint_X h|\nu|$.

LEMMA 1.F.

Suppose that $\phi : X \longrightarrow X'$ is a smooth (submersive) map of F-manifolds. Suppose that X, h and ν satisfy (1.9) - (1.11) with the further constraint on the coordinates $u_1, \ldots, u_n$:

(1.16) there are coordinates $v_1, \ldots, v_r$ on $\phi(\mathbf{U})$ such that $u_j = v_j \circ \phi$, $j = 1, \ldots, r$.

Let ν' be a differential form on X' given on $\phi(\mathbf{U})$ by

$$\alpha' \prod_{j=1}^{r} v_j^{d_j} \frac{dv_1}{v_1} \wedge \ldots \wedge \frac{dv_r}{v_r} , \tag{1.17}$$

where $|\alpha'|$ is constant and d_j is rational, $1 \leq j \leq r$. Then for $x' \in X'(F)$ the principal value integral

$$H(x') = \oint h \frac{|\nu|}{|\phi^*(\nu')|} ,$$

taken over the fiber above x' in X, is well-defined outside a locally finite family of divisors. Moreover, $\oint_{X'} H|\nu'|$ is well-defined and

$$\oint_X h|\nu| = \oint_{X'} H|\nu'| .$$

Proof: We may assume that the support of h is contained in a neighborhood $\mathbf{U}$ as in the statement of the lemma. Let $x' \in \phi(\mathbf{U})$ have coordinates $v_1, \ldots, v_r$. The fiber integral

$$H(x') = \gamma|\alpha||\alpha'|^{-1} \prod_{j=1}^{r} \theta_j(u_j)|u_j|^{c_j - d_j} \oint \prod_{j=r+1}^{n} \theta_j(u_j)|u_j|^{c_j} \frac{du_{r+1}}{|u_{r+1}|} \cdots \frac{du_n}{|u_n|}$$

is well-defined provided none of $v_1, \ldots, v_r$ vanish at x'. Then

$$\oint H|\nu'| = \gamma|\alpha| \oint \prod_{j=r+1}^{n} \theta_j(u_j)|u_j|^{c_j} \frac{du_{r+1}}{|u_{r+1}|} \cdots \frac{du_n}{|u_n|} \oint \prod_{j=1}^{r} \theta_j(u_j)|u_j|^{c_j} \frac{du_1}{|u_1|} \cdots \frac{du_r}{|u_r|} .$$

is well-defined and coincides with $\oint h|\nu|$. Thus the lemma is proved.

Proof of Lemma 1.C:

Let p be the point in $\mathbf{P}^n(F)$ where $U_0 = U_1 = \ldots = U_{n-1} = 0$. Suppose that we blow up $\mathbf{P}^n$ at p to obtain the smooth variety Q over F. The local conditions of Lemma 1.D are met since

$$\sum_{j=0}^{n-1} (t_j+c_j) = -(t_n+c_n) \quad \text{and} \quad \prod_{j=0}^{n-1} \theta_j = \theta_n^{-1}$$

(cf. (1.12)). Corollary 1.E then implies that $\oint_{\mathbf{P}^n(F)} h|\nu| = \oint_{Q(F)} \bar{h}|\bar{\nu}|$. We define a smooth map $\phi^0 : \mathbf{P}^n - \{p\} \longrightarrow \mathbf{P}^{n-1}$ by mapping the point with homogeneous coordinates $U_0, \ldots, U_n$ in $\mathbf{P}^n$ to the point with homogeneous coordinates $U_0, \ldots, U_{n-1}$ in $\mathbf{P}^{n-1}$. There is a smooth extension $\phi : Q \longrightarrow \mathbf{P}^{n-1}$ of ϕ^0, with fiber $\mathbf{P}^1$. An easy calculation verifies that the conditions of Lemma 1.F are met and that the integral $H(x')$ over the fiber above $x' \in \mathbf{P}^{n-1}(F)$ takes the form (1.15). But then $H \equiv 0$. We conclude that $\oint_{Q(F)} \bar{h}|\bar{\nu}| = 0$, and the lemma is proved.

Finally, there are two remarks which will be useful for the proof of Lemma 4.B. We state them only in the generality needed for that lemma.

REMARK 1.G.

Let $L \subset \bar{F}$ be a quadratic extension of F. Denote the natural action of the nontrivial element σ of $\mathrm{Gal}(L/F)$ by $^-$. We define a twisted form S of $\mathbf{P}^1$ by requiring that σ act on the homogeneous coordinates U_0, U_1 by $U_0 \longrightarrow \bar{U}_1$, $U_1 \longrightarrow \bar{U}_0$. Then $S(F)$ is contained in the affine patch $U_1 \neq 0$ and is given by $u\bar{u} = 1$ if we require $U_1 = 1$ and set $U_0 = u$. The form ν on $\mathbf{P}^1(L) = S(L)$ given by (1.13) with $c_1 = c_2 = 0$ is preserved by the Galois action of S; $|\nu| = \frac{du}{|u|}$ is a Haar measure on $S(F)$. Thus, for any character θ on $\{u \in L^\times : u\bar{u} = 1\}$, $\int_{S(F)} \theta(u) \frac{du}{|u|}$ exists as an ordinary integral and is zero unless θ is trivial.

REMARK 1.H.

Again L will be a quadratic extension of F. We regard $\mathbf{P}^1(L)$ as the F-rational points on a twisted form R of $\mathbf{P}^1 \times \mathbf{P}^1$, as follows: $R(L) = \mathbf{P}^1(L) \times \mathbf{P}^1(L)$ and σ acts by $(p, q) \longrightarrow (\bar{q}, \bar{p})$, so that $R(F) = \{(p, \bar{p}) : p \in \mathbf{P}^1(L)\}$. Define a form on $R(L)$ by $\nu = \frac{du \wedge dv}{uv}$, where u (respectively, v) denotes the coordinate U_0 on $U_1 = 1$ in the first (respectively, second) copy of $\mathbf{P}^1(L)$. At a point of $R(F)$ on $(U_1 = 1) \times (U_1 = 1)$ we have $v = \bar{u}$. Let h be given at such a point by $\theta(u\bar{u})|u\bar{u}|^t$, where θ is a character on $F^\times$ and t is a real number such that $t \neq 0$ if $\theta^2 \equiv 1$. Observe that, in general, h and ν do not satisfy the conditions of (1.9) - (1.11). We may, however, blow up R at $u = v = 0$ to obtain a variety $\bar{R}$ over F and projection $\pi : \bar{R} \longrightarrow R$. Set $\bar{h} = h \circ \pi$ and $\bar{\nu} = \pi^*(\nu)$. Let $\bar{N}$ be the inverse image in $\bar{R}(F)$ of the neighborhood $|u|_L \leq 1$ of $u = v = 0$ in $R(F)$. Then a calculation with coordinates shows that

$$\text{(1.18)} \qquad \oint_{\bar{N}} \bar{h}|\bar{\nu}| \text{ is well-defined}$$

(here $t \neq 0$ if $\theta^2 \equiv 1$ is needed) and

$$\text{(1.19)} \qquad \oint_{N} \bar{h}|\bar{\nu}| = \oint_{|u|_L \leq 1} \theta \circ \mathrm{Nm}(u)|u|_L^t \frac{d_L u}{|u|_L},$$

where the subscript L indicates that we are computing on the L-manifold $|u|_L \leq 1$ in $\mathbf{P}^1(L)$. Observe that if θ is trivial on $\mathrm{Nm}_F^L L^\times$ and $t = -1$ then $\oint_{\bar{R}(F)-\bar{N}} \bar{h}|\bar{\nu}| = \oint_{R(F)-N} h|\nu|$ is well-defined and equals $\oint_{|u|_L<1} d_L u = \oint_{|u|_L>1} \frac{d_L u}{|u|_L^2}$.
Thus, in this case, we have

$$\text{(1.20)} \qquad \oint_{\bar{R}(F)} \bar{h}|\bar{\nu}| = \oint_{\mathbf{P}^1(L)} \frac{d_L u}{|u|_L^2} = 0 \quad \text{(Lemma 1.C)} .$$

§2. IGUSA THEORY.

Recall the setting of [L, §1]: Y is a smooth variety over F, $\phi : Y \longrightarrow C$ is an Igusa fibering of Y over a smooth curve C over F, ω is an Igusa form on Y, and f an Igusa integrand (the definitions will be reviewed presently). There is a distinguished point c_0 on $C(F)$ and ϕ is smooth except on the special fiber $\phi^{-1}(c_0)$. Choose an F-coordinate λ around c_0 on C; assume $\lambda(c_0) = 0$. Then Igusa's theory establishes the existence of an asymptotic expansion

$$\sum_{(\theta,\beta,r)} \theta(\lambda)|\lambda|^{\beta-1}(-\log_q|\lambda|)^{r-1}F_r(\theta,\beta,f)$$

near $\lambda = 0$ for the integral

$$F(\lambda) = \int f \frac{|\omega|}{|\phi^*(d\lambda)|}$$

over the fiber in $Y(F)$ above the point in C with coordinate λ. Here θ denotes a character on $F^\times$, β a real number and r a positive integer. The coefficients $F_r(\theta, \beta, f)$ are the principal value integrals of [L, Proposition 1.2]. Under an assumption we will make (2.9), only $r = 1$ occurs and $F_1(\theta, \beta, f)$ is an integral of the type considered in the last section. In this paragraph we will relax the constraints on the form ω and integrand f. The fiber integral $F(\lambda)$ may then exist only as a principal value integral, but it will still have an asymptotic expansion. The coefficients are again given by [L, Proposition 1.2], i.e., by (related) principal value integrals.

For the rest of this section we require the following of Y, C, ϕ, ω and f:

(2.1) Y is a smooth variety over F, C is a smooth curve over F with distinguished point $c_0 \in C(F)$, $\phi : Y \longrightarrow C$ is an F-morphism smooth except over c_0, ω is a differential form of maximal degree on Y, f is a $\mathbf{C}$-valued function supported on a compact open subset of $Y(F)$, <u>and</u>

(2.2) if $y_0 \in Y(F)$ lies over the coordinate patch for λ, a fixed local F-coordinate around c_0 on C, then there exist local F-coordinates $\mu_1, \ldots, \mu_n$ around y_0 on Y such that:

(2.3) if $y_0 \in \phi^{-1}(c_0)$ then ϕ is given near y_0 by $\lambda = \alpha\mu_1^{a_1} \ldots \mu_n^{a_n}$, where α is regular and invertible at y_0 and $a_1, \ldots, a_n$ are nonnegative integers; if $y_0 \notin \phi^{-1}(c_0)$ and λ_0 is the coordinate of $\phi(y_0)$ then $\mu_1 = \lambda-\lambda_0$,

(2.4) ω is given near y_0 by

$$W \prod_{j=1}^{n} \mu_j^{b_j} \frac{d\mu_1}{\mu_1} \wedge \ldots \wedge \frac{d\mu_n}{\mu_n},$$

where W is regular and invertible at y_0 and $b_1, \ldots, b_n$ are rational numbers; if $y_0 \notin \phi^{-1}(c_0)$ then $b_1 = 1$,

(2.5) f is given on points of $Y(F)$ near y_0 by $\gamma K_1(\mu_1) \ldots K_n(\mu_n)$, where γ is locally constant around y_0 and $K_1, \ldots, K_n$ are quasicharacters on $F^\times$ such that:

(2.6) if $\mu_j = 0$ is the branch of a divisor E in $\phi^{-1}(c_0)$ through y_0 then K_j depends only on E, if $y_0 \notin \phi^{-1}(c_0)$ then $K_1 \equiv 1$, and

(2.7) if $K_j = \kappa_j|.|^{t_j}$ with κ_j unitary and t_j real then either $t_j + b_j \neq 0$ or $\kappa_j \not\equiv 1$, $1 \leq j \leq n$. For Lemma 2.A, (2.7) need only be satisfied for $y_0 \notin \phi^{-1}(c_0)$.

REMARK. These are the conditions of [L, §1] for $\phi : Y \longrightarrow C$ to be an Igusa fibering; ω is an Igusa form if $b_1,\ldots,b_n$ are positive integers, i.e., ω has no singularities, and the zeros ω lie on the special fiber, i.e., $b_j = 0$ unless $\mu_j = 0$ is the branch of a divisor in $\phi^{-1}(c_0)$; f is an Igusa integrand if $K_1, \ldots, K_n$ are unitary and $K_j \equiv 1$ unless $\mu_j = 0$ is the branch of a divisor in $\phi^{-1}(c_0)$.

Let $\mathcal{E}$ be the set of all divisors in $\phi^{-1}(c_0)$ meeting the support of f. If $\mu_j = 0$ is the branch of $E \in \mathcal{E}$ through y_0 then $a_j = a(E)$, the multiplicity of E in $\phi^{-1}(c_0)$. We then also set $b_j = b(E)$, $K_j = K(E)$, $\kappa_j = \kappa(E)$ and $t_j = t(E)$, as our assumptions allow.

Let

$$F(\lambda) = \oint f \frac{|\omega|}{|\phi^*(d\lambda)|} ,$$

the integral being taken over the fiber in $Y(F)$ above the point on $C(F)$ with coordinate $\lambda \neq 0$. Then $F(\lambda)$ is a well-defined principal value integral of the type studied in the last section. To check this we may assume that f is supported on a neighborhood $|\mu_j| \leq \varepsilon_j$, $1 \leq j \leq n$, in a coordinate patch (2.2) around $y_0 \in \phi^{-1}(c_0)$. We may also assume $|W|$ and γ constant. Then

$$(2.8) \qquad F(\lambda) = \gamma|W| \oint_{\substack{|u_j|<\varepsilon_j \\ (j>1)}} \prod_{j>1} K_j(\mu_j)|\mu_j|^{b_j} \frac{d\mu_2}{|\mu_2|} \cdots \frac{d\mu_n}{|\mu_n|} ,$$

and we are done.

The data for the asymptotic expansion of $F(\lambda)$ will be a slight modification of that of [L, Proposition 1.1]. Consider pairs (θ, β), where θ is a character on $F^\times$ and β is a real number. Let $\mathcal{E}(\theta, \beta)$ be the set of those $E \in \mathcal{E}$, i.e., of those divisors E in $\phi^{-1}(c_0)$ meeting the support of f, such that $\kappa(E) = \theta^{a(E)}$ and

$$\beta(E) \overset{\text{def.}}{=} \frac{b(E) + t(E)}{a(E)} = \beta .$$

Let $e(\theta, \beta)$ be the maximum number of branches of divisors in $\mathcal{E}(\theta, \beta)$ meeting at a point. For the purposes of this paper it will be sufficient to consider the case:

$$(2.9) \qquad e(\theta, \beta) \leq 1 .$$

Then:

LEMMA 2.A.

<u>For</u> $|\lambda|$ <u>sufficiently small</u>,

$$F(\lambda) = \sum_{(\theta,\beta)} \theta(\lambda)|\lambda|^{\beta-1} F_1(\theta, \beta, f)$$

where $F_1(\theta, \beta, f)$ is the constant of [L, Proposition 1.2].

If $e(\theta, \beta) = 0$ then $F_1(\theta, \beta, f) = 0$. Otherwise, let E be a divisor in $\mathcal{E}(\theta, \beta)$. Suppose that $y_0 \in E(F)$. Choose coordinates $\mu_1, \ldots, \mu_n$ as in (2.2) and assume that $\mu_1 = 0$ is a branch of E through y_0. Following [L, Proposition 1.2] we define h and ν near y_0 by

$$h = h(\mu_2, \ldots, \mu_n) = \frac{\gamma(0, \mu_2, \ldots, \mu_n)}{\theta^{\beta}(\alpha(0, \mu_2, \ldots, \mu_n))} \prod_{j=2}^{n} K_j(\mu_j)\theta(\mu_j^{-a_j})|\mu_j|^{-\beta a_j} ,$$

where $\theta^{\beta} = \theta|.|^{\beta}$, and

$$\nu = W(0, \mu_2, \ldots, \mu_n) \prod_{j=2}^{n} \mu_j^{b_j} \frac{d\mu_2}{\mu_2} \wedge \ldots \wedge \frac{d\mu_n}{\mu_n} .$$

Then

$$F_1(\theta, \beta, f) = \sum_{E} \oint_{E(F)} h|\nu| , \tag{2.10}$$

hese integrals to be calculated by the methods of §1.

Proof of Lemma 2.A:

We may assume that f is supported on a coordinate patch (2.2) around $y_0 \in \phi^{-1}(c_0)$. Then f and ω, and hence $F(\lambda)$, come with the parameters $t = (t_1, \ldots, t_n)$ and $b = (b_1, \ldots, b_n)$. We write $F(\lambda) = F(\lambda, t, b)$.

If $t_j + b_j \geq 1$ $(1 \leq j \leq n)$ then arguments of [L, Propositions 1.1 and 1.2] carry through without modification, for $F(\lambda, t, b)$ is an ordinary integral. Thus the lemma is proved in this case.

We now relax this condition on t and b. Let $t' = (t'_1, \ldots, t'_n) \in \mathbb{R}^n$. It is convenient to assume that

$$\frac{t'_i}{a_i} = \frac{t'_j}{a_j} \quad \text{if } a_i, a_j \neq 0 .$$

Suppose that $E \in \mathcal{E}$ has data (θ, β) with respect to (t, b), i.e., with respect to f and ω. If $\mu_j = 0$ is a branch of E through y_0 then E has data $\theta' = \theta$ and

$$\beta' = \frac{b_j + t_j + t'_j}{a_j} = \beta + \frac{t'_j}{a_j}$$

with respect to $(t+t', b)$. Thus (2.9) is satisfied by $(t+t', b)$. If $t'_j >> 0$, $1 \leq j \leq n$, then $t_j + t'_j + b_j \geq 1$, $1 \leq j \leq n$, and there exists $\varepsilon > 0$ independent of t' such that

$$(2.11) \qquad F(\lambda, t+t', b) = \sum_{(\theta',\beta')} \theta'(\lambda)|\lambda|^{\beta'-1} F_1(\theta', \beta', f)$$

for $|\lambda| < \varepsilon$. One verifies easily that $F(\lambda, t, b)$ is the value at $t' = 0$ of $F(\lambda, t+t', b)$. At $t' = 0$ the right side of (2.11) has the value

$$\sum_{(\theta,\beta)} \theta(\lambda)|\lambda|^{\beta-1} F_1^0(\theta, \beta', f)$$

where $F_1^0(\theta, \beta', f)$ is the value of $F_1(\theta, \beta', f)$ at $t' = 0$. This is readily seen to be $F_1(\theta, \beta, f)$, and the lemma is proved.

The following remarks will not be needed in this paper.

Let $\varepsilon > 0$, $I(\varepsilon) = \{c \in C(F) : |\lambda| = |\lambda(c)| \leq \varepsilon\}$ and $Y(\varepsilon)$ be the inverse image of $I(\varepsilon)$ in $Y(F)$. The asymptotic expansion for $F(\lambda)$ allows us to define the principal value integral $\oint_{I(\varepsilon)} F(\lambda)d\lambda$ as the value of $\int_{I(\varepsilon)} F(\lambda)|\lambda|^s d\lambda$ at $s = 0$ provided $F_1(1, 0, f) = 0$, i.e., provided there is no contribution from the pair $\theta \equiv 1$, $\beta = 0$ to the expansion. On the other hand our initial assumptions ensure that $\oint_{Y(\varepsilon)} f|\omega|$ is well-defined (in the sense of §1).

LEMMA 2.B.

<u>Assume</u> $F_1(1, 0, f) = 0$. <u>Then</u>

$$\oint_{Y(\varepsilon)} f|\omega| = \oint_{I(\varepsilon)} F(\lambda)d\lambda .$$

<u>Proof</u>: We may assume that f is supported on a neighborhood in a coordinate patch (2.2) around $y_0 \in \phi^{-1}(c_0)$. Suppose that $\mu_1 = 0, \ldots, \mu_r = 0$ are branches of divisors in $\phi^{-1}(c_0)$, and that $\mu_{r+1} = 0, \ldots, \mu_n = 0$ are not. Then for $\mathrm{Re}\, s >> 0$ $\oint_{Y(\varepsilon)} f|\lambda|^s|\omega|$ is the value at $s_{r+1} = \ldots = s_n = 0$ of $\int_{Y(\varepsilon)} f|\lambda|^s|\mu_{r+1}|^{s_{r+1}} \ldots |\mu_n|^{s_n}|\omega|$. Since $\lambda = \alpha\mu_1^{a_1} \ldots \mu_r^{a_r}$, where $|\alpha| \neq 0$, on the support of f it follows that $\lim_{\varepsilon'\to 0} \oint_{Y(\varepsilon')} f|\lambda|^s|\omega| = 0$. Since ϕ is smooth away from the special fiber we have (cf. Lemma 1.F) that for $\varepsilon' < \varepsilon$ and $\mathrm{Re}\, s >> 0$

$$\oint_{Y(\varepsilon)-Y(\varepsilon')} f|\lambda|^s|\omega| = \int_{I(\varepsilon)-I(\varepsilon')} F(\lambda)|\lambda|^s d\lambda .$$

The asymptotic expansion for $F(\lambda)$ implies that

$$\lim_{\varepsilon'\to 0} \int_{I(\varepsilon')} F(\lambda)\,|\lambda|^s d\lambda = 0 \quad \text{for} \quad \operatorname{Re} s >> 0 \quad .$$

Thus

$$\oint_{Y(\varepsilon)} f\,|\lambda|^s\,|\omega| = \int_{I(\varepsilon)} F(\lambda)\,|\lambda|^s d\lambda \quad .$$

Since the value of the left side at $s = 0$ is $\oint_{Y(\varepsilon)} f|\omega|$ the lemma is proved.

LEMMA 2.C.

If $F_1(\theta, \beta, f) = 0$ for all $\beta \leq 0$ then

$$\lim_{\varepsilon\to 0} \oint_{Y(\varepsilon)} f|\omega| = 0 \quad .$$

Proof: Under this assumption the asymptotic expansion involves only positive exponents β. Then $\lim_{\varepsilon\to 0} \oint_{I(\varepsilon)} F(\lambda)d\lambda = 0$. Hence, by the last lemma, $\lim_{\varepsilon\to 0} \oint_{Y(\varepsilon)} f|\omega| = 0$.

§3. SOME IGUSA DATA.

Following [L, §§2-5] we now construct a smooth variety Y over F, an Igusa fibering $\phi : Y \longrightarrow C$ of Y over a curve C, a differential form ω of maximal degree on Y, and an integrand f_K (notation of [L, §2]) on $Y(F)$.

Fix an inner form G of $SL(2)$ and a maximal torus T over F in G. Let c_0 be a point in the center of G. For the curve C we take T with the other central point removed.

The construction of Y starts with the variety S of stars ([L, §2]). Here S is just $\mathcal{B} \times \mathcal{B}$, $\mathcal{B}$ denoting the variety of Borel subgroups of G. Let $B_\infty \in \mathcal{B}$. Then $S(B_\infty)$ is $\mathcal{B} - \{B_\infty\} \times \mathcal{B} - \{B_\infty\}$. Let $B_0 \in \mathcal{B} - \{B_\infty\}$. Then $S(B_\infty, B_0)$ consists of the pairs (B_+, B_-) in $S(B_\infty)$ with $B_+ = B_0$. If $N(.)$ indicates unipotent radical and B^g the Borel subgroup $g^{-1}Bg$, $g \in \mathcal{B}$, we have

$$\begin{aligned} S(B_\infty) &= \{(B_0^{n_1}, B_0^{nn_1}); \ n, n_1 \in N(B_\infty)\} \\ &\cong S(B_\infty, B_0) \times N(B_\infty) \\ &\cong N(B_\infty) \times N(B_\infty) \quad . \end{aligned}$$

Coordinates for $S(B_\infty)$ are evident, but the demands for Galois action require that a little care be taken in the choice.

First, and for the rest of the paper, we fix data as in [L, §2]: $G^* = SL(2)$, $\mathbf{B}^*$ is the upper triangular subgroup of G^*, $\mathbf{B}_*$ the lower triangular subgroup, $\mathbf{T}^*$ the diagonal subgroup; $\psi : G \longrightarrow G^*$ is an inner twist such that $\psi : T \longrightarrow G^*$ is defined over F, T^* denotes $\psi(T)$, $\eta^* : G^* \longrightarrow G^*$ is a diagonalization of T^* and, finally, η denotes $\eta^* \circ \psi$.

By means of ψ we identify G with G^* as a group over $\overline{F}$, and hence $\mathcal{B}$ with $\mathcal{B}^*$ and S with $\mathcal{B}^* \times \mathcal{B}^*$, where $\mathcal{B}^*$ is the variety of Borel subgroups of $SL(2)$. View $\mathcal{B}^*$ as the variety $\mathbf{P}^1$ of lines through the origin in $\mathbf{A}^2$ via

$$(\mathbf{B}^*)^g \longleftrightarrow [0, 1] \cdot g \quad .$$

Write a for $[a, 1]$.

Returning to B_∞ and B_0, now elements of $\mathcal{B}^*$, we choose $h \in G^*$ such that

$$(3.1) \qquad (B_0)^h = \mathbf{B}^* \quad \text{and} \quad (B_\infty)^h = \mathbf{B}_* \quad .$$

Then h allows us to identify $S(B_\infty)$ with $S(\mathbf{B}_*)$. If $n = \begin{bmatrix} 1 & 0 \\ x & 1 \end{bmatrix}$ and $n_1 = \begin{bmatrix} 1 & 0 \\ y & 1 \end{bmatrix}$ we have

$$(3.2) \qquad S(\mathbf{B}_*) \ni ((\mathbf{B}^*)^{n_1}, (\mathbf{B}^*)^{nn_1}) \longleftrightarrow (y, x+y) \in \mathbf{P}^1 \times \mathbf{P}^1 \quad .$$

Thus h provides coordinates, informally denoted x and y, on $S(B_\infty)$.

The variety $S_1(B_\infty)$ of [L, §3] is naturally identified with $S(B_\infty)$; S_1 is then obtained by gluing together the $S(B_\infty)$, $B_\infty \in \mathcal{B}^*$, according to the rules of [L, (3.7) and (3.8)]. But these are the rules for the natural gluing of open subsets of $\mathcal{B}^* \times \mathcal{B}^* = \mathbf{P}^1 \times \mathbf{P}^1$, and so $S_1 = S = \mathbf{P}^1 \times \mathbf{P}^1$ (cf. [L, Lemma 3.10(a)]).

To describe the Galois action on S and at the same time maintain our identification of $G(\overline{F})$ with $G^*(\overline{F})$ and of S with $\mathcal{B}^* \times \mathcal{B}^*$ we equip $G^*(\overline{F})$ with the Galois action $\sigma_G = \psi \circ \sigma \circ \psi^{-1}$, $\sigma \in \mathrm{Gal}(\overline{F}/F)$. Recall that the identification $\psi : T \longrightarrow T^*$ is over F. Let $L \subset \overline{F}$ be a quadratic extension of F. Write $\mathcal{J}_L$ for the set of tori in G defined over F (i.e., in G^* and preserved by σ_G, $\sigma \in \mathrm{Gal}(\overline{F}/F)$) which are anisotropic over F and split over L. We allow also $L = F$, then meaning by $\mathcal{J}_L$ the set of F-split tori in G.

The action of $\sigma \in \mathrm{Gal}(\overline{F}/F)$ on $S = \mathcal{B}^* \times \mathcal{B}^*$ will be denoted $\sigma_{(G,T)}$. From [L, §2 and §4] we get

(3.3) $$\sigma_{(G,T)}((B_+, B_-)) = (\sigma_G(B_-), \sigma_G(B_+))$$

if $T \in \mathcal{T}_L$, $L \neq F$, and σ is nontrivial on L, and

(3.4) $$\sigma_{(G,T)}((B_+, B_-)) = (\sigma_G(B_+), \sigma_G(B_-))$$

otherwise.

The following elaborate remark will be helpful later on.

(3.5) If $\sigma_G(\mathbf{B}^*) = \mathbf{B}^*$ and $\sigma_G(\mathbf{B}_*) = \mathbf{B}_*$, as we may assume if G is split over F, then S is covered by patches $S(B_\infty)$, where $\sigma_G(B_\infty) = B_\infty$ and $\sigma_G(B_0) = B_0$, $\sigma \in \mathrm{Gal}(\bar{F}/F)$, for some $B_0 \neq B_\infty$. For example, $S = S(\mathbf{B}_*) \cup S(\mathbf{B}^*)$. Each such patch $S(B_\infty)$ is preserved by $\mathrm{Gal}(\bar{F}/F)$. The element h of (3.1) can be chosen so that $\sigma_G(h)h^{-1}$ is central, $\sigma \in \mathrm{Gal}(\bar{F}/F)$. Then the identification of $S(B_\infty)$ with $S(\mathbf{B}_*)$ provided by h respects Galois action.

(3.6) Suppose that L is a quadratic extension of F. Assume, as we may if $T \in \mathcal{T}_L$, that $\sigma_G(\mathbf{B}^*) = \mathbf{B}_*$ for σ nontrivial on L and $\sigma_G(\mathbf{B}^*) = \mathbf{B}^*$, $\sigma_G(\mathbf{B}_*) = \mathbf{B}_*$ otherwise. Then S is covered by coordinate patches $S(B_\infty)$ where for some $B_0 \neq B_\infty$, $\sigma_G(B_\infty) = B_0$ for σ nontrivial on L and $\sigma_G(B_\infty) = B_\infty$, $\sigma_G(B_0) = B_0$ otherwise. Again $S = S(\mathbf{B}_*) \cup S(\mathbf{B}^*)$ will do. Now, however, $\sigma_{(G,T)}$ preserves only $S(B_\infty) \cap S(B_0) = S(B_\infty) - \{(B_0, B_0)\}$ if σ is nontrivial on L. The element h of (3.1) may be chosen so that $\sigma_G(h)h^{-1}$ is central, $\sigma \in \mathrm{Gal}(\bar{F}/F)$. Then the identification of $S(B_\infty) \cap S(B_0)$ with $S(\mathbf{B}_*) \cap S(\mathbf{B}^*)$ provided by h respects Galois action.

Returning to the construction of Y we find it convenient to make yet another identification, that of T and T^* with $\mathbf{T}^*$ using the diagonalization η^*. We equip $\mathbf{T}^*(\bar{F})$ with the action $\sigma_T = \sigma_{T^*} = \eta^* \circ \sigma \circ (\eta^*)^{-1}|_{\mathbf{T}^*}$, $\sigma \in \mathrm{Gal}(\bar{F}/F$, and regard C as a curve in $\mathbf{T}^*$ preserved by this action. Note that $\eta^*(c_0) = c_0$.

A star $s = (B_+, B_-)$ is regular in the sense of [L, §2] if and only if $B_+ \neq B_-$. For the variety X_1 of [L, §4] we take the closure in $G^* \times S$ of $\{(g, s = (B_+, B_-)) : g, s \text{ regular}, g \in B_+ \cap B_-\}$; X_1 is defined over F for the Galois action given by $\sigma_G \times \sigma_{(G,T)}$, $\sigma \in \mathrm{Gal}(\bar{F}/F)$. There are maps defined over F:

$$\begin{array}{ccc} X_1 & \xrightarrow{\pi} & G^* \\ \downarrow{\scriptstyle \phi_1} & & \\ \mathbf{T}^* & & \end{array}$$

where $\sigma \in \mathrm{Gal}(\bar{F}/F)$ acts on G^* by σ_G and on $\mathbf{T}^*$ by σ_T. The horizontal arrow is projection on the first component. To define ϕ_1, note that X_1 is contained in $\{(g,\ s = (B_+,\ B_-)) \in G^* \times S : g \in B_+ \cap B_-\}$. Thus if $(g,\ s) \in X_1$ we may choose $h \in G^*$ such that $B_+^h = \mathbf{B}^*$. Then $\phi_1((g,\ s))$ is the image of $h^{-1}gh \in \mathbf{B}^*$ under the projection $\mathbf{B}^* = \mathbf{T}^* N(\mathbf{B}^*) \longrightarrow \mathbf{T}^*$.

The variety Y will be the intersection of $\phi_1^{-1}(C)$ with the closure in X_1 of $\phi_1^{-1}(C - \{c_0\})$. By restriction we have:

$$\begin{array}{ccc} Y & \xrightarrow{\ \pi\ } & G^* \\ \Big\downarrow{\scriptstyle \phi} & & \\ C & & \end{array}$$

Let M be the Springer-Grothendieck variety $\{(g,\ b) : g \in B\} \subset G^* \times \mathcal{B}^*$, with the usual maps:

$$\begin{array}{ccc} M & \xrightarrow{\ \pi_M\ } & G^* \\ {\scriptstyle \phi_M}\Big\downarrow & & \\ \mathbf{T}^* & & \end{array}$$

Define $\xi : Y \longrightarrow M$ by $(g,\ (B_+,\ B_-)) \longrightarrow (g,\ B_+)$. Then $\phi = \phi_M \circ \xi$ and $\pi = \pi_M \circ \xi$. If M' is M with the fibers over the central points removed then $\phi^{-1}(C - \{c_0\}) \xrightarrow{\ \xi\ } M'$ is an isomorphism of varieties over $\bar{F}$. In particular, $\phi^{-1}(C - \{c_0\})$ is smooth.

To examine the special fiber $\phi^{-1}(c_0)$ we introduce coordinates as in [L, §3]. Let $Y \subset G^* \times S \longrightarrow S$ be projection on the second factor. Let $Y(B_\infty)$ be the inverse image of $S(B_\infty)$, $B_\infty \in \mathcal{B}^*$. Identify $S(B_\infty)$ with $S(\mathbf{B}_*)$ by means of some h as in (3.1). We may then work with the coordinates x, y of (3.2) on $S(\mathbf{B}_*)$ and with $Y(\mathbf{B}_*)$.

Let λ be a local F-coordinate around c_0 in C. Recall that $C \subset \mathbf{T}^*$ and that $\sigma \in \mathrm{Gal}(\bar{F}/F)$ acts by σ_T. Assume that $\lambda = 0$ at c_0. If α is the root of $\mathbf{T}^*$ in $\mathbf{B}^*$ then we may write $1 - \alpha^{-1}$ as $\lambda b(\lambda)$ near c_0, with b regular and invertible near $\lambda = 0$. Suppose that $(g,\ s) \in Y(\mathbf{B}_*)$. As in (3.2) write s as $((\mathbf{B}^*)^{n_1},\ (\mathbf{B}^*)^{nn_1})$, with $n = \left[\begin{smallmatrix} 1 & 0 \\ x & 1 \end{smallmatrix}\right]$ and $n_1 = \left[\begin{smallmatrix} 1 & 0 \\ y & 1 \end{smallmatrix}\right]$. Note that x is the coordinate $z(W_+,\ \alpha)$ from [L, §3]. Write

$$g = n_1^{-1} t \begin{bmatrix} 1 & u \\ 0 & 1 \end{bmatrix} n_1, \quad \text{with } t \in \mathbf{T}^*,\ u \in \bar{F} . \tag{3.7}$$

Assume $x \neq 0$. Then $g \in (B^*)^{nn_1}$ is equivalent to

$$1 - \alpha(t)^{-1} = xu$$

or, if (g, s) is near $\phi^{-1}(c_0)$ and we pull back λ to Y, to

$$(3.8) \qquad \lambda b(\lambda) = xu \quad .$$

As a consequence, u, x and y serve as coordinates on $Y(\mathbf{B}_*)$, and $Y(\mathbf{B}_*)$ is smooth. Then each $Y(B_\infty)$ is smooth, $B_\infty \in \mathcal{B}^*$. Hence Y is a smooth variety.

Near $\phi^{-1}(c_0)$ on $Y(\mathbf{B}_*)$, ϕ is given by

$$(3.9) \qquad \lambda = Axu$$

with A regular and invertible near $\lambda = 0$. Thus $u = 0$ is the branch of a divisor E_1 of $\phi^{-1}(c_0)$. This branch consists of the pairs (c_0, s), $s \in S(\mathbf{B}_*)$, and so E_1 must be $\{c_0\} \times S = \{c_0\} \times \mathbf{P}^1 \times \mathbf{P}^1$; E_1 maps under $\pi : Y \longrightarrow G$ to $\{c_0\}$. On the other hand, $x = 0$ is the branch $\{(g, (B, B)) : B \neq B_*,\ g \in G,\ c_0 g$ unipotent$\}$ of a divisor E_2. Thus $E_2 = \{(g, (B, B));\ B \in \mathcal{B},\ g \in B,\ c_0 g$ unipotent$\}$. For convenience we call g c_0-unipotent if $c_0 g$ is unipotent. Then π maps $E_2 - E_1$ isomorphically to the orbit of regular c_0-unipotent elements in G. Note that the two divisors E_1 and E_2 cover $\phi^{-1}(c_0)$, and that E_2 has no F-rational points unless G is split over F.

The relation of Y to the Springer-Grothendieck variety M is now evident. Under $Y \xrightarrow{\xi} M$ the divisor E_2 is mapped isomorphically to the fiber over c_0; Y is obtained from M with the fiber over $-c_0$ removed by blowing up along the subvariety $\{c_0\} \times \mathcal{B}^* = \xi(E_1)$ of the fiber over c_0.

To verify that $\phi : Y \longrightarrow C$ is an Igusa fibering it remains only to check that one of E_1, E_2 is defined over F. For then both divisors are defined over F and we may apply (3.9) and Hilbert's Theorem 90 (for the field of functions regular and invertible near a point) to replace around each F-rational point $y_0 \in \phi^{-1}(c_0)$ the coordinates u, x and y with F-coordinates μ_1, μ_2, μ_3 such that:

(3.10) $\mu_i = 0$ is a branch of E_i if y_0 lies on E_i $(i = 1, 2)$.

(3.11) λ is given near y_0 by $\lambda = \alpha\mu_1^{a_1}\mu_2^{a_2}$, where α is regular and invertible at y_0, and $a_i = 1$ if y_0 lies on E_i and $a_i = 0$ otherwise $(i = 1, 2)$.

Since $E_1 = \{c_0\} \times S$ is clearly defined over F, we are done.

The indices $a(.)$ of §2 are:

(3.12) $$a(E_1) = a(E_2) = 1 \quad \text{if } G \text{ is split over } F,$$
$$a(E_1) = 1 \qquad \text{otherwise.}$$

The next step is to define an Igusa form ω. Let ω_T be the (right) invariant form on $\mathbf{T}^*$ equal to $d\lambda$ at c_0. Let $a \in \bar{F}$ be such that $\bar{\omega} = a\omega_T$ is defined over F for the Galois action on $\mathbf{T}^*$ as F-split torus. Let $H \in \mathrm{Lie}(\mathbf{T}^*)$ be such that $\bar{\omega}(H) = 1$. Choose $X_+ \in \mathrm{Lie}(N(\mathbf{B}^*))$, $X_- \in \mathrm{Lie}(N(\mathbf{B}_*))$ and right invariant 1-forms $\omega_0, \omega_+, \omega_-$ on G defined over F so that $\langle \omega_0, \omega_+, \omega_- \rangle$ is dual to $\langle H, X_+, X_- \rangle$. Then $\omega_G = \omega_0 \wedge \omega_+ \wedge \omega_-$ is a (right) invariant form of maximal degree on G defined over F. The form ω_M on M associated to ω_G (more precisely, to $\nu_1 = \omega_0$, $\nu_2 = \omega_+$, $\omega_1 = \omega_-$) in [L, Lemma 2.8] is $\pi_M^*(\omega_G)$. We set $\omega_Y = \xi^*(\omega_M) = \pi^*(\omega_G)$ and $\omega = a^{-1}\omega_Y$.

The form ω is regular; it is nonvanishing off the special fiber. The discussion of [L, §2] implies that locally $\omega = W'\omega'$, where W' is a regular invertible function and ω' is defined over F. This ensures that the measure $|\omega|$ is well-defined.

Suppose that $y_0 \in Y(\mathbf{B}_*)$ is near but not on $\phi^{-1}(c_0)$. We may as well take $X_+ = \begin{bmatrix} 0 & 1 \\ 0 & 0 \end{bmatrix}$ and $X_- = \begin{bmatrix} 0 & 0 \\ 1 & 0 \end{bmatrix}$. Then it may be shown that ω is given near y_0 by $\phi^*(d\lambda) \wedge du \wedge dy = W(\lambda)d(xu) \wedge du \wedge dy = W(\lambda)u\, dx \wedge du \wedge dy$, where W is regular and invertible near $\lambda = 0$, with $W(0) = (ab(0))^{-1}$ (cf. (3.8)). From this it follows that

(3.13) $$\omega = W(\lambda)u^2 x \,\frac{dx}{x} \wedge \frac{du}{u} \wedge dy$$

around a point of $Y(\mathbf{B}_*) \cap \phi^{-1}(c_0)$.

Note that ω may be expressed in terms of the coordinates μ_1, μ_2, μ_3 of (3.10), but that the coordinates u and x will do just as well to compute the indices $b(.)$ of §2:

(3.14) $$b(E_1) = 2, \quad b(E_2) = 1 \quad \text{if } G \text{ is split over } F$$
$$b(E_1) = 2 \qquad \text{otherwise.}$$

It remains to define the Igusa integrand. Let κ be a character on $\mathfrak{D}(T)$, the definition of which will be noted in (3.15). Recall that $T(\bar{F})\backslash \mathfrak{A}(T, F)$ is the set of F-rational points in $T(\bar{F})\backslash G(\bar{F}) = (T\backslash G)(\bar{F})$. If $\gamma \in T(F) - \{c_0\}$ then $\pi : Y \longrightarrow G$ induces an F-isomorphism from the fiber ϕ_γ^{-1} over γ in Y to

$T\backslash G$ (cf. [L, Lemma 2.1]). We have therefore:

$$\phi_\gamma^{-1}(F) \longrightarrow T(\bar{F})\backslash \mathfrak{A}(T,F) \longrightarrow \mathfrak{B}(T,F) = T(\bar{F})\backslash \mathfrak{A}(T,F)/G(F) \cong H^1(\mathrm{Gal}(\bar{F}/F), T(\bar{F})) \quad , \tag{3.15}$$

allowing us to regard κ as a function m_κ on $\phi_\gamma^{-1}(F)$. If κ is trivial then $m_\kappa \equiv 1$. Suppose then that $T \in \mathcal{J}_L$, $L \neq F$, and κ is nontrivial. We will need an explicit formula for m_κ near an F-rational point y_0 on the special fiber. Proposition 5.1 of [L] shows that m_κ depends locally only on the coordinate x, at least if G is split over F, but for the formula we will need an F-coordinate.

Suppose that $S(B_\infty)$ is a coordinate patch as in (3.5). We may as well take $\sigma_G = \sigma_{G^*}$, $\sigma \in \mathrm{Gal}(\bar{F}/F)$, or $G = SL(2)$. Identify $S(B_\infty)$ with $S(\mathbf{B}_*)$ using h as in (3.5). Recall that this identification respects the Galois action on S. The formulas (3.2), (3.3) and (3.4) imply that the coordinates x, y on $S(\mathbf{B}_*)$ satisfy $\sigma(y) = x + y$ if σ is nontrivial on L and that $\sigma(y) = y$, $\sigma(x+y) = x + y$ otherwise. Then $\sigma(x) = \sigma(x+y-y) = y - (x+y) = -x$ for σ nontrivial on L. Fix $\tau \in L - F$ such that $\tau^2 \in F$. Then $\mu = \tau x$ is an F-coordinate (and will serve as μ_2 in (3.10)). Let $(g, s) \in \phi_\gamma^{-1}(F)$ lie in $Y(B_\infty)$ which we have identified with $Y(\mathbf{B}_*)$. The coordinate μ then being F-valued, we have that

$$\sigma \longrightarrow \begin{bmatrix} \mu & 0 \\ 0 & \mu^{-1} \end{bmatrix} \quad \text{if } \sigma|_L \not\equiv 1, \ \sigma \longrightarrow 1 \ \text{otherwise} \quad ,$$

represents an element of $H^1(T)$ which we denote μ_σ. Let ε_σ denote the image of (g, s) under (3.15). Then Proposition 5.2 of [L] implies that there is an element t_σ of $H^1(T)$ independent of (g, s) such that

$$\varepsilon_\sigma = \mu_\sigma t_\sigma \tag{3.16}$$

(see the Appendix to this section). Thus

$$m_\kappa((g, s)) = \kappa(\varepsilon_\sigma) = \kappa(\mu_\sigma)\kappa(t_\sigma) = \kappa(\mu)\kappa(t_\sigma) \quad ,$$

where κ now also denotes the quadratic character on $F^\times$ attached to L/F.

By requiring that $S(B_\infty)$ be as in (3.5) we have excluded the case G anisotropic over F. This is of no consequence, for then if $(g, s) \in Y(F)$ the star $s = (B_+, B_-)$ must be regular ($\sigma_G(B_+ \cap B_-) = B_+ \cap B_-$ implies $B_+ \cap B_-$ is a torus so that $B_+ \neq B_-$). From Lemma 2.10 of [L] we conclude that m_κ is locally constant on $Y(F)$.

Finally, fix $f \in C_c^\infty(G(F))$. The Igusa integrand will be:

$$f_\kappa((g, s)) = m_\kappa((g, s))(f \circ \pi)(g, s)$$

$$= m_\kappa((g, s))f(g), \qquad (g, s) \in Y(F) - \phi^{-1}(c_0) \quad .$$

The characters $\kappa(.)$ of §2 are:

$$(3.17) \qquad \begin{cases} \kappa(E_1) \equiv 1 \text{ and } \kappa(E_2) = \kappa & \text{if } G \text{ is split over } F \\ \kappa(E_1) \equiv 1 & \text{otherwise} \end{cases}$$

Appendix

Here we note the explicit calculation of ε_σ in (3.16) and another local expression for m_κ which applies to anisotropic groups as well.

For (3.16) recall that we have assumed that G is $SL(2)$ (and $\psi \equiv 1$, $T = T^*$). We refrain from identifying $T(\bar{F})$ with $\mathbf{T}^*(\bar{F})$. Then ε_σ is the class of $\sigma \longrightarrow \sigma(h_1)h_1^{-1}$, where $h_1 \in G(L)$ satisfies:

$$h_1 g h_1^{-1} \in T, \; B_+^{h_1} = \eta^{-1}(\mathbf{B}^*) \text{ and } B_-^{h_1} = \eta^{-1}(\mathbf{B}_*)$$

if $s = (B_+, B_-)$. Write η as $t \longrightarrow h_2 t h_2^{-1}$, $h_2 \in G(L)$. For h_1 we can take $h_2^{-1}h_3$ if $h_3 \in G(L)$ satisfies:

$$h_3 g h_3^{-1} \in \mathbf{T}^*, \; B_+^{h_3} = \mathbf{B}^* \text{ and } B_-^{h_3} = \mathbf{B}_* \quad .$$

On $Y(\mathbf{B}_*)$ we have

$$g = \begin{bmatrix} 1 & 0 \\ -y & 1 \end{bmatrix} t \begin{bmatrix} 1 & u \\ 0 & 1 \end{bmatrix} \begin{bmatrix} 1 & 0 \\ y & 1 \end{bmatrix} \; ,$$

with $t \in \mathbf{T}^*$ and $ux = 1 - \alpha(t)^{-1}$. It is easily checked that

$$h_3' = \begin{bmatrix} 1 & 1/x \\ 0 & 1 \end{bmatrix} \begin{bmatrix} 1 & 0 \\ y & 1 \end{bmatrix}$$

will do;

$$\sigma(h_3)h_3^{-1} = \begin{bmatrix} 0 & -1/x \\ x & 0 \end{bmatrix} = \begin{bmatrix} \mu^{-1} & 0 \\ 0 & \mu \end{bmatrix}\begin{bmatrix} 0 & -1/\tau \\ \tau & 0 \end{bmatrix}$$

or σ nontrivial on L. Then

$$\sigma(h_1)h_1^{-1} = \sigma(h_2^{-1})\begin{bmatrix} \mu^{-1} & 0 \\ 0 & \mu \end{bmatrix}\begin{bmatrix} 0 & -1/\tau \\ \tau & 0 \end{bmatrix} h_2$$

$$= \eta^{-1}\left(\begin{bmatrix} \mu & 0 \\ 0 & \mu^{-1} \end{bmatrix}\right)\sigma(h_2^{-1})\begin{bmatrix} 0 & -1/\tau \\ \tau & 0 \end{bmatrix} h_2 ,$$

so that (3.16) holds with t_σ the class of

$$\sigma \longrightarrow \sigma(h_2^{-1})\begin{bmatrix} 0 & -1/\tau \\ \tau & 0 \end{bmatrix} h_2 .$$

Suppose now that L is a quadratic extension of F and that

(3.18)
$$\sigma_G = \begin{cases} \operatorname{ad}\begin{bmatrix} 0 & 1 \\ \zeta & 0 \end{bmatrix} \circ \sigma_{G^*} & \text{if } \sigma|_L \not\equiv 1 , \\ \sigma_{G^*} & \text{otherwise} , \end{cases}$$

where $\zeta \in F^\times$. Note that G is split over F if and only if $\zeta \in \mathrm{Nm}_F^L L^\times$. Assume that $T \in \mathcal{J}_L$. Now σ_G satisfies the conditions of (3.6). The coordinates x and y on $S(\mathbf{B}_*)$ satisfy $\sigma(x+y) = \zeta/y$, $\sigma(y) = \zeta/(x+y)$ if $\sigma|_L \not\equiv 1$ and $\sigma(x+y) = x + y$, $\sigma(y) = y$ otherwise. Then $x/(x+y) = 1 - y/(x+y)$ is defined over F. It serves as a coordinate around an F-rational point (g, s) of $Y(\mathbf{B}_*)$ near $\pi^{-1}(c_0)$. A calculation as in the last paragraph shows that:

(3.19)
$$m_\kappa((g, s)) = \kappa(x/(x+y)) ,$$

where κ now denotes the quadratic character of $F^\times$ attached to L/F if κ is nontrivial, and the trivial character otherwise.

4. APPLICATION

Continuing from the last section, we have $f \in C_c^\infty(G(F))$, Haar measures $|\omega_G|$ on G(F) and $|\omega_T|$ on T(F), and a character κ on $\mathfrak{K}(T)$. For γ regular in T(F), form the κ-orbital integral

$$\Phi^{\kappa}(\gamma, f) = \Phi_T^{\kappa}(\gamma, f, |\omega_T|, |\omega_G|)$$

$$= \sum_{\delta} \kappa(\delta) \int_{T^h(F)\backslash G(F)} f(g^{-1}h^{-1}\gamma h g) \frac{|\omega_G|}{|\omega_T|^h} ,$$

where $h \in \mathfrak{A}(T, F)$ represents $\delta \in \mathfrak{D}(T, F)$, and then the normalized integral

$$F^{\kappa}(\gamma, f) = |1 - \alpha(\gamma^{-1})| \Phi^{\kappa}(\gamma, f) .$$

Recall that T has been identified with $\mathbf{T}^*$ by means of η; α is the root of $\mathbf{T}^*$ in $\mathbf{B}^*$.

Assume that γ lies in $C(F)$ near c_0 and has coordinate λ. The Igusa data of the last section was chosen so that

$$F^{\kappa}(\gamma, f) = \int_{\phi_\gamma^{-1}(F)} f_\kappa \frac{|\omega|}{|d\lambda|} .$$

Thus for $|\lambda|$ sufficiently small we have that

(4.1) $$F^{\kappa}(\gamma, f) = |\lambda| \Lambda_1 + \kappa(\lambda)\Lambda_2$$

where, in the notation of Lemma 2A,

(4.2) $$\Lambda_1 = F_1(1, 2, f)$$

and

(4.3) $$\Lambda_2 = \begin{cases} F_1(\kappa, 1, f) & \text{if } G \text{ is split over } F \\ 0 & \text{otherwise.} \end{cases}$$

On the right side of (4.1) we have, as in §3, regarded κ as a character on $F^\times$, trivial if κ is trivial on $\mathfrak{D}(T)$ and the character on $F^\times$ attached to the quadratic splitting field L of T otherwise.

The term Λ_1 is the contribution from the divisor E_1 which maps to $\{c_0\}$ under $\pi : Y \longrightarrow G$, while Λ_2 is the contribution from E_2; under π, E_2 maps (isomorphically) to the conjugacy class of regular c_0-unipotent elements in G.

Thus (4.1) assumes a familiar form, but with the coefficients now expressed as principal value integrals.

If T is split over F then blowing up the Springer-Grothendieck variety M to obtain Y is unnecessary, and as a consequence we have introduced the spurious term Λ_1. It is quickly dismissed, for if T is split over F then $E_1 = \{c_0\} \times S$ is F-isomorphic to $\mathbf{P}^1 \times \mathbf{P}^1$ (cf. (3.4)) and by (2.10) Λ_1 is given, up to a constant, by

$$\oint_{\mathbf{P}^1(F)\times\mathbf{P}^1(F)} \frac{da\,db}{|a-b|^2}$$

where a, b each denote the coordinate U_0 on $U_1 = 1$ in $\mathbf{P}^1$. We apply Lemma 1.F to this integral and the fibering $\phi : \mathbf{P}^1 \times \mathbf{P}^1 \longrightarrow \mathbf{P}^1$ given by projection on the first component. The fiber integral $H(x')$, $x' \in \mathbf{P}^1(F)$, is seen to be an integral over $\mathbf{P}^1(F)$ of the form (1.15). Thus it is zero. We conclude then from Lemma 1.F that $\Lambda_1 = 0$.

LEMMA 4.A.

If κ is nontrivial then $\Lambda_1 = 0$.

Proof: If G is anisotropic over F then this is immediate from the definition of κ-orbital integral (see also the remark following the proof of Lemma 4.B). Suppose then that G is split over F. We may as well assume that $G = G^* = SL(2)$. Since κ is nontrivial $T \in \mathcal{J}_L$, some $L \neq F$; E_1 is the variety $R = \mathrm{Res}_F^L \mathbf{P}^1$ of Remark 1.H, i.e., $E_1(L) = \mathbf{P}^1(L) \times \mathbf{P}^1(L)$ and $E_1(F) = \{(p, \bar{p}) : p \in \mathbf{P}^1(L)\}$, where $^-$ denotes the action of the nontrivial element of Gal(L/F) (cf. (3.3), (3.4)). Then by (2.10), (3.2), (3.5) and (3.16) Λ_1 is, up to a constant,

$$\oint_{R(F)} \frac{\kappa(\frac{b-a}{\tau})\,da\,db}{|b-a|^2} \tag{4.4}$$

where a, b each denote the coordinate U_0 on $U_1 = 1$ in $\mathbf{P}^1(L)$. The element τ of L - F was fixed for (3.16); $\frac{b-a}{\tau}$ lies in $F^\times$ if $b = \bar{a} \neq 0$.

Abbreviate the point $U_0 = a$, $U_1 = 1$ in $\mathbf{P}^1$ by a, and $U_0 \neq 0$, $U_1 = 0$ by ∞. We define a smooth morphism $\phi^0 : R - \{(\infty, \infty)\} \longrightarrow \mathbf{P}^1$ by $(a, b) \longrightarrow \frac{b-a}{\tau}$ and (a, ∞), $(\infty, b) \longrightarrow \infty$; ϕ^0 is defined over F. We blow up R at (∞, ∞) to obtain the variety $\hat{R}$ over F. The fiber over (∞, ∞) in $\hat{R}$ meets the proper inverse image of the divisor $a = b$ at a single point p_0. Blow up $\hat{R}$ at this point, which is F-rational, to obtain the variety $\hat{\hat{R}}$ over F. A calculation with coordinates shows that ϕ^0 extends to an Igusa fibering

$\phi : \hat{\hat{R}} \longrightarrow \mathbf{P}^1$ with distinguished point $c_0 = \infty$; moreover, the fiber over ∞ is the union of three divisors, each occurring with multiplicity one. Only one of the divisors has F-rational points. We conclude then that on $\hat{\hat{R}}(F)$ the map ϕ is smooth.

To compute (4.4) by lifting to $\hat{\hat{R}}(F)$ we must check the conditions of Lemma 1.D at (∞, ∞) on $R(F)$ and at p_0 on $\hat{R}(F)$. We find that $\theta_1 = \kappa$, $c_1 = -1$ and $\theta_2 \equiv 1$, $c_2 = 1$ for a suitable choice of local coordinates around (∞, ∞) on $R(F)$. Thus $\theta_1\theta_2 = \kappa$ and $c_1 + c_2 = 0$, and so to lift to $\hat{R}(F)$ it is crucial that κ be nontrivial. On $\hat{R}(F)$ at p_0 we find $\theta_1 = \kappa$, $c_1 = 0$ and $\theta_2 = \kappa$, $c_2 = -1$. The conditions of Lemma 1.D are met and we may apply Corollary 1.E to rewrite (4.4) as a principal value integral I over $\hat{\hat{R}}(F)$. We compute I by applying Lemma 1.F to the fibering ϕ. For any $p \in \mathbf{P}^1(F) - \{0, \infty\}$ the corresponding fiber integral is seen immediately to be a constant times $\int_{\mathbf{P}^1(F)} da = 0$ (cf. (1.15)). Thus Lemma 1.F implies that $I = 0$, and Lemma 4.A is proved.

Suppose now that $T \in \mathfrak{J}_L$, $L \neq F$; κ may be either character on $\mathfrak{G}(T)$. Recall that the maximal torus $T^* = \psi(T)$ in G^* is defined over F, as is the map $\psi : T \longrightarrow T^*$. Let κ^* be the character on $\mathfrak{G}(T^*)$ associated to κ by ψ. We indicate by Λ_1^* the contribution (4.2) for the data G^*, T^*, κ^* and $f^* \in C_c^\infty(G^*(F))$. It may be written as $M_1^* f^*(c_0)$. Similarly the contribution (4.2) for the data G, T, κ and f can be written as $M_1 f(c_0)$.

LEMMA 4.B.

$$M_1 = \varepsilon(\kappa, G) M_1^*$$

where $\varepsilon(\kappa, G) = 1$ if κ is nontrivial and

$$\varepsilon(1, G) = \begin{cases} 1 & \text{if } G \text{ is split over } F \\ -1 & \text{otherwise.} \end{cases}$$

Proof: We may assume that G satisfies (3.18), i.e.,

$$\sigma_G = \begin{cases} \mathrm{ad}\begin{bmatrix} 0 & 1 \\ \zeta & 0 \end{bmatrix} \circ \sigma_{G^*} & \text{if } \sigma|_L \not\equiv 1 \\ \sigma_{G^*} & \text{otherwise} \end{cases},$$

where $\zeta \in F^\times$; G is split over F if and only if $\zeta \in \mathrm{Nm}_F^L L^\times$. Write $M_1(\zeta)$ for he term M_1. Then from (2.10), (3.2), (3.6) and (3.19) we find that $M_1(\zeta)$ is given up to a constant independent of G (i.e., of ζ) by:

$$(4.5)\qquad \oint_{Q_\zeta(F)} \frac{\kappa(1-\frac{a}{b})\ da\ db}{|b-a|^2}$$

where Q_ζ is the form of $\mathbf{P}^1 \times \mathbf{P}^1$ on which $1 \neq \sigma \in \mathrm{Gal}(L/F)$ acts on the homogeneous coordinates U_0, U_1 (on the first copy of $\mathbf{P}^1$) and V_0, V_1 (on the second copy) by $U_0 \longrightarrow \zeta V_1$, $U_1 \longrightarrow V_0$, $V_0 \longrightarrow \zeta U_1$, $V_1 \longrightarrow U_0$. Also a denotes U_0 on $U_1 = 1$ and $b = V_0$ on $V_1 = 1$.

We define now a smooth variety Y and an Igusa figering $\phi : Y \longrightarrow \mathbf{A}^1$ with distinguished point zero on $\mathbf{A}^1$ such that if ζ is the coordinate on $\mathbf{A}^1$ then (4.5) is the fiber integral $F(\zeta)$, $\zeta \neq 0$. The asymptotic expansion for $F(\zeta)$ at $\zeta = 0$ will be seen to have the one term, that corresponding to $\theta = \theta_L\kappa$, where θ_L is the quadratic character of $F^\times$ attached to L/F, and $\beta = 1$. Then in the notation of Lemma 2.A we have

$$(4.6)\qquad F(\zeta) = \theta_L(\zeta)\kappa(\zeta)F_1(\theta_L\kappa,\ 1,\ *)$$

for $|\zeta|$ sufficiently small, where * is the Igusa integrand yet to be defined. This will prove the lemma.

We start with a variety $Y_1 \subset (\mathbf{P}^1)^4 \times \mathbf{A}^1$. Let a, b, a_1, b_1 each denote the coordinate U_0 on $U_1 = 1$ in $\mathbf{P}^1$, and ζ be the coordinate on $\mathbf{A}^1$. On $(U_1 = 1)^4 \times \mathbf{A}^1$, Y_1 is given by $ab_1 = a_1b = \zeta$. Let a', b', a_1', b_1' each denote U_1 on $U_0 = 1$. On $(U_0 = 1) \times (U_1 = 1)^3 \times \mathbf{A}^1$, Y_1 is given by $a_1b = \zeta$ and $b_1 = a'\zeta$; on $(U_0 = 1)^2 \times (U_1 = 1)^2 \times \mathbf{A}^1$ by $b_1 = a'\zeta$, $a_1 = b'\zeta$, and so on. We define Y_1 over F by twisting the natural F-structure by $\sigma(a) = a_1$, $\sigma(b) = b_1$, $\sigma(a') = a_1'$, $\sigma(b') = b_1'$, σ being the nontrivial element of $\mathrm{Gal}(L/F)$.

The variety Y_1 is smooth except at the point y_1 given by $a = a_1 = b = b_1 = 0$. Let ϕ_1 be the projection of $Y_1 \subset (\mathbf{P}^1)^4 \times \mathbf{A}^1$ onto $\mathbf{A}^1$; $\phi_1 : Y_1 - \{y_1\} \longrightarrow \mathbf{A}^1$ is an Igusa fibering with distinguished point zero.

For $\zeta \neq 0$ the projection of Y_1 onto the product of the first and second copies of $\mathbf{P}^1$ yields an F-isomorphism of the fiber $\phi_1^{-1}(\zeta)$ with Q_ζ. Let $Y_1' = Y_1 - \phi_1^{-1}(0)$. We define a form ω_1 on Y_1' by $\dfrac{da \wedge db \wedge d\zeta}{(b-a)^2}$ and an integrand $f_1 = \kappa(1-\frac{a}{b})$ on $Y_1'(F)$. The fiber integral

$$(4.7)\qquad \oint_{\phi_1^{-1}(\zeta)(F)} f_1 \frac{|\omega_1|}{|d\zeta|}$$

is (4.5).

We now attend to the fiber in $Y_1 - \{y_1\}$ over $\zeta = 0$. It is the union of divisors $E_1', \ldots, E_4'$. Their branches on $((U_1 = 1)^4 \times \mathbf{A}^1) \cap (Y_1 - \{y_1\})$ are:

$$(E_1') \quad b = 0,\ b_1 = 0 \qquad (E_2') \quad a = 0,\ a_1 = 0$$

$$(E_3') \quad a = 0,\ b = 0 \qquad (E_4') \quad a_1 = 0,\ b_1 = 0 \ .$$

Note that E_3', E_4' have no F-rational points; E_1', E_2' are each defined over F.

The point y_1 on Y_1 is F-rational. Blow up Y_1 at this point to obtain the variety Y over F and projection $\pi : Y \longrightarrow Y_1$. Set $\phi = \phi_1 \circ \pi$, $\omega = \pi^*(\omega_1)$ and $f_Y = f_1 \circ \pi$. Then Y, $C = \mathbf{A}^1$, $c_0 = 0$, ϕ, ω and f_Y satisfy the conditions of (2.1) - (2.7) (i.e., are "generalized" Igusa data), as well as (2.9). The proof is routine. We will include as much of it as will be needed to write down the asymptotic expansion for the fiber integral which, by construction, coincides with the integral (4.7).

The fiber $\phi^{-1}(0)$ is the union of five divisors $E_0, E_1, \ldots, E_4$, where E_i is the proper inverse image of E_i' $(i = 1, \ldots, 4)$. Let $u_1 = a$, $u_2 = b$, $u_3 = a_1$, $u_4 = b_1$; let $U_1 = A$, $U_2 = B$, $U_3 = A_1$, $U_4 = B_1$ be homogeneous coordinates on $\mathbf{P}^3$. Then Y is given near $\pi^{-1}(y_1)$ by $u_iU_j = u_jU_i$ $(i, j = 1, \ldots, r)$. The divisor E_0 is given by $a = b = a_1 = b_1 = 0$ and $AB_1 = BA_1$ (homogeneous coordinates); on $E_0 \cap E_1$ we have $B = 0$, $B_1 = 0$; on $E_0 \cap E_2$, $A = 0$ and $A_1 = 0$, and so on. The divisors E_0, E_1 and E_2 are each defined over F, while E_3 and E_4 have no F-rational points and so may be ignored for the asymptotic expansion. Also $E_1 \cap E_2$ is empty and $E_i \cap E_j$ $(i = 1, 2$ and $j = 3, 4)$ consists of a single point on E_0 which is not F-rational.

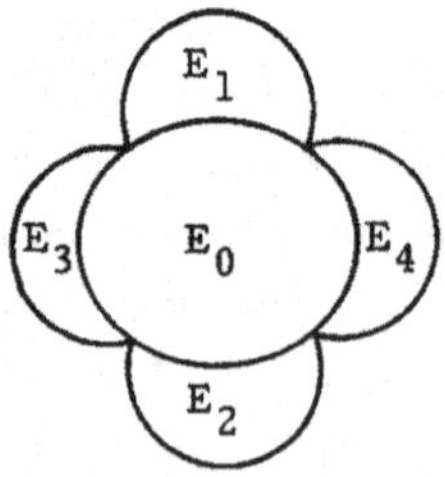

The variety E_0 is a form of $\mathbf{P}^1 \times \mathbf{P}^1$, the natural projections being given by

$(A, B, A_1, B_1) \longrightarrow \frac{B}{A} = \frac{B_1}{A_1}, \frac{A_1}{A} = \frac{B_1}{B}$ (where we allow the value ∞ and ignore quotients of the form $\frac{0}{0}$). For these to be defined over F, the first $\mathbf{P}^1$ has to be provided with its natural F-structure and the second with the structure of Remark 1.G. The variety E_1 is the blow-up $\overline{R}$ of the twisted form R of $\mathbf{P}^1 \times \mathbf{P}^1$ described in Remark 1.H (cf. also (4.4)). To see this, we note that the projection of $Y_1 \subset (\mathbf{P}^1)^4 \times \mathbf{A}^1$ onto the product of the first and third copies of $\mathbf{P}^1$ yields an F-isomorphism of E_1' with $R - \{r_0\}$, where r_0 is given by $a = a_1 = 0$. Then E_1 is the blow-up $\overline{R}$ of R at r_0, and $E_0 \cap E_1$ is the inverse image of r_0 in $\overline{R}$. The divisor E_2 is described similarly.

Suppose that $y_0 \in E_0(F)$ and that $A \neq 0$ at y_0. We may assume $A = 1$. Then $t = a_1, A_1, B$ serve as coordinates on Y near y_0; $a_1 = tA_1$ and $b = tB$. For $1 \neq \sigma \in \mathrm{Gal}(L/F)$ we have $\sigma(B) = B$, $A_1\sigma(A_1) = 1$ and $\sigma(t) = tA_1$. To obtain F-coordinates we may take B, r and s with $t = st_0$, $t_0 \neq 0$, $\sigma(t_0)/t_0 = A_1$ and $t_0 = 1 + \tau r$, where $\tau \in L - F$ and $\tau^2 \in F$. Also, $t = 0$ is a branch of E_0, $A_1 \neq 0$ since $A_1\sigma(A_1) = 1$, and $B = 0$ is a branch of E_1. Finally, $\zeta = a_1 b = t^2 A_1 B$,

$$\omega = \frac{da \wedge db \wedge d\zeta}{(b-a)^2} = tB\,\frac{dt \wedge dB \wedge dA_1}{(B-1)^2}$$

and $f_Y = \kappa(1 - 1/B) = \kappa(B)\kappa(B-1)$. We conclude that

$$\beta(E_0) = \frac{b(E_0)}{a(E_0)} = \frac{2}{2} = 1, \ \kappa(E_0) \equiv 1 \quad .$$

Similarly,

$$\beta(E_1) = \frac{2}{1} = 2, \ \kappa(E_1) = \kappa$$

and

$$\beta(E_2) = \frac{2}{1} = 2, \ \kappa(E_2) \equiv 1 \quad .$$

This implies that (2.9) is satisfied (i.e., $e(\theta, \beta) \leq 1$ for all (θ, β)) since E_1 and E_2 do not intersect.

The asymptotic expansion for the fiber integral, i.e., for the integral (4.5), is then

(4.8) $$\sum_{\theta} \theta(\zeta)F_1(\theta, 1, f_Y) + |\zeta|\kappa(\zeta)F_1(\kappa, 2, f_Y) + |\zeta|F_1(1, 2, f_Y)$$

if κ is nontrivial, or

$$(4.9) \qquad \sum_{\theta} \theta(\zeta) F_1(\theta, 1, f_Y) + |\zeta| F_1(1, 2, f_Y)$$

if κ is trivial. The summation is over characters θ of $F^\times$ for which $\theta^2 = 1$.

The integrals $F_1(\kappa, 2, f_Y)$ and $F_1(1, 2, f_Y)$ of (4.8) and the two integrals contributing to $F_1(1, 2, f_Y)$ in (4.9) (cf. (2.10)) are each of the form (1.20) and hence vanish.

Since $\zeta = s^2(t_0^2 A_1 B)$ and $t_0^2 A_1 B = t_0 \sigma(t_0) B$ the formula (2.10) yields

$$(4.10) \qquad F_1(\theta, 1, f_Y) = \oint_{E_0(F)} \frac{\kappa(B)}{\theta(B)} \frac{\kappa(B-1)}{|B-1|^2} \frac{1}{\hat{\theta}(A_1)} \frac{dA_1}{|A_1|} dB ,$$

where $\hat{\theta}(A_1) = \theta(t_0 \sigma(t_0))$. Recall that A_1 ranges over $\{x \in L^\times : Nm_F^L x = 1\}$. The character $\hat{\theta}$ is trivial if and only if $\theta \equiv \kappa$ or $\theta \equiv \kappa\theta_L$. In the case $\hat{\theta} \not\equiv 1$,

$$\int \frac{dA_1}{\hat{\theta}(A_1)|A_1|} = 0 \quad \text{(cf. Remark 1.G)}$$

and so (4.10) vanishes. In the case $\theta \equiv \kappa$,

$$\oint_{\mathbf{P}^1(F)} \frac{\kappa(B-1)}{|B-1|^2} dB = \oint_{\mathbf{P}^1(F)} \frac{\kappa(B)}{|B|^2} dB = 0 \quad \text{(Lemma 1.C)}$$

and again (4.10) is zero. Thus only $\theta \equiv \kappa\theta_L$ may give a nonzero contribution. The expansions (4.8) and (4.9) therefore take the form (4.6), and Lemma 4.B is proved.

Lemma 4.B and Lemma 4.A in the case G split over F imply Lemma 4.A for G anisotropic over F.

REFERENCES

[L-L] J.-P. Labesse and R. Langlands, L-indistinguishability for SL(2), Canad. J. Math., vol. 31 (1979), pp. 726-785.

[L] R. Langlands, Orbital integrals on forms of SL(3), I, Amer. J. Math., vol. 105 (1983), to appear.

[L_1] ——————, Les débuts d'une formule des traces stable, Publ. Math. Univ. Paris VII, vol. 13 (1983).

[R] J. Repka, Shalika's germs for p-adic GL(n) II: the subregular term, preprint.

[V] M.-F. Vignéras, Caractérisation des integrales orbitales sur un groupe réductif p-adique, J. Fac. Sci. Univ. Tokyo, vol. 28 (1982), pp. 945-961.

School of Mathematics
The Institute for Advanced Study
Princeton, NJ 08540

WORK OF WALDSPURGER

Ilya Piatetski-Shapiro
Yale University
and
University of Tel Aviv

§0. INTRODUCTION

The work of Waldspurger [1-5] is devoted to a very deep study of the automorphic forms on $\overline{SL}_2$. The main tool for such a study is the correspondence between automorphic forms on $\overline{SL}_2$ and automorphic forms on PGL_2. This correspondence was first discovered by Shintani and Niwa using the Weil representation. An earlier approach to this correspondence, based on L-functions, was suggested by Shimura [10]. Indeed, Shimura's work seemed to stimulate Shintani's and Niwa's work on the subject.

R. Howe has outlined a general theory of duality correspondence based on the use of the Weil representation. He has introduced the general notion of a dual reductive pair, and has defined both a local and global duality correspondence. R. Howe has obtained many deep results in the general situation; but many important problems remain [6].

A systematic study of the duality correspondence for the simplest dual reductive pair $\overline{SL}_2$, PGL_2 from the point of view of representation theory has been carried out by Rallis and Schiffmann [8]. In his work, Waldspurger refers in many places to Rallis and Schiffmann, and in a way, Waldspurger's work is a continuation of that of Rallis and Schiffmann. However, I would like to emphasize that Waldspurger's work contains many fundamental new ideas especially in the global case.

Flicker has studied a correspondence between the automorphic forms of GL_2 and those of $\overline{GL}_2$ using the trace formula [16]. He has in fact obtained a complete description of this correspondence. Since $\overline{SL}_2$ is a subgroup of $\overline{GL}_2$, there is a close connection between the automorphic forms of these two groups. Waldspurger has used Flicker's results in a substantial way to obtain his own results. However, let me say that Waldspurger's results for $\overline{SL}_2$ are quite surprising and were not predicted from the results for $\overline{GL}_2$. It remains a mystery to me why the automorphic forms on $\overline{SL}_2$ and $\overline{GL}_2$ behave so differently. For example, strong multiplicity one is true for $\overline{GL}_2$ but not for $\overline{SL}_2$. Also, the descent (correspondence) of automorphic forms from GL_2 to $\overline{SL}_2$ has only a local obstruction, while the correspondence from PGL_2 to $\overline{SL}_2$ has a global obstruction, but no local obstruction.

Let me also mention work [13], [14] which deals with L-functions for $\overline{GL}_2$. This work can be considered as an adelization of Shimura's work. It establishes an injection of the automorphic representations of $\overline{GL}_2$ into those of GL_2.

In this talk, I would like to explain Waldspurger's work in the framework of representation theory. I will explain all of Waldspurger's work except [2], which deals with the Fourier coefficients of automorphic forms of half-integral weight. This latter work, which is based on the material explained here, is very important for number theory, but lies outside the framework of this talk. Despite the fact that I have omitted many local proofs, I hope this talk will be useful to the mathematical community. A beautiful exposition of Waldspurger's work from the classical point of view has been given in a talk bv Marie-France Vigneras [9].

§1. AUTOMORPHIC FORMS ON $\overline{SL_2(\mathbb{A})}$

Let k be a global field. The adele group $SL_2(\mathbb{A})$ has a unique non-trivial two-fold covering $\overline{SL_2(\mathbb{A})}$:

$$1 \to \{\pm 1\} \to \overline{SL_2(\mathbb{A})} \to SL_2(\mathbb{A}) \to 1.$$

There is a unique embedding of $SL_2(k)$ into $\overline{SL_2(\mathbb{A})}$ such that the following diagram commutes.

$$\begin{array}{ccc} & \nearrow & \overline{SL(\mathbb{A})} \\ & & \downarrow \\ SL_2(k) & \rightarrow & SL_2(\mathbb{A}) \end{array}$$

This means the covering splits over $SL_2(k)$. Similarly, there is an embedding of $Z(\mathbb{A})$ into $\overline{SL_2(\mathbb{A})}$, where Z is the upper unipotent subgroup of SL_2.

Let A_0 denote the space of genuine cuspidal functions on $\overline{SL_2(\mathbb{A})}$. In particular, if $f \in A_0$, then

i) $f(\xi\gamma g) = \xi f(g)$ $\qquad (\xi \in \{\pm 1\},\ \gamma \in SL_2(k),\ g \in \overline{SL_2(\mathbb{A})})$,

ii) $\displaystyle\int_{k\backslash\mathbb{A}} f(\begin{pmatrix}1 & z\\ 0 & 1\end{pmatrix}g)\,dz = 0.$

Under right translation, A_0 decomposes discretely into a countable number of irreducible subspaces. An irreducible representation of $\overline{SL_2(\mathbb{A})}$ which occurs in A_0 is called a genuine automorphic cuspidal representation. Let A_{00} denote the subspace of forms in A_0 orthogonal to the Weil representations of $\overline{SL_2(\mathbb{A})}$.

Theorem 1.1 (Multiplicity One) [1]. The multiplicity of an irreducible genuine automorphic cuspidal representation in A_{00} is one.

Remark. If σ is a genuine irreducible automorphic cuspidal representation lying in a Weil representation of $\overline{SL_2(\mathbb{A})}$, then multiplicity one is obvious.

If ψ is a character of $k\backslash\mathbb{A}$, and $f \in A_{00}$, the ψ-Fourier coefficient of f is defined to be

$$f_\psi(g) = \int_{k\backslash\mathbb{A}} \psi(z) f(\begin{pmatrix}1 & z\\ 0 & 1\end{pmatrix}g)\,dz \qquad (g \in \overline{SL_2(\mathbb{A})}).$$

The multiplicity result follows from the uniqueness of Whittaker models for $\overline{SL_2(\mathbb{A})}$, and the following result of Waldspurger.

heorem 1.2 [2,5]. Let (σ,V) be a genuine irreducible automorphic uspidal representation of $\overline{SL_2(\mathbb{A})}$. If $v \to \varphi(v)$ $(v \in V,\ \varphi(v) \in A_{00})$ s an embedding of (σ,V) into A_{00}, then the vanishing of the -Fourier coefficient $\varphi(v)_\psi$ depends only on (π,V) as an abstract epresentation, and not on the embedding φ.

roof of the multiplicity one. Suppose $v \to \varphi'(v)$ and $v \to \varphi''(v)$ $(v \in V)$ are two distinct embeddings of an irreducible genuine auto- orphic cuspidal representation (σ,V) into A_0. We may select a char- cter ψ of $k\backslash\mathbb{A}$ so that the ψ-Fourier coefficient $\varphi'(v)_\psi$ does ot vanish for some $v \in V$. Let us consider the ψ-Fourier coefficient $''(v)_\psi$. If $\varphi''(v)_\psi$ vanishes, then Theorem 1.2 says $\varphi'(v)_\psi$ must also anish, a contradiction. If $\varphi''(v)_\psi$ does not vanish, then the unique- ess of Whittaker models for $\overline{SL_2(\mathbb{A})}$ tells us that $\varphi''(v)_\psi = c\varphi'(v)_\psi$ or some constant c. Since φ' and φ'' are assumed to be distinct mbeddings of (σ,V) into A_{00}, the map $w \to \varphi''(w) - c\varphi'(w)$ is a on-trivial embedding of (π,V) into A_{00}. The ψ-Fourier coefficient f $\varphi''(v) - c\varphi'(v)$ vanishes. This again contradicts Theorem 1.2; herefore (σ,V) must occur in A_{00} with multiplicity one.

Two irreducible genuine automorphic cuspidal representations of $\overline{L_2(\mathbb{A})}$, $\sigma = \underset{v}{\otimes}\,\sigma_v$ and $\varphi' = \underset{v}{\otimes}\,\sigma'_v$, are said to be nearly equivalent f $\sigma_v \simeq \sigma'_v$ for almost all places v. Let $\ell(\sigma)$ denote the set of rreducible genuine automorphic cuspidal representations nearly equiv- lent to σ. $\ell(\sigma)$, of course, just measures departure from strong ultiplicity one. In order to determine the set $\ell(\sigma)$, Waldspurger as defined an involution $\sigma \to \sigma^W$ whenever σ is a discrete series epresentation of $\overline{SL_2(k_v)}$. If $\sigma = \underset{v}{\otimes}\,\sigma_v \in A_{00}$, define

$$\Sigma = \{v \mid \sigma_v \text{ is a discrete series representation}\}\ .$$

f $M \subset \Sigma$, and $|M|$ is even, put

$$\sigma^M = \otimes \sigma_v^M \quad \text{where} \quad \sigma_v^M = \begin{cases} \sigma_v & \text{if } v \notin M \\ \sigma_v^W & \text{if } v \in M. \end{cases}$$

The relationship of the σ^M's and $\ell(\sigma)$ is given in the following theorem.

<u>Theorem 1.3</u> [3]. Any representation in $\ell(\sigma)$ is of the form σ^M for some $M \subseteq \Sigma$.

<u>Corollary 1.4</u>. $|\ell(\sigma)| = 2^{|\Sigma|-1}$.

Remark: Recall that $\begin{pmatrix}-1 & 0\\ 0 & -1\end{pmatrix}$ lies in the center of $\overline{SL_2(k_v)}$. Waldspurger has shown that $\sigma_v^W\begin{pmatrix}-1 & 0\\ 0 & -1\end{pmatrix} = -\sigma_v\begin{pmatrix}-1 & 0\\ 0 & -1\end{pmatrix}$. Since $\begin{pmatrix}-1 & 0\\ 0 & -1\end{pmatrix} \in SL_2(k)$, it follows that if $M \subseteq \Sigma$ has an odd number of elements, σ^M cannot be an automorphic representation.

§2. THE OSCILLATOR REPRESENTATION OVER A LOCAL FIELD

Let k be a local field, and let X be a 2n-dimensional vector space over k with a symplectic form $<\,,\,>$. If $X = X_1 \oplus X_2$ is a polarization of X, let P be the subgroup of $Sp(<,>)$ which preserves X_2. If ψ is a non-trivial character of k, let ω_ψ be the oscillator representation of $\overline{Sp_{2n}(k)} = \overline{Sp(<,>)}$, the double cover of $Sp_{2n}(k) = Sp(<,>)$. ω_ψ acts on the Schwartz-Bruhat space $S(X_1)$.

Let us now consider the 3-dimensional vector space $M = \{m \in M_2(k) | tr(m) = 0\}$. PGL_2 acts on M by conjugation:

$$m \mapsto g^{-1}mg \qquad (g \in PGL_2,\ m \in M).$$

This conjugation action preserves the symmetric form $q(x) = -\det(x)$. Let Y be a 2-dimensional vector space over k with a symplectic form $<,>$. Define a symplectic vector space X by $X = M \otimes_k Y$, $<m_1 \otimes y_1, m_2 \otimes y_2> = (m_1,m_2)<y_1,y_2>$. Since PGL_2 and SL_2 preserve the forms $(\,,\,)$ and $<\,,\,>$ respectively, there is a natural embedding of $PGL_2 \times SL_2$ into $Sp(<,>) = Sp_6$. Our aim is to use the oscillator representation of $\overline{Sp_6}$ to define a correspondence between certain irreducible representations of PGL_2 and certain irreducible representations of $\overline{SL_2}$. Waldspurger has given a different definition of the correspondence based on explicit integral formulas. These integral formulas, though complicated and defined only for the case PGL_2, $\overline{SL_2}$, yield much more information about the correspondence.

Let T be a subgroup of $G = PGL_2$ and let N be a subgroup of $H = \overline{SL_2}$. Let α and β be characters of T and N respectively. Let $X = X_1 \oplus X_2$ be a polarization of X such that $T \times N \subset P$. Let us suppose that $x_1 \in X_1$ is a vector such that the linear functional

$$\phi \to \phi(x_1) \qquad\qquad (\phi \in S(X_1))$$

transforms under $T \times N$ by $\alpha \times \beta$, i.e.,

$$\omega_\psi(t,n)\cdot\phi(x_1) = \alpha(t)\beta(n)\phi(x_1).$$

Let (π,V) be an irreducible admissible representation of PGL_2 and let us assume that ℓ is a linear functional on V such that $\ell(\pi(t)v) = \alpha^{-1}(t)\ell(v)$ $(t \in T)$. If the integral

$$F(h) = \int_{T\backslash G} (\omega_\psi(g,h)\cdot\phi)(x_1)\,\ell(\pi(g)v)\,dg \qquad\qquad (h \in H)$$

converges, then $F(nh) = \beta(n)F(h)$ $(n \in N)$. Let W be the space of all the functions F obtained in this fashion by varying ϕ and v. $\overline{SL_2}$ acts on W by right translation. We shall denote this representation by $\theta(\pi,\psi)$. Conversely, given an irreducible admissible genuine representation σ of $\overline{SL_2}$ it is possible to define a representation $\theta(\sigma,\psi)$ of PGL_2, which may be a zero representation.

In order to explain Waldspurger's integral formulas for the correspondence, we have to consider two polarizations of X. For the first polarization, let $y_1, y_2 \in Y$ be a symplectic basis, i.e., $\langle y_1,y_2\rangle = 1$, and put $X_1 = M \otimes y_1$, $X_2 = M \otimes y_2$. Let m_1 be an element of M such that $\det m_1 \neq 0$ and let $T = \operatorname{Stab} m_1$. T is a torus in G. Let N

be the unipotent subgroup of SL_2 which preserves y_2. Let α be the trivial character, and β the character $\beta\begin{pmatrix}1 & n\\ 0 & 1\end{pmatrix} = \psi(q(m_1)n)$. We shall now describe the second polarization which has the property that the unipotent subgroups of PGL_2 and $\overline{SL}_2$ both lie in P. Let e_1, e_2, e_3 be a basis of M such that the matrix of the symmetric form is

$$\begin{pmatrix} 0 & 0 & 1\\ 0 & 1 & 0\\ 1 & 0 & 0 \end{pmatrix}$$

Define $X_1 = e_1 \otimes Y + e_2 \otimes ky_1$, and $X_2 = e_3 \otimes Y + e_2 \otimes ky_2$. It is clear that the unipotent subgroup of $G = PGL_2$ which preserves e_3 also preserves X_2. We shall denote this subgroup by T. Similarly, the unipotent subgroup N of SL_2 which preserves y_2 preserves X_2. Let $x_1 = e_1 \otimes y_2 + \lambda e_2 \otimes y_1$ and define α and β by

$$\alpha\begin{pmatrix}1 & t\\ 0 & 1\end{pmatrix} = \psi(-\lambda t)$$

$$\beta\begin{pmatrix}1 & n\\ 0 & 1\end{pmatrix} = \psi(\lambda^2 n).$$

Waldspurger has proved the following theorems:

<u>Theorem 2.1</u> [1]. Let T and N be as above. If (π, V) (respectively (σ, V)) is an irreducible admissible representation of PGL_2 (respectively $\overline{SL}_2$), then the representation of $\overline{SL}_2$ (respectively PGL_2) obtained from the above integral formulas is irreducible admissible and depends only on the additive character ψ. It is independent of the choice of the subgroups T and N and the characters α and β.

<u>Theorem 2.2</u> [3]. Let $\xi \in k^\times$, and let χ_ξ be the quadratic character of $k^\times$ associated to $k(\sqrt{\xi})$. If $\theta(\sigma, \psi^\xi)$ and $\theta(\sigma, \psi)$ are both non-zero representations of PGL_2, then $\theta(\sigma, \psi^\xi) = \theta(\sigma, \psi) \otimes \chi_\xi$.

Remarks: $\theta(\pi, \psi)$ is non-zero for any irreducible admissible representation π of PGL_2 and any ψ. It follows from this that any irreducible admissible representation of PGL_2 admits a linear functional

which is invariant with respect to the split torus. $\theta(\sigma,\psi)$ is non-
ero if and only if σ admits a linear functional which transforms
under N by ψ^{-1}.

Let us now make a few remarks about a similar construction for the
uaternion algebra D over k. Let M' be the elements of trace zero
n D, and let q be the symmetric form on M' given by $q(m) =$
norm(m). $PD^{\times}$ acts on M' by conjugation, and this action preserves
he form q. We can introduce a symplectic space $X' = M' \otimes_k Y$ and as
bove, we have an embedding $PD^{\times} \times SL_2 \hookrightarrow Sp_6$. In an analogous fashion,
e can also introduce integral formulas to describe a correspondence
etween some of the irreducible admissible representations of $PD^{\times}$ and
ome of the irreducible genuine admissible representations of $\overline{SL}_2$.
he analogues of Theorems 2.1 and 2.2 are also true for the quaternion
lgebra. If σ (respectively π) is an irreducible admissible re-
resentation of $\overline{SL}_2$ (respectively $PD^{\times}$), we shall denote the corres-
onding representation of PGL_2 (respectively $\overline{SL}_2$) by $\theta'(\sigma,\phi)$
respectively $\theta'(\pi,\psi)$).

From the explicit integral formulas, it is easy to show that
$'(\pi,\psi)$ does not admit a linear functional which transforms under N
y ψ^{-1}. This together with the remark after Theorem 2.2 implies that
he representations $\theta(\sigma,\psi)$ and $\theta'(\sigma,\psi)$ cannot both be non-zero
epresentations. However, Waldspurger has the following result.

heorem 2.3 [3]. One of the representations $\theta(\sigma,\psi)$ and $\theta'(\sigma,\psi)$ is
lways non-zero.

laim: $\theta'(\pi',\psi')$ is non-zero if and only if π' is a spherical re-
resentation, i.e., π' possesses a T-invariant vector for some torus
$\subset PD^{\times}$.

roof (Waldspurger). Consider for the moment, an irreducible admis-
ible representation π of PGL_2. If χ is a quadratic character of
$^{\times}$, we define the Waldspurger symbol as follows. Let $\varepsilon(\pi,s,\psi)$ be
he ε-factor introduced in [17]. It is easy to check that $\varepsilon(\pi,\frac{1}{2},\psi) =$

± 1 does not depend on ψ. Let $\varepsilon(\pi,\frac{1}{2})$ denote $\varepsilon(\pi,\frac{1}{2},\psi)$. We then define $(\frac{\chi}{\pi})$ by

$$\varepsilon(\pi \otimes \chi,\tfrac{1}{2}) = (\tfrac{\chi}{\pi})\chi(-1)\varepsilon(\pi,\tfrac{1}{2}).$$

$(\frac{\chi}{\pi}) = \pm 1$, and if χ is the trivial character, then $(\frac{\chi}{\pi}) = 1$. It is easy to see that if π is an irreducible principal series representation, then $(\frac{\chi}{\pi}) = 1$ for all χ. On the other hand, if π is a discrete series representation, then there exists a χ such that $(\frac{\chi}{\pi}) = -1$ [3]. Let us now return to the proof of the claim. Let π be the discrete series representation of PGL_2 associated to π' under the Jacquet-Langlands map. Let χ be a quadratic character of $k^{\times}$ such that $(\frac{\chi}{\pi}) = -1$, and denote by $K = k(\sqrt{\xi})$ the field corresponding to χ. Put

$$\sigma_1 = \theta(\pi \otimes \chi,\psi^{\xi}), \qquad \sigma = \theta(\pi,\psi).$$

Waldspurger has proved that

$$\sigma_1\begin{pmatrix}-1 & 0\\ 0 & -1\end{pmatrix} = (\tfrac{\chi}{\pi})\sigma\begin{pmatrix}-1 & 0\\ 0 & -1\end{pmatrix}.$$

Since $(\frac{\chi}{\pi}) = -1$, $\sigma_1 \neq \sigma$. This means that σ does not admit a ψ^{ξ}-linear functional, for if it did, then $\theta(\sigma,(\psi^{\xi})^{-1}) \neq 0$, so $\theta(\sigma_1(\psi^{\xi})^{-1}) = \theta(\sigma,\psi^{-1}) \otimes \chi_{\xi}$ which would imply $\sigma_1 = \sigma$, a contradiction. Now, Theorem 2.3 tells us that $\theta'(\sigma,(\psi^{\xi})^{-1}) \neq 0$ and so $\theta(\pi',\psi^{\xi}) \not\equiv 0$ which means π' is spherical.

The next theorem defines Waldspurger's involution.

Theorem 2.4 [3]. Let σ be an irreducible representation of the discrete series of $\overline{SL}_2$, and let ψ be a character of k such that $\theta(\sigma,\psi) \not\equiv 0$.

The composition of the following 3 maps

$$\sigma \to \theta(\sigma,\psi) = \pi \xrightarrow{JL} \pi' \to \theta(\pi',\psi^{-1})$$

(where JL means the Jacquet-Langlands map) is independent of ψ and defines an involution.

Finally, it is not difficult to prove

Theorem 2.5. If $\pi = \theta(\sigma,\psi) \not\equiv 0$, then

$$\theta(\pi \otimes \chi_\xi, \psi^\xi) = \begin{cases} \sigma & \text{if } (\frac{\chi_\xi}{\pi}) = 1 \\ \sigma^W & \text{if } (\frac{\chi_\xi}{\pi}) = -1 \end{cases}$$

where χ_ξ is the character associated to $k(\sqrt{\xi})$.

§3. THE θ-CORRESPONDENCE

Let k be a global field. We shall use the same notation globally as was previously introduced locally. The global Weil (oscillator) representation w_ψ acts on $S(X_1(\mathbb{A}))$. It is easy to see that it is the tensor product of the local Weil representations. Let $\mathbb{X} = X_1 \oplus X_2$ be the standard polarization of X, and identify X_1 with M. For $\phi \in S(X_1(\mathbb{A}))$,

$$\vartheta^\phi_\psi(g,h) = \sum_{x\in X_1(k)} \omega_\psi(g,h)\cdot\phi(x) \qquad (g \in G(\mathbb{A}),\ h \in \overline{SL_2(\mathbb{A})}).$$

Here, G is either PGL_2 or $PD^\times$. It is well known that ϑ^ϕ_ψ is an automorphic function on $G(\mathbb{A}) \times \overline{SL_2(\mathbb{A})}$ of moderate growth.

The theta function ϑ^ϕ_ψ's can be used to define a correspondence between the automorphic representations of $G(\mathbb{A})$ and those of $\overline{SL_2(\mathbb{A})}$. To describe this correspondence, let π be an irreducible automorphic cuspidal representation of $G(\mathbb{A})$. If $f \in \pi \subset A_0$, put

$$\varphi(h) \overset{\text{definition}}{=} \int_{G(k)\backslash G(\mathbb{A})} \vartheta^\phi_\psi(g,h)f(g)\,dg.$$

In the case $G = PD^\times$, we assume that $\int_{G(k)\backslash G(\mathbb{A})} f(g)\,dg = 0$. The fact that ϑ^ϕ_ψ is a function of moderate growth on $G(k) \times SL_2(k)\backslash G(\mathbb{A}) \times \overline{SL_2(\mathbb{A})}$ means that the integral is well-defined, and that φ is a function on $SL_2(k)\backslash \overline{SL_2(\mathbb{A})}$.

Claim: φ is a cusp form.

Proof. It is enough to show that $\int_{k\backslash\mathbb{A}} \varphi\begin{pmatrix}1 & z\\ 0 & 1\end{pmatrix}dz = 0.$

$$\int_{k\backslash\mathbb{A}} \varphi\begin{pmatrix}1 & z\\ 0 & 1\end{pmatrix}dz = \int_{k\backslash\mathbb{A}}\int_{G(k)\backslash G(\mathbb{A})} \sum_{x\in X(k)} \omega_\psi\left(g\begin{pmatrix}1 & z\\ 0 & 1\end{pmatrix}\right)\cdot\phi(x)f(g)\,dg\,dz$$

$$= \int_{G(k)\backslash G(\mathbb{A})} \sum_{x\in X(k)} \omega_\psi(g)\cdot\phi(x)f(g)\int_{k\backslash\mathbb{A}} \psi(zq(x))\,dz\,dg.$$

The inner integral $\int_{k\backslash\mathbb{A}} \psi(zq(x))dz$ is zero unless $q(x) = 0$. If $G = PD^\times$, then $q(x) = 0$ if and only if $x = 0$, and the integral becomes

$$\int_{k\backslash\mathbb{A}} \varphi\begin{pmatrix}1 & z\\ 0 & 1\end{pmatrix}dz = \int_{G(k)\backslash G(\mathbb{A})} \phi(0)f(g)dg = 0.$$

If $G = PGL_2$, then $q(x) = 0$ means either $x = 0$ or x is a non-zero nilpotent element of $M_2(k)$. The integral in this situation is

$$\int_{k\backslash\mathbb{A}} \varphi\begin{pmatrix}1 & z\\ 0 & 1\end{pmatrix}dz = \int_{G(k)\backslash G(\mathbb{A})} \phi(0)f(g)dg$$

$$+ \int_{G(k)\backslash G(\mathbb{A})} \sum_{N(k)\backslash G(k)} \phi\left(g^{-1}\gamma^{-1}\begin{pmatrix}0 & 1\\ 0 & 0\end{pmatrix}\gamma g\right)f(g)dg$$

$$= 0 + \int_{N(k)\backslash G(\mathbb{A})} \phi(g^{-1}xg)f(\gamma^{-1}g)dg$$

$$= \int_{N(\mathbb{A})\backslash G(\mathbb{A})} w_\psi(g)\cdot\phi(x)\int_{N(k)\backslash N(\mathbb{A})} f(ng)\,dn\,dg = 0$$

Here, N = centralizer in G of $\begin{pmatrix}0 & 1\\ 0 & 0\end{pmatrix}$, and $\int_{N(k)\backslash N(\mathbb{A})} f(ng)dn = 0$, $\int_{N(\mathbb{A})\backslash G(\mathbb{A})} f(g)dg = 0$ since f is a cusp form.

Let $\theta(\pi,\psi)$ denote the representation of $\overline{SL_2(\mathbb{A})}$ spanned by the φ's $(\phi \in S(X_1(\mathbb{A})),\ f \in \pi)$. $\theta(\pi,\psi)$ is a genuine automorphic cuspidal representation of $\overline{SL_2(\mathbb{A})}$.

Theorem 3.1 [1]. The θ-correspondence $\pi \to \theta(\pi,\psi)$ is compatible with the local correspondence introduced in §2.

roof. Let π be an irreducible automorphic cuspidal representation f $G(\mathbb{A})$. For $f \in \pi$, and $\phi \in S(X,\mathbb{A}))$, let φ again be the cusp orm

$$\varphi(h) = \int_{G(k)\backslash G(\mathbb{A})} \vartheta_\psi^\phi(g,h)f(g)dg.$$

f $a \in k^\times$, then a calculation similar to the one used to show is a cusp form shows

$$\text{(F)}\quad \varphi_a(1) =^{\text{definition}} \int_{k\backslash\mathbb{A}} \varphi\begin{pmatrix}1 & z\\ 0 & 1\end{pmatrix}\overline{\psi(az)}dz$$

$$= \int_{T^a(\mathbb{A})\backslash G(\mathbb{A})} \omega_\psi(g)\cdot\phi(x_a) \int_{T^a(k)\backslash T^a(\mathbb{A})} f(tg)dtdg.$$

ere, x_a is any element in X, such that $q(x_a) = a$ (if $G = PD^\times$, e assume a is representable by q), and T^a is the stabilizer f x_a. T^a is a torus in G. Put

$$U(f,g) =^{\text{definition}} \int_{T^a(t)\backslash T^a(\mathbb{A})} f(tg)dt \qquad (g \in G(\mathbb{A}),\ f \in \pi).$$

he function $U(f,-)$ satisfies the property $U(f,tg) = U(f,g)$ for $\in T^a(\mathbb{A})$, and the linear function $\ell\colon f \to U(f,1)$ is a linear functional on $\{f \in \pi\}$ for which $\ell(\pi(t)f) = \ell(f)$ $(t \in T^a(\mathbb{A}))$. Locally, uch a linear functional is unique; hence ℓ is globally unique and

$$U(f,-) = \bigotimes_v U_v(-),$$

here U_v is a function on G_v such that $U_v(t_v g_v) = U_v(g_v)$ $t_v \in T^a(k_v)$, $g_v \in G(k_v)$). Under right translations by G_v on $^a_v\backslash G_v$, U_v generates a representation equivalent to π_v. In analogy ith the global formula

$$\varphi_a(h) = \int_{T^a(\mathbb{A})\backslash G(\mathbb{A})} \omega_\psi(g,h)\cdot\phi(x_a)U(f,g)dg,$$

f U is an element in the space generated by U_v, and if

$$W_{\psi^a}(h) \overset{\text{definition}}{=} \int_{T^a \backslash G_v} \omega_{\phi,v}(g,h)\cdot\phi(x_a)U(g)\,dg,$$

then $W_{\psi^a}(\begin{pmatrix}1 & z\\ 0 & 1\end{pmatrix}h) = \psi_v(za)W_{\psi^a}(h)$.

Theorem 3.2 [1]. The θ-correspondence is a 1-1 correspondence between certain automorphic cuspidal irreducible representations of $G(\mathbb{A})$ and certain genuine automorphic cuspidal irreducible representations of $\overline{SL_2(\mathbb{A})}$.

Theorem 3.3 [1][7]. Let $G = PGL_2$. Suppose $\sigma \in A_\infty$, and π is an automorphic cuspidal representation of $PGL_2(\mathbb{A})$. Then

i) $\theta(\sigma,\psi^{-1}) \not\equiv 0$ if and only if σ possesses a nonvanishing ψ-Fourier coefficient.

ii) $\theta(\pi,\psi) \not\equiv 0$ if and only if $L(\pi,\frac{1}{2}) \neq 0$.

Proof. In order to prove this theorem, we must use a polarization for which the usual unipotent subgroups of $PGL_2(\mathbb{A})$ and $\overline{SL_2(\mathbb{A})}$ lie inside P. As before, let M be the elements of $M_2(k)$ of trace zero, and let $q(m) = -\det(m)$. Let Y be a 2-dimensional simplectic vector space over k with form $<\ ,\ >$ and symplectic basis y_1, y_2. Let e_1, e_2, e_3 be a basis of M such that q has the matrix $\begin{pmatrix}0&0&1\\0&1&0\\1&0&0\end{pmatrix}$. Put $X_1 = e_1 \otimes Y + e_2 \otimes ky_1$, $X_2 = e_3 \otimes Y + e_2 \otimes ky_2$. Suppose σ is an irreducible genuine automorphic representation of $\overline{SL_2(\mathbb{A})}$ lying in A_{00}. If $\varphi \in \sigma$, let $f(g) = \int_{SL_2(k)\backslash SL_2(\mathbb{A})} \vartheta_\psi^\phi(g,h)\varphi(h)\,dh$. We can identify X_1 with $Y \oplus k$, and we can choose ϕ in the form $\phi = \phi_1\cdot\phi_2$, where $\phi_1 \in S(Y(\mathbb{A}))$, $\phi_2 \in S(\mathbb{A})$. In this situation,

$$\vartheta_\psi^\phi(1,h) = F_1(h)F_2(h),$$

where

$$F_1(h) = \sum_{y\in Y_k} \phi_1(yh) = \phi_1(0) + \sum_{\gamma\in B_k\backslash SL_2(k)} \phi_1(y_2\gamma)$$

$$F_2(h) = \sum_{t\in k} \omega'_\psi(h)\cdot\phi_2(t).$$

In the formula for F_2, ω'_ψ is the 1-dimensional Weil representation.

$$f(1) = \int_{SL_2(T)\backslash SL_2(\mathbb{A})} \phi_1(0)F_2(h)\varphi(h)dh$$

$$+ \int_{N_k\backslash SL_2(\mathbb{A})} \phi_1(y_2h)F_2(h)\varphi(h)dh.$$

Since $\sigma \in A_{00}$, and F_2 lies in the space of the Weil representation of $\overline{SL_2(\mathbb{A})}$, the first integral is zero. It follows that $\theta(\sigma,\psi^{-1}) \not\equiv 0$ if and only if the second integral does not vanish identically.

$$f(1) = \int_{N_k\backslash SL_2(\mathbb{A})} \phi_1(y_2h)F_2(h)\varphi(h)dh$$

$$= \sum_{t\in k} \int_{N_k\backslash SL_2(\mathbb{A})} \phi_1(y_2h)\omega'_\psi(h)\phi_2(t)\varphi(h)dh.$$

Since $\phi_1(y_2nh) = \phi_1(y_2h)$ and $\omega'_\psi(nh)\cdot\phi_2(t) = \psi(t^2n)\phi_2(t)$ $(n = \begin{pmatrix}1 & n\\ 0 & 1\end{pmatrix}$ $\in N_{\mathbb{A}})$ it follows that

$$f(1) = \sum_{t\in k} \int_{N_{\mathbb{A}}\backslash SL_2(\mathbb{A})} \phi_1(y_2h)\,\omega'_\psi(h)\cdot\phi_2(t)\varphi_{\psi_{t^2}}(h)dh.$$

Thus, if $\theta(\sigma,\psi^{-1}) \neq 0$, then there exists a t for which $\varphi_{\psi_{t^2}}$ is non-zero. This means σ possesses a non-zero ψ-Fourier coefficient. Conversely, now suppose σ possesses a non-vanishing ψ-Fourier coefficient. Let

$$f_t(1) = \int_{N_{\mathbb{A}}\backslash SL_2(\mathbb{A})} \phi_1(y_2h)\omega'_\psi(h)\phi_2(t)\varphi_{\psi_{t^2}}(h)dh$$

$$= \int_{N_{\mathbb{A}}\backslash SL_2(\mathbb{A})} \omega_\psi(h)\phi(y_2,t)\varphi_{\psi_{t^2}}(h)dh.$$

The latter formula allows us to define $f_t(1)$ for arbitrary ϕ. In this situation, we still have $f(1) = \sum_{t\in k^\times} f_t(1)$. ($f_0(1) = 0$, since f is a cusp form). Let Z be the upper unipotent subgroup of PGL_2. For $z \in Z$

$$\omega_\psi(z)\cdot\phi(y_2,t) = \psi(tz)\phi(y_2,t).$$

It follows from this formula that $f_t(1)$ is a Fourier coefficient of f. Therefore, if $\varphi_\psi \not\equiv 0$, then $f_t(1) \neq 0$ and so $\theta(\sigma,\psi^{-1}) \not\equiv 0$. To prove the second part of the theorem, we use the standard polarization. If $\sigma = \theta(\pi,\psi) \not\equiv 0$, then $\theta(\sigma,\psi^{-1})$ equals π. This means by part i) that σ possesses a non-zero ψ-Fourier coefficient. If T is the split torus in PGL_2, then formula (F) in the proof of Theorem 3.1 tells us that

$$\int_{T_k\backslash T_{\mathbb{A}}} f(t)dt \neq 0.$$

From the Jacquet-Langlands theory of L-functions, it is known that for an appropriate choice of f,

$$L(\pi,s) = \int_{T_k\backslash T_{\mathbb{A}}} f(t)|t|^{s-\frac{1}{2}}dt.$$

In particular,

$$L(\pi,\tfrac{1}{2}) = \int_{T_k\backslash T_{\mathbb{A}}} f(t)dt \neq 0.$$

Conversely, if $L(\pi,\frac{1}{2}) \neq 0$, then it is clear that $\int_{T_k\backslash T_{\mathbb{A}}} f(t)dt \neq 0$, and hence that $\theta(\pi,\psi) \not\equiv 0$.

§4. NON-VANISHING OF A FOURIER COEFFICIENT

In this section we will prove Theorem 1.2 that the non-vanishing of the ψ-Fourier coefficient of σ depends on σ only as an abstract representation. Let $\sigma \in A_{00}$, and let ψ be a non-trivial character of $k\backslash\mathbb{A}$. There exists a $\xi \in k^\times$ such that $\theta(\sigma,\psi^\xi) \not\equiv 0$. Define $W(\sigma,\psi)$ to be $\theta(\sigma,\psi^\xi) \otimes \chi_\xi$. By Theorem 2.3, $Wd(\sigma,\psi)$ depends only on ψ. It does not depend on ξ. Define $L_\psi(\sigma,s)$ to be $L(Wd(\sigma,\psi),s)$.

Theorem 4.1. Let $\sigma = \underset{v}{\otimes} \sigma_v \subseteq A_{00}$. σ admits a non-zero ψ-Fourier coefficient if and only if

i) at each place v, there is a linear functional ℓ_v on the space W of σ_v such that

$$\ell_v(\sigma\begin{pmatrix}1 & t\\ 0 & 1\end{pmatrix}w) = \psi_v(t)\ell_v(w) \qquad (w \in W),$$

ii) $L_\psi(\sigma,\frac{1}{2}) \neq 0$.

Proof. If σ admits a non-zero ψ-Fourier coefficient, it is clear that i) is satisfied. ii) follows from Theorem 3.3. In order to prove the converse state, Waldspurger developed a remarkable method, based on the generalization of the Siegel-Weil formula. We shall now describe Waldspurger's generalization of the Siegel-Weil formula. The Siegel-Weil formula for the simplest dual reductive pair Sp_{2n} and O_m expresses the integral $\int_{O_m(k)\backslash O_m(\mathbb{A})} \vartheta_\psi^\phi(g)dg$ in terms of an Eisenstein series on Sp_{2n}, when m is sufficiently large compared to n. Waldspurger's generalization of the Siegel-Weil formula considers the case when m is small. Let T be an anisotropic form of SO_2; thus, T is isomorphic to the norm one elements of some quadratic extension K of k. Let χ be the idèle class character associated to K. Let X be the 2-dimensional space on which T acts, and let Y be 2-dimensional with a symplectic form $<\,,\,>$. Put $Z = X \otimes_k Y$, $<x_1 \otimes y_1, x_2 \otimes y_2> = (x_1,x_2)<y_1,y_2>$. As usual, $SO_2 \times SL_2 \hookrightarrow Sp_4$. For $h_v \in SL_2(k_v)$, we have an Iwasawa decomposition $h_v = \begin{pmatrix}\alpha & *\\ 0 & \alpha^{-1}\end{pmatrix}u$, with $u \in SL_2(\mathcal{O}_v)$ if k_v is non-archimedean, and $u \in SO_2(\mathbb{R})$ or $SU_2(\mathbb{C})$ in the archimedean case. Define $A_v(h_v)$ to be $|\alpha|$, and if $h \in SL_2(\mathbb{A})$ put

$$A(h) = \prod_v A_v(h_v).$$

We define Eisenstein series by

$$E^{\phi}(h,s) \;=\; L(\chi,s+\tfrac{1}{2}) \sum_{\gamma\in B_k\backslash SL_2(k)} A^{s-\frac{1}{2}}(\gamma h)\,\omega_{\psi}(\gamma h)\,\phi(0)$$

where ϕ is a Schwartz-Bruhat function on $X_{\mathbb{A}}$. Using the standard theory of Eisenstein series, it is easy to show that this Eisenstein series converges absolutely in some half-plane and admits a meromorphic continuation to the entire plane. We have the following Siegel-Weil-Waldspurger identity

$$E^{\phi}(h,\tfrac{1}{2}) \;=\; c\int_{T_k\backslash T_{\mathbb{A}}} \vartheta_{\psi}^{\phi}(g,h)\,dg,$$

where c is a constant depending only on K/k. This identity can be proved by Poisson summation. We now return to the proof of Theorem 4.1. According to Theorem 3.3, it is sufficient to prove for the dual reductive pair PGL_2, $\overline{SL_2}$ that

$$\zeta \;=\; \int_{SL_2(k)\backslash SL_2(\mathbb{A})} \varphi(h)\,\vartheta_{\psi}^{\phi}(g,h)\,dh$$

is non-zero for some choice of ϕ, φ, and g. Suppose $\zeta \equiv 0$. Since $\varphi \neq 0$, there is an $a \in k^{\times}$ such that $\varphi_{\psi^a} \neq 0$. If $a \in (k^{\times})^2$, then since φ_{ψ} and $\varphi_{\psi^{\lambda^2}}$ are related in an elementary fashion, our statement is true. Thus, we may assume that $a \notin (k^{\times})^2$. Let x_a be an element of X so that $q(x_a) = a$, and decompose X into the line $(x_a) = kx_a$ generated by x_a and the orthogonal complement X_a'. We may take a ϕ of the form $\phi(\lambda x_a+x') = \phi_1(\lambda x_a)\phi_2(x')$ $(x' \in X_a')$. For $g \in T = \mathrm{Stab}\; x_a$, we have

$$0 \equiv \zeta \;=\; \int_{SL_2(k)\backslash SL_2(\mathbb{A})} \varphi(h)\,\vartheta_{\psi}^{\phi_1}(h)\,\vartheta_{\psi}^{\phi_2}(g,h)\,dh$$

Let $K = k(\sqrt{a})$, and let T be the anisotropic torus of norm one elements in $K^{\times}$. We can integrate with respect to $g \in T_k\backslash T_{\mathbb{A}}$. Since $T_k\backslash T_{\mathbb{A}}$ is compact, we can change the order of integration to obtain

$$0 \equiv \int_{SL_2(k)\backslash SL_2(\mathbb{A})} \varphi(h)\vartheta_\psi^{\phi_1}(h)E^{\phi_2}(h,\tfrac{1}{2})\,dh.$$

Let

$$\zeta(s) = \int_{SL_2(k)\backslash SL_2(\mathbb{A})} \varphi(h)\vartheta_\psi^{\phi_1}(h)E^{\phi_2}(h,s)\,dh.$$

For s sufficiently large, the Eisenstein series converges absolutely, and hence we can write

$$\zeta(s) = \int_{B_k\backslash SL_2(\mathbb{A})} \varphi(h)\vartheta_\psi^{\phi_1}(h)L(\chi,s+\tfrac{1}{2})A^{s-\frac{1}{2}}(h)\,\omega_\psi(h)\phi_2(0)\,dh$$

$$= \int_{N_k\backslash SL_2(\mathbb{A})} \varphi(h)\omega_\psi'(h)\cdot\phi_1(x_a)L(\chi,s+\tfrac{1}{2})A^{s-\frac{1}{2}}(h)\,\omega_\psi(h)\phi_2(0)\,dh$$

$$= L(\chi,s+\tfrac{1}{2})\int_{N_{\mathbb{A}}\backslash SL_2(\mathbb{A})} \varphi_{\psi^0}(h)\omega_\psi'(h)\cdot\phi_1(x_a)\,\omega_\psi(h)\cdot\phi_2(0)A^{s-\frac{1}{2}}(h)\,dh$$

$$= L(\chi,s+\tfrac{1}{2})\int_{N_{\mathbb{A}}\backslash SL_2(\mathbb{A})} \varphi_{\psi^0}(h)\omega_\psi(h)\cdot\phi(x_a)A^{s-\frac{1}{2}}(h)\,dh.$$

Each function in the integral factorizes as a product of local factors so

$$\zeta(s) = L(\chi,s+\tfrac{1}{2})\prod_v \int_{N_v\backslash SL_2(k_v)} \ell_v(\sigma(h_v)w)\,\omega_\psi(h_v)\phi_v(x_a)A^{s-\frac{1}{2}}(h_v)\,dh_v.$$

By §2, we know that the local integral $\int_{N_v\backslash SL_2(k_v)} \ell_v(\sigma(h_v)w)\omega_\psi(h_v)$ $\cdot\phi_v(t_a)A^{s-\frac{1}{2}}(h_v)\,dh_v$ does not vanish identically if and only if $\theta(\sigma_v,\psi_v^{-1}) \not\equiv 0$, which in turn is equivalent to the existence of a linear functional ℓ_v, which transforms under N_v by ψ_v^a. We have

$$\frac{\zeta(s)}{L_\psi(\sigma,s)} = \prod R_v(s),$$

where $R_v(s) \equiv 1$ for almost all v, and $R_v(\frac{1}{2}) \neq 0$ for all v. Since $L_\psi(\sigma,\frac{1}{2}) \neq 0$, we obtain $0 \neq \zeta(\frac{1}{2}) = \zeta$, a contradiction. Thus $\zeta \neq 0$.

Thus, there exists a $\phi = \prod_v \phi_v$ and a $w = \otimes_v w_v$ such that for v, we have

$$\int_{N_v \backslash S_{L_2}(k_v)} \ell_v(\sigma(h_v)w_v)\omega_{\phi_v}(h_v)\phi_v(x_a)dh_v \neq 0.$$

Proof of Theorem 1.3. Let $\sigma_1 = \otimes_v \sigma_{1,v}, \sigma_2 = \otimes \sigma_{2,v} \subseteq A_{00}$, and assume that they are nearly equivalent. In this situation, $\pi_1 = \mathrm{Wd}(\sigma_1,\psi)$ and $\pi_2 = \mathrm{Wd}(\sigma_2,\phi)$ will have the same local components at almost all places. By the strong multiplicity theorems for PGL_2, it follows that $\pi_1 \simeq \pi_2$. This means that $\sigma_{1,v} \simeq \sigma_{2,v}$ at all places v for which $\sigma_{1,v}$ is not a discrete series representation. Furthermore, at the places v for which $\sigma_{1,v}$ is in the discrete series, it follows from Theorem 2.5 and the local Waldspurger involution, that either $\sigma_{2,v} = \sigma_{1,v}$ or $\sigma_{2,v} = \sigma^W_{1,v}$. Since

$$\sigma^W_{1,v}(\begin{pmatrix}-1 & 0\\ 0 & -1\end{pmatrix}) = -\sigma_{1,v}(\begin{pmatrix}-1 & 0\\ 0 & -1\end{pmatrix}),$$

and $\begin{pmatrix}-1 & 0\\ 0 & -1\end{pmatrix} \in SL_2(k)$, the number of places for which $\sigma_{2,v} = \sigma^W_{1,v}$ is even. From this, we conclude that the representations in A_{00} nearly equivalent to σ_1 must be of the form σ_1^M (see §1 for the definition of σ_1^M). To complete the proof of Theorem 1.3, we must show that every σ^M (with M even) lies in A_{00}. To do this, Waldspurger used a result of Flicker [16] which we shall now describe. Flicker established a correspondence between the representations of $\overline{GL_2}$ and GL_2 $(\rho \to \pi)$. A representation $\pi = \otimes_v \pi_v$ of GL_2 lies in the image of the Flicker correspondence if and only if at each place v for which π_v is a principal series representation $\pi_v = \pi_v(\mu_v^1, \mu_v^2)$ with $\mu_v^1(-1) = \mu_v^2(-1) = 1$. It is known [14] [3] that π is in the Waldspurger correspondence if and only if there is an idèle character ω such that $\pi \otimes \omega$ is in the Flicker correspondence. Waldspurger used this fact to prove that σ^M is automorphic.

APPENDIX: A CONJECTURE OF HOWE.

R. Howe introduced in his Corvallis talk "θ-series and invariant theory" (1977) the notion of a dual reductive pair and defined a duality correspondence between the irreducible admissible representations of the members of a dual reductive pair. Howe also conjectured the following: Let (G,H) be a dual reductive pair over a global field k. Suppose that $\pi = \otimes_v \pi_v$ is an automorphic representation of $G(\mathbb{A})$, and suppose that locally at each place v, σ_v is the associated representation of $H(k_v)$ under the local duality correspondence. Then, $\sigma = \otimes_v \sigma_v$ is an automorphic representation of $H(\mathbb{A})$.

The pair $G = PGL_2$, $H = \overline{SL_2}$ is one of the simplest examples of a dual reductive pair. If D is a quaternion algebra over k, then the pair $(PD^\times, \overline{SL_2})$ is also a dual reductive pair. We shall see that if σ is an automorphic representation of $\overline{SL_2(\mathbb{A})}$, then the associated representation π of $PGL_2(A)$ is automorphic. However, we will give an example which the correspondence in the opposite direction does not send an automorphic representation of $PGL_2(\mathbb{A})$ to an automorphic representation of $\overline{SL_2(\mathbb{A})}$. Finally, we shall show that Howe's conjecture in weak form is true for $(PD^\times, \overline{SL_2})$, i.e. there exists a nearly equivalent automorphic representation.

By the definition of the Waldspurger map, we have that $Wd(\sigma,\psi) = \theta(\sigma_v,\psi_v)$, where $\sigma_v \to \theta(\sigma_v,\psi_v)$ is the local Waldspurger map. The Waldspurger map is always defined; thus, the Howe conjecture is true in the direction from $\overline{SL_2}$ to PGL_2 or $PD^\times$.

Let us now consider the other direction. If π is an automorphic cuspidal representation of $PGL_2(\mathbb{A})$, and denote by $H(\pi,\psi)$ the corresponding representation of $\overline{SL_2(\mathbb{A})}$ under the Howe correspondence. The following theorems are a consequence of Waldspurger's work.

Theorem A.1. $\sigma = H(\pi,\psi)$ is an automorphic representation of $\overline{SL_2(\mathbb{A})}$ if and only if there is a quadratic character χ_ξ such that 1) $L(\pi\otimes\chi_\xi,\frac{1}{2}) \neq 0$, and 2) $(\frac{\chi_{\xi,p}}{\pi_p}) = 1$ $\forall$ local place p. If (2) is not satisfied than $\exists\sigma'$ which is an automorphic and nearly equivalent to σ.

<u>Proof</u>. Assume that σ is automorphic. The L-function $L_\psi(\sigma,\omega,s) = L(\pi\otimes\omega,s)$ (ω any idèle class character) is entire since π is automorphic cuspidal for GL_2. This means, since σ is automorphic, that it must be cuspidal and in fact $\sigma \in A_{00}$. π is equal to $Wd(\sigma,\psi^{-1})$. If ψ^ξ is a character for which σ possesses a non-zero ψ^ξ-Fourier coefficient, then $\pi\otimes\chi_\xi = \theta(\sigma,\psi^\xi)$. By Theorem 3.3, we have $L(\pi\otimes\chi_\xi,\frac{1}{2}) \neq 0$. Conversely, if $L(\pi\otimes\chi_\xi,\frac{1}{2}) \neq 0$, then $\sigma' = \theta(\pi\otimes\chi_\xi,(\psi^\xi)^{-1})$ and so $\sigma' \in A_{00}$. It is easy to see that σ' is nearly equivalent to σ and $\sigma' \cong \sigma$ iff $(\frac{\chi_{\xi,p}}{\pi_p}) = 1$ $\forall p$.

<u>Theorem A.2</u>. Let $\pi = \underset{v}{\otimes}\, \pi_v$ be an automorphic cuspidal representation of $PGL_2(\mathbb{A})$. If either of the following conditions is satisfied:

i) there is a v for which π_v lies in the discrete series

ii) $\varepsilon(\pi,\frac{1}{2}) = 1$ (see §2),

then there is a χ_ξ such that $L(\pi\otimes\chi_\xi,\frac{1}{2}) \neq 0$. Also, if $L(\pi\otimes\chi_\xi,\frac{1}{2}) \neq 0$, then π satisfies one of the above two conditions.

<u>Proof</u>. We shall show that $L(\pi\otimes\chi_\xi,\frac{1}{2}) = 0$ for all χ_ξ is equivalent to all the π_v's being in the principal series, and $\varepsilon(\pi,\frac{1}{2}) = -1$. If $\varepsilon(\pi,\frac{1}{2}) = -1$ and all the π_v's are principal series, then $\varepsilon(\pi\otimes\chi_\xi,\frac{1}{2}) = -1$ for any χ_ξ. This means $L(\pi\otimes\chi_\xi,\frac{1}{2}) = 0$. The converse result was proved by Waldspurger using the result of Flicker [16] formulated in §4.

We shall now construct a counterexample to Howe's conjecture. Let $\pi = \underset{v}{\otimes}\, \pi_v$ be an automorphic representation of $PGL_2(\mathbb{A}_\mathbb{Q})$ for which π_∞ lies in the holomorphic discrete series, and π_v for v finite is unramified. In classical language, such a representation corresponds to a holomorphic modular form with respect to the full modular group $PSL_2(\mathbb{Z})$. Let K be any imaginary quadratic extension of $\mathbb{Q}$, and denote by Π, the base change lift of π to $PGL_2(\mathbb{A}_K)$. Π_v lies in the principal series for all v, and $\varepsilon(\Pi,\frac{1}{2}) = -1$. Thus, $\varepsilon(\Pi\otimes\chi_\xi,\frac{1}{2}) = -1$ for all χ_ξ. By Theorem A.1, $\sigma = H(\pi,\psi)$ is not an automorphic representation of $\overline{SL_2(\mathbb{A}_K)}$.

Let us now consider the dual-reductive pair $(PD^{\times}, \overline{SL_2})$. Let π' be an infinite dimensional automorphic representation of $PD^{\times}$. Denote by $\pi = \underset{v}{\otimes}\, \pi_v$ the automorphic cuspidal representation of PGL_2 associated to π' by the Jacquet-Langlands correspondence. For some place v, π_v will lie in the discrete series. It follows from Theorem A.2 that there is a χ_ξ for which $L(\pi \otimes \chi_\xi, \frac{1}{2}) \neq 0$. This means that there exists $\sigma \in A_{00}$, which is nearly equivalent to $H(\pi', \psi)$.

<u>Additional note on Theorem A.1</u>

Waldspurger's results [3] imply that the assumptions (1) and (2) of A.1 are equivalent to saying that $\varepsilon(\pi, \frac{1}{2}) = 1$, where $\varepsilon(\pi, \frac{1}{2})$ was defined above after Theorem 2.3. For all p such that π_p is unramified, we have $\varepsilon(\pi_p, \frac{1}{2}) = 1$. Hence, in order to verify this assumption, we have to check that

$$\prod_{p \in S} \varepsilon(\pi_p, \tfrac{1}{2}) = 1,$$

where S is a certain finite set.

I want to thank J. Waldspurger for conversations in which he explained his results to me. Finally I thank A. Moy for help in the preparation of this manuscript.

REFERENCES

[1] J.-L. Waldspurger, "Correspondance de Shimura," J. Math. Pures Appl. 59 (1980), 1-133.

[2] ______, "Sur les coefficients de Fourier des formes modulaires de poids demi-entier," J. Math. Pures Appl. 60 (1981), 375-484.

[3] ______, "Correspondance de Shimura et quaternions," preprint.

[4] ______, "Engendrement par des séries thêta de certains espaces de formes modulaires," Invent. Math. 50 (1979), 135-168.

[5] ______, "Correspondance de Shimura," Seminaire de Théorie de Nombres Paris, 1979-80, Progress in Math., Vol. 12, Birkhäuser, Boston, 1981, pp. 357-369.

[6] R. Howe, "θ-series and invariant theory," in Proceeding of Symposi in Pure Mathematics, A.M.S. vol. XXXIII, p. 275-285.

[7] R. Howe and I. Piatetski-Shapiro, "Some examples of automotphic forms on Sp_4," to appear in Duke Math. J.

[8] S. Rallis and Schiffmann, G., "Représentations supercuspidales du groupe métaplectique," J. Math. Kyoto. Univ. 17 (1977), 567-603.

[9] M.-F. Vignéras, "Valeur au centre de symétrie des fonctions L associées aux formes modulaires," Seminaire de Théorie de Nombres, Paris, 1979-80, Progress in Math., vol. 12, Birkhäuser, Boston, 1981, pp. 331-356.

[10] G. Shimura, "On modular forms of half-integral weight," Ann. of Math. (2) 97 (1973), 440-481.

[11] S. Rallis, "On the Howe duality conjecture," preprint.

[12] S. Gelbart, "Weil's representation and the spectrum of the metaplectic group," Lecture Notes in Math., vol. 530, Springer-Verlag, Berlin-New York, 1976.

[13] S. Gelbart and I. Piatetski-Shapiro, "On Shimura's correspondence for modular forms of half-integral weight," in Automorphic forms, Representation theory and Arithmetic, Tata Institute of Fundamental Research, Bombay, 1979.

[14] ______, "Some remarks on metaplectic cusp forms and the correspondence of Shimura and Waldspurger," Israel J. Math., to appear.

[15] A. Weil, "Sur la formule de Siegel dans la théorie des groupes classiques," Acta Math. 113 (1965), 1-87.

[16] Y. Flicker, "Automorphic forms on covering groups of GL(2)," Invent. Math. 57 (1980), 119-182.

[17] H. Jacquet and R. P. Langlands, "Automorphic forms on GL(2)," Lecture Notes in Math., vol. 114, Springer-Verlag, Berlin-New York, 1970.

The Fourier Transform of Orbital Integrals on SL_2 over a p-adic Field

by

Paul J. Sally, Jr.* and Joseph A. Shalika*

Table of Contents

*Both authors supported by the National Science Foundation

§1. Introduction.

Let k be a p-adic field (that is, a locally compact, totally disconnected, non-discrete field). We denote by R the ring of integers in k, and by P the prime ideal in R. We assume that $q = |R/P|$ is odd. Let dx be a Haar measure on k^+ normalized so that $\int_R dx = 1$. A valuation is defined on k by the equations $d(ax) = |a|\, dx$, $a \in k^x$, $|0| = 0$. Let $U = \{x \mid |x| = 1\}$, the units in R, and $U_n = 1 + P^n = \{x \in U \mid |x-1| \leq q^{-n}\}$, $n \geq 1$. Throughout this paper, we use τ to denote a prime element, that is, a generator of P, and ε to denote a fixed primitive (q-1)st root of unity in U.

Let $G = SL(2,k)$, and denote by $C_c^\infty(G)$ the space of compactly supported, locally constant, complex valued functions on G. If $\gamma \in G'$, the set of regular elements in G, or γ is a unipotent element in G, we set

$$(1.1) \qquad I_f(\gamma) = \int_{G/G_\gamma} f(x\gamma x^{-1})\, d\dot{x}, \quad f \in C_c^\infty(G),$$

where G_γ is the centralizer of γ in G, and $d\dot{x}$ is a G-invariant measure on G/G_γ. It is known (see [Sk2]) that the integrals in (1.1) converge. Such integrals are called orbital integrals on G.

If $\gamma \in G'$, then $G_\gamma = T$ is a maximal torus in G, and the invariant integral of f relative to T is defined by

$$(1.2) \qquad F_f^T(t) = |D(t)|^{1/2} \int_{G/T} f(xtx^{-1})\, d\dot{x}, \quad t \in T' = T \cap G',$$

where D(t) is the usual Weyl discriminant. For SL(2,k), $D(x) = (\lambda - \lambda^{-1})^2$, where λ, λ^{-1} are the eigenvalues of x, and $G' = \{x \in G \mid D(x) \neq 0\}$.

If γ is fixed, the map $f \longmapsto I_f(\gamma)$ defines an invariant distribution on G which we denote by Λ_γ. For $f \in C_c^\infty(G)$ and $\Pi \in \hat{G}$, the unitary dual of G, we set

$$(1.3) \qquad \hat{f}(\Pi) = \operatorname{tr} \Pi(f),$$

where $\Pi(f) = \int_G f(x)\Pi(x)dx$, an operator of finite rank.

The purpose of this paper is to derive the Fourier transform of the distribution Λ_γ where γ is a regular or unipotent element in G. Thus, we determine a linear functional $\hat{\Lambda}_\gamma$ on $\hat{G}$ such that

$$(1.4) \qquad \hat{\Lambda}_\gamma(\hat{f}) = \Lambda_\gamma(f), \quad f \in C_c^\infty(G) .$$

In §2, we give explicit formulas for the germs of F_f^T on the compact Cartan ubgroups of G. This is an easy computation following from the results in [Sk2]. brief summary of the (tempered) representation theory and character theory of G is resented in §3. For $\gamma \in G'$, a formula for $\hat{\Lambda}_\gamma$ is derived in §4 and §5. The lancherel formula for G is then obtained in §6 by using Shalika's germ expansion or F_f^T. In §7, we find $\hat{\Lambda}_\gamma$ in the case when γ is a non-trivial unipotent element n G. In an appendix, we outline the methods used in computing the kernel which is iven in §4. We assume throughout that measures are normalized as in [Sk2].

The results in this paper were completed a number of years ago. In the intervening period, many authors have studied orbital integrals on p-adic groups and heir applications. We give a fairly complete list of references after the ibliography. To our knowledge, the only other case in which the full Fourier ransform of elliptic orbital integrals has been computed is that of GL(2,k) [Sc]. n [Sc], the characteristic of k is assumed to be zero, but there is no restriction n residual characteristic.

The authors would like to thank the University of Maryland, and, in particular, . Herb, R. Lipsman and J. Rosenberg for their gracious hospitality and stimulating athematics during 1982-1983.

2. Germs for SL_2.

In this section, we give explicit formulas for the germs associated to the ompact Cartan subgroups of $G = SL(2,k)$. We set $N = \{\begin{pmatrix} 1 & x \\ 0 & 1 \end{pmatrix} \mid x \in k\}$, $^- = \{\begin{pmatrix} 1 & 0 \\ x & 1 \end{pmatrix} \mid x \in k\}$, and $K = SL(2,R)$.

Let T be a compact Cartan subgroup of G and $\mathcal{O}$ a non-trivial unipotent orbit in . If $\gamma \in \mathcal{O}$, we write $\Lambda_{\mathcal{O}}(f)$ for $\Lambda_\gamma(f)$. Then, according to [Sk2], Theorem 2.2.2, nere is a subset $T_{\mathcal{O}}$ of T' such that

2.1) $$\{F_f^T\} = -A_T\{|D|^{1/2}\}f(1) + B_T \sum_{\dim \mathcal{O} > 0} \{C_{\mathcal{O}}\}\Lambda_{\mathcal{O}}(f),$$

here $f \in C_c^\infty(G)$, $C_{\mathcal{O}}$ is the characteristic function of $T_{\mathcal{O}}$, and A_T, B_T are constants o be determined below. This is the germ expansion for F_f^T.

To obtain explicit formulas for the functions $C_{\mathcal{O}}$, we assume that the compact Cartan subgroup T is written in the form

$$(2.2)\qquad T = \left\{ \begin{pmatrix} \alpha & \beta\omega_1 \\ \beta\omega_2 & \alpha \end{pmatrix} \;\middle|\; \alpha,\beta \in k \right\} \quad \text{(see (3.3))}.$$

Let $V = V_T$ be the quadratic extension of k generated by the eigenvalues of T, and denote by sgn_V the character of order two on $k^\times$ associated to V. Each nontrivial unipotent class $\mathcal{O}$ contains an element of the form $\begin{pmatrix} 1 & \zeta \\ 0 & 1 \end{pmatrix}$ with ζ unique modulo $(k^\times)^2$. Thus, $\mathrm{sgn}_V(\mathcal{O}) = \mathrm{sgn}_V(\zeta)$ is defined for any quadratic extension V of k.

We define

$$(2.3)\qquad \varepsilon(t) = \mathrm{sgn}_V(\beta\omega_1),\quad \beta \neq 0,\quad t = \begin{pmatrix} \alpha & \beta\omega_1 \\ \beta\omega_2 & \alpha \end{pmatrix} \in T.$$

Lemma 2.4. For $t \in T'$,

$$C_{\mathcal{O}}(t) = \frac{1}{2}\{1 + \varepsilon(t)\,\mathrm{sgn}_V(\mathcal{O})\}.$$

Proof. By definition, $C_{\mathcal{O}}$ is the characteristic function of $T' \cap \Omega$ where $\Omega = \Omega(x_0)$ for some $x_0 \in \mathcal{O}$ of the form $\begin{pmatrix} 1 & \zeta \\ 0 & 1 \end{pmatrix}$, and $\Omega(x_0) = \{x x_0 n x^{-1} \mid x \in G,\ n \in N^-\}$ (see [Sk2], §2.2). Let $t = \begin{pmatrix} \alpha & \beta\omega_1 \\ \beta\omega_2 & \alpha \end{pmatrix} \in T'$ and $g = \begin{pmatrix} a & b \\ * & * \end{pmatrix} \in G$. Then $gtg^{-1} = \begin{pmatrix} * & \beta(a^2\omega_1 - b^2\omega_2) \\ * & * \end{pmatrix}$.

If $t \in \Omega$, then $gtg^{-1} = x_0 n$ for some $g \in G$, $n \in N^-$. It follows that

$$\mathrm{sgn}_V(\mathcal{O}) = \mathrm{sgn}_V(\zeta) = \mathrm{sgn}_V[\beta\omega_1(a^2 - b^2\omega_2\omega_1^{-1})] = \mathrm{sgn}_V(\beta\omega_1) = \varepsilon(t).$$

Conversely, if $\mathrm{sgn}_V\,\mathcal{O} = \varepsilon(t)$, a simple computation shows that we may choose $g \in G$ so that $gtg^{-1} = x_0 n$ for some $n \in N^-$.

We now turn to the constants A_T and B_T which appear in (2.1).

Lemma 2.5. Let T be a compact Cartan subgroup of G, and let V_T denote the uadratic extension generated by the eigenvalues of T. Then

$$A_T = \mu(K)\mu(T)^{-1}(q - 1)^{-1},$$

$$B_T = \mu(T)^{-1}\mu(N \cap K)^{-1}q(q - 1)^{-1}\kappa_T ,$$

here

$$\kappa_T = \begin{cases} (q+1)/q & \text{if } V_T \text{ is unramified,} \\ 2q^{-1/2} & \text{if } V_T \text{ is ramified .} \end{cases}$$

emark. Here, and in the remainder of this paper, $\mu(\cdot)$ denotes the measure of a set ith respect to an appropriate Haar measure.

Proof. Let T be a compact Cartan subgroup of G, and let $\{K_m \mid m \geq 1\}$ be the sual filtration in K. Thus $K_m = \{x \in K \mid x \equiv I \pmod{P^m}\}$.
et f_m be the characteristic function of K_m, and suppose that T is ramified. It ollows from (2.1) and Lemma 2 of [SS2] that, for $t \in T$ close to 1,

$$(q-1)^{-1}\mu(K)\mu(T)^{-1}[q^{-m+1/2} - |D(t)|^{1/2}]$$

$$= -A_T|D(t)|^{1/2} + B_T \sum_{\dim \mathcal{O} > 0} c_{\mathcal{O}}(t)\Lambda_{\mathcal{O}}(f_m) .$$

t is then clear that $A_T = \mu(K)\mu(T)^{-1}(q - 1)^{-1}$. If T is unramified, a similar omputation yields the same value for A_T. The value of B_T given above will be erived in §7.

3. The representations of SL_2 and their characters.

Let V be any quadratic extension of k and $N_{V/k}$ the norm from V to k. V may be ritten in one of the forms $k(\sqrt{\tau})$, $k(\sqrt{\varepsilon\tau})$ (the ramified extensions) or $k(\sqrt{\varepsilon})$ the unramified extension). For a valuation on V, we choose the unique valuation hat extends the given valuation on k.

For any quadratic extension $V = k(\sqrt{\theta})$, let C_θ denote the kernel of $N_{V/k}$, and $_\theta$ the prime ideal in V. If V is unramified, we set $C_\varepsilon^{(h)} = (1 + P_\varepsilon^h) \cap C_\varepsilon$, $h \geq 1$.

If V is ramified, we set $C_\theta^{(h)} = (1 + P_\theta^{2h+1}) \cap C_\theta$, $h \geq 0$. In any case the collection $\{C_\theta^{(h)}\}$ is a neighborhood basis for 1 in C_θ. If $\psi \in \hat{C}_\theta$, the character group of C_θ, we denote the conductor of ψ by cond ψ. This is the largest subgroup in the filtration $\{C_\theta^{(h)}\}$ on which ψ is trivial. On each C_θ, there is a unique character of order two denoted by ψ_0.

If $\theta = \tau$ or $\varepsilon\tau$, we can write $C_\theta = C_\theta^{(0)} \cup (-1)C_\theta^{(0)}$. Clearly $[C_\theta^{(h)} : C_\theta^{(h+1)}] = q$, $h \geq 0$, so that

$$(3.1) \qquad [C_\theta : C_\theta^{(h)}] = 2q^h, \quad h \geq 0, \ \theta = \tau, \ \varepsilon\tau \ .$$

When $\theta = \varepsilon$, we have $[C_\varepsilon : C_\varepsilon^{(1)}] = q+1$ and $[C_\varepsilon^{(h)} : C_\varepsilon^{(h+1)}] = q$, $h \geq 1$, so that

$$(3.2) \qquad [C_\varepsilon : C_\varepsilon^{(h)}] = (q+1)q^{h-1}, \quad h \geq 1 \ .$$

Now consider the subgroups of $G = SL(2,k)$ defined by

$$A = \left\{ \begin{pmatrix} \lambda^{-1} & 0 \\ 0 & \lambda \end{pmatrix} : \lambda \in k^\times \right\},$$

$$(3.3) \qquad \begin{aligned} T_\tau &= \left\{ \begin{pmatrix} x & y \\ \tau y & x \end{pmatrix} : x,y \in k \right\}, & T_\tau^\# &= \left\{ \begin{pmatrix} x & \varepsilon y \\ \tau\varepsilon^{-1} y & x \end{pmatrix} : x,y \in k \right\}, \\ T_{\varepsilon\tau} &= \left\{ \begin{pmatrix} x & y \\ \varepsilon\tau y & x \end{pmatrix} : x,y \in k \right\}, & T_{\varepsilon\tau}^\# &= \left\{ \begin{pmatrix} x & \varepsilon y \\ \tau y & x \end{pmatrix} : x,y \in k \right\}, \\ T_\varepsilon &= \left\{ \begin{pmatrix} x & y \\ \varepsilon y & x \end{pmatrix} : x,y \in k \right\}, & T_\varepsilon^\# &= \left\{ \begin{pmatrix} x & \tau y \\ \varepsilon\tau^{-1} y & x \end{pmatrix} : x,y \in k \right\}. \end{aligned}$$

If $-1 \in (k^\times)^2$, the collection (3.3) is a complete set (up to conjugation) of Cartan subgroups of G, that is, maximal abelian subgroups in which every element is semisimple. If $-1 \notin (k^\times)^2$, T_θ and $T_\theta^\#$ are conjugate for $\theta = \tau, \varepsilon\tau$.

The group A is naturally isomorphic to $k^\times$, and we denote by A_d the image of U_d under this isomorphism. For any θ, T_θ and $T_\theta^\#$ are naturally isomorphic to C_θ, and we denote by $(T_\theta)_d$ or $(T_\theta^\#)_d$ the image of $C_\theta^{(d)}$ under this isomorphism. For any subset M of G, we set $M^G = \{gmg^{-1} : g \in G, m \in M\}$. For any subset M of G, we set $M' = M \cap G'$. The set of elliptic elements in G is defined by

$$(3.4) \qquad G_e = \bigcup_T (T')^G ,$$

here T runs through the collection $\{T_\theta, T^\#_\theta : \theta = \tau, \varepsilon\tau, \varepsilon\}$ of compact Cartan ubgroups of G.

The representations of G with which we are concerned fall into two classes. hese are the tempered representations of G.

i) <u>The principal series</u> ([GG], [Sa2]).

These representations are induced from the one-dimensional unitary epresentations of

$$B = \left\{ \begin{pmatrix} \lambda^{-1} & \mu \\ 0 & \lambda \end{pmatrix} : \lambda \in k^\times, \mu \in k \right\},$$

nd are in one-to-one correspondence with the unitary characters on $k^\times$. These epresentations are all irreducible except those corresponding to the characters of rder two on $k^\times$ and, if $\pi \in \hat{k}^\times$, the representations corresponding to π and π^{-1} are nitarily equivalent.

For any $g \in G'$, we write the eigenvalues of g as λ, λ^{-1}. For the character of ne representation of the principal series indexed by $\pi \in \hat{k}^\times$, we have

$$(3.5) \qquad \Theta_\pi(g) = \begin{cases} \dfrac{\pi(\lambda) + \pi(\lambda^{-1})}{|\lambda - \lambda^{-1}|} & , \quad g \in (A')^G, \\ 0 & , \quad g \in G_e . \end{cases}$$

his formula can be computed easily by the usual method for computing such induced aracters (see e.g. [GG]).

There are three characters of order two on $k^\times$, corresponding to the three adratic extensions, denoted sgn_θ, $\theta = \tau, \varepsilon\tau, \varepsilon$, where sgn_θ is the character of $k^\times$ whose kernel is the image of $k(\sqrt{\theta})^\times$ under the norm map. The representations of e principal series corresponding to these characters on $k^\times$ split into two irre-ucible components. The characters of these irreducible components do not vanish on , and they play a central role in our development. The formulas for the characters these representations are given below.

(ii) The discrete series

(iia) The special representation ([GG], [Sa2], [Sk3], [Ca]).

The special representation can be obtained from the principal series by analytic continuation. This representation is irreducible, unitary and square integrable. The character of the special representation is given by

$$\Theta_0(g) = \begin{cases} \dfrac{|\lambda| + |\lambda|^{-1}}{|\lambda - \lambda^{-1}|} - 1 \ , & g \in (A')^G \ , \\ -1 \ , & g \in G_e \ . \end{cases} \tag{3.6}$$

(iib) The supercuspidal representations ([GG], [Sk1], [T]). The supercuspidal representations may be indexed by a non-trivial character $\Phi \in \hat{k}^+$ and a nontrivial character $\psi \in \hat{C}_\theta$, $\theta = \tau, \varepsilon\tau, \varepsilon$. The corresponding representation is denoted $\Pi(\Phi, \psi, V)$. For any $b \in k^x$, set $\Phi_b(x) = \Phi(bx)$.

Most of the following facts are proved in [Sk1] (~ denotes unitary equivalence).

(1) For fixed V, $\Pi(\Phi,\psi,V) \sim \Pi(\Phi_b, \psi, V)$ iff $b \in N_{V/k}(V^x)$.

(2) $\Pi(\Phi, \psi, V) \sim \Pi(\Phi, \psi', V)$ iff $\psi' = \psi$ or ψ^{-1}.

(3) If V and V' are distinct quadratic extensions and $\psi^2 \not\equiv 1$, $(\psi')^2 \not\equiv 1$, then $\Pi(\Phi, \psi, V) \not\sim \Pi(\Phi, \psi', V')$.

(4) If $\psi \neq \psi_0$, $\Pi(\Phi, \psi, V)$ is irreducible.

(5) $\Pi(\Phi, \psi_0, V)$ splits into two inequivalent irreducible components denoted $\Pi^+(\Phi, V)$ and $\Pi^-(\Phi, V)$. From (1) it follows that we get four inequivalent irreducible representations corresponding to ψ_0. For any two distinct quadratic extensions these representations are pairwise equivalent in some order.

(6) Each of the representations $\Pi(\Phi, \psi, V)$, $\psi \neq \psi_0$, along with the irreducible components of $\Pi(\Phi, \psi_0, V)$, is induced from an irreducible representation of some maximal compact subgroup of G.

(7) All the above representations have compactly supported matrix coefficients in some orthonormal basis.

(8) Any irreducible unitary representation of a maximal compact subgroup K occurs with finite multiplicity in the restriction to K of the direct sum of the representations $\Pi(\Phi, \psi, V)$, $\psi^2 \not\equiv 1$, $\Pi^+(\Phi, V)$, $\Pi^-(\Phi, V)$.

The characters of the representations $\Pi(\Phi, \psi, V)$, $\psi^2 \not\equiv 1$, $\Pi^+(\Phi, V)$, $\Pi^-(\Phi,V)$ re given in [SS1]. The character of the representation $\Pi(\Phi, \psi, V)$ will be ritten $\Theta_{(\Phi,\psi)}$, and the characters of the representations $\Pi^{\pm}(\Phi, V)$ will be ritten $\Theta^{\pm}_{\Phi}$, respectively.

As stated in [SS1], if $g \in G'$ and $\psi \in \hat{C}_\theta$, $\theta = \tau, \varepsilon\tau, \varepsilon$, then

3.7) $$\Theta_{(\Phi,\psi)}(-g) = \mathrm{sgn}_\theta(-1)\psi(-1)\Theta_{(\Phi,\psi)}(g).$$

We take this opportunity to correct a discrepancy in the character tables. or $Z = x + \sqrt{\theta}\, y$ in $k(\sqrt{\theta})$, we set $\mathrm{tr}\, Z = Z + \bar{Z} = 2x$. In [SS1], the fourth ormula from the bottom on page 1234 should read

$$\frac{q^{-1/2}}{|\lambda - \lambda^{-1}|} \sum_{\gamma \in C_\tau^{(h-1)}/C_\tau^{(h)}} \mathrm{sgn}_\tau(\mathrm{tr}\,\lambda - \mathrm{tr}\,\gamma).$$

iii) <u>The reducible principal series.</u>

It can be shown that the irreducible components of the reducible principal eries are unitarily equivalent to representations $\Pi(\Phi, 1, V)$ where $\Phi \in \hat{k}^+$, $= k(\sqrt{\theta})$, $\theta = \tau, \varepsilon\tau, \varepsilon$, and 1 denotes the trivial character on C_θ. The repre- entations $\Pi(\Phi, 1, V)$ can be obtained from the construction of Weil (see [T]), nd, as in the case of the supercuspidal representations, $\Pi(\Phi, 1, V) \sim \Pi(\Phi_b, 1, V)$ f and only if $b \in N_{V/k}(V^\times)$. Let $\Theta_{(\Phi,V)}$ denote the character of $\Pi(\Phi, 1, V)$. Then, f $V = k(\sqrt{\theta})$, we have

3.8) $$\Theta_{(\Phi,V)}(g) = \begin{cases} \dfrac{\mathrm{sgn}_\theta(\lambda)}{|\lambda - \lambda^{-1}|} & \text{if } g \in (A')^G, \\ \kappa(\Phi,V)\varepsilon(t)|D(t)|^{-1/2}, & \text{if } g \text{ is conjugate to } t \in T'_\theta \cup (T^\#_\theta)', \\ 0 & , \text{if } g \in G_e,\ g \notin [T'_\theta \cup (T^\#_\theta)']^G . \end{cases}$$

(See the Appendix or [SS1] for the definition of $\kappa(\Phi,V)$.) These character formulas are computed in [Fr]. In the remainder of the paper, we denote the representations of the reducible principal series associated to V by RPS_V.

§4. An orthogonality relation on the dual of G.

In this section, we give a formula for summing the characters of the discrete series which is analogous to a well-known orthognality relation for finite groups. This formula allows us to compute explicitly the Fourier transform of a distribution on G with support in a closed conjugacy class. We are interested in the sums $\sum_{\Pi} \Theta_{\Pi}(t_1)\overline{\Theta_{\Pi}(t_2)}$, where $t_1 \in G'$, $t_2 \in G_e$, and Π runs over a finite number of members of the discrete series. More precisely, for any positive integer d, we set

$$K_d(t_1,t_2) = \Theta_0(t_1)\overline{\Theta_0(t_2)} + \sum_{\Phi} [\Theta_{\Phi}^{+}(t_1)\overline{\Theta_{\Phi}^{+}(t_2)} + \Theta_{\Phi}^{-}(t_1)\overline{\Theta_{\Phi}^{-}(t_2)}]$$

$$(4.1) \qquad + \frac{1}{2}\sum_{\Phi}\sum_{\theta=\tau,\varepsilon\tau}\sum_{h=1}^{d}\sum_{\substack{\psi\in\hat{C}_\theta \\ \text{cond } \psi=C_\theta^{(h)}}} \Theta_{(\Phi,\psi)}(t_1)\overline{\Theta_{(\Phi,\psi)}(t_2)}$$

$$+ \frac{1}{2}\sum_{\Phi}\sum_{h=1}^{d}\sum_{\substack{\psi\in\hat{C}_\varepsilon \\ \text{cond } \psi=C_\varepsilon^{(h)} \\ \psi\neq\psi_0}} \Theta_{(\Phi,\psi)}(t_1)\overline{\Theta_{(\Phi,\psi)}(t_2)}$$

In (4.1), $\sum_{\Phi}$ is used to denote the sum over the characters of the inequivalent representations $\Pi(\Phi,\psi,V)$, $\Pi(\Phi_b,\psi,V)$, $b \notin N_{V/k}(V^{x})$. This notation will be retained throughout.

It follows immediately from (3.6) and (3.7) that

$$(4.2) \qquad K_d(-t_1, -t_2) = K_d(t_1,t_2) \quad \text{for all } t_1\in G',\ t_2\in G_e.$$

Thus, if λ_2, λ_2^{-1} are the eigenvalues of t_2, it is sufficient for computational purposes to assume that $|1+\lambda_2| = |1+\lambda_2^{-1}| = 1$. It is also clear that

$$(4.3) \qquad K_d(t_2,t_1) = \overline{K_d(t_1,t_2)}\ .$$

We now define two functions on the Cartan subgroups (3.3) of G. These functions enter in a significant way into our computations.

Suppose that h is an integer greater than zero. We define

$$(4.4)\qquad \Delta_h(\lambda) = \begin{cases} 1, & \lambda \in U_h, \\ 0, & \lambda \in k^x, \ \lambda \notin U_h, \\ 1, & \lambda \in C_\theta^{(h)}, \ \theta = \tau, \varepsilon\tau, \varepsilon, \\ 0, & \lambda \in C_\theta \backslash C_\theta^{(h)}, \ \theta = \tau, \varepsilon\tau, \varepsilon. \end{cases}$$

Observe that we can also define $\Delta_0(\lambda)$ if $\lambda \in C_\theta$, $\theta = \tau, \varepsilon\tau$. The function Δ_h carries over to a function on the Cartan subgroups of G via the natural isomorphism mentioned in §3.

The second function is defined for t a regular element in the union of the compact Cartan subgroups of G. If $\varepsilon(t)$ is defined as in §2, we set, for $\theta = \tau, \varepsilon\tau, \varepsilon$,

$$(4.5)\qquad \sigma(t) = \begin{cases} \varepsilon(t)|D(t)|^{-1/2}, & t \in T_\theta', \\ -\mathrm{sgn}_\theta(-1)\varepsilon(t)|D(t)|^{-1/2}, & t \in (T_\theta^{\#})'. \end{cases}$$

We can extend $\sigma(t)$ to all of G_e by conjugation.

If

$$t = \begin{pmatrix} \alpha & \beta \\ \theta\beta & \alpha \end{pmatrix} \in T_\theta \quad \text{or} \quad t = \begin{pmatrix} \alpha & \omega\beta \\ \omega^{-1}\theta\beta & \alpha \end{pmatrix} \in T_\theta^{\#},$$

it is clear that

$$\sigma(t) = \pm \frac{\mathrm{sgn}_\theta[(\lambda - \lambda^{-1})/2\sqrt{\theta}]}{|\lambda - \lambda^{-1}|},$$

where $\lambda = \alpha + \sqrt{\theta}\,\beta$ in either case. This is the form in which $\sigma(t)$ appears in the character tables [SS1].

The computation of the kernel K_d is straightforward but long. We summarize all the computations in the following theorem. Some details of the computation are presented in the Appendix. Note that, if $T = T_\theta$ or $T_\theta^{\#}$, then $[1 + \mathrm{sgn}_\theta(-1)]/2 = |W_T| - 1$, where $W_T = N(T)/T$ is the Weyl group of T.

Theorem 4.6. Suppose that t_1 and t_2 are regular elements in $A \cup \bigcup_{\theta=\tau,\varepsilon\tau,\varepsilon} (T_\theta \cup T_\theta^\#)$. If λ is an eigenvalue of t_2, assume that $|1+\lambda| = 1$. Take d a positive integer. Then

(i) If $t_1 \in T_\theta$ or $T_\theta^\#$ and $t_2 \in T_{\theta'}$ or $T_{\theta'}^\#$, $\theta \neq \theta'$,

$$K_d(t_1,t_2) = \left(\frac{q+1}{q}\right)q^{3d}\Delta_d(t_1)\Delta_d(t_2) .$$

(ii) If $t_1 \in T_\theta$ and $t_2 \in T_\theta^\#$, and T_θ and $T_\theta^\#$ are not conjugate

$$K_d(t_1,t_2) = \left(\frac{q+1}{q}\right)q^{3d}\,\Delta_d(t_1)\Delta_d(t_2) + \sigma(t_1)\sigma(t_2)[1 - |\theta|^{-1/2}\kappa_T\Delta_d(t_1)\Delta_d(t_2)].$$

(iii) If t_1 and t_2 are both in T_θ or $T_\theta^\#$,

$$K_d(t_1,t_2) = \left(\frac{q+1}{q}\right)q^{3d}\Delta_d(t_1)\Delta_d(t_2) - \sigma(t_1)\sigma(t_2)$$
$$+\ \kappa_T|\theta|^{-1/2}q^d\sigma(t_1)\sigma(t_2)\begin{cases}\Delta_d(t_1t_2^{-1}) + (|W_T| - 1)\Delta_d(t_1t_2), & \Delta_d(t_1)\Delta_d(t_2) = 0, \\ 1, & \Delta_d(t_1)\Delta_d(t_2) = 1,\end{cases}$$

where $T = T_\theta$ or $T_\theta^\#$ as the case may be.

(iv) If $t_1 \in A$ and $t_2 \in \bigcup_{\theta=\tau,\varepsilon\tau,\varepsilon} (T_\theta \cup T_\theta^\#)$, and the eigenvalues $\lambda_2, \lambda_2^{-1}$ of t_2 lie in $C_\theta^{(h_2)} \setminus C_\theta^{(h_2+1)}$ ($h_2 = 0$ if $\theta = \varepsilon$ and $|1-\lambda_2| = 1$), then, for $h_2 < d$

$$K_d(t_1,t_2) = \begin{cases}\dfrac{-|\lambda_1|}{|\lambda_1 - \lambda_1^{-1}||1-\lambda_1|^2}, & \lambda_1 \notin U_{h_2+1}, \\[2ex] \dfrac{-1}{|\lambda_1 - \lambda_1^{-1}||\lambda_2 - \lambda_2^{-1}|^2}, & \lambda_1 \in U_{h_2+1} .\end{cases}$$

Remark: The value of $K_d(t_1,t_2)$ in other cases can be obtained readily from (4.2) and (4.3).

In §5, we consider the particular case when d is large and t_1 is "far" from 1. In this case, the formula for K_d simplifies considerably. More precisely, if $t_1 \in T_\theta \setminus (T_\theta)_d$ or $t_1 \in T_\theta^\# \setminus (T_\theta^\#)_d$, $\theta = \tau, \varepsilon\tau, \varepsilon$, we have

$$K_d(t_1,t_2) = \begin{cases} 0, & \text{if } t_2 \in T_{\theta'} \;\; T^{\#}_{\theta'},\;\; \theta \neq \theta', \\ \sigma(t_1)\sigma(t_2), & \text{if } t_1 \text{ and } t_2 \text{ lie in two non-conjugate} \\ & \text{tori associated to } \theta, \\ -\sigma(t_1)\sigma(t_2) + \kappa_T|\theta|^{-1/2}q^d\sigma(t_1)\sigma(t_2)[\Delta_d(t_1t_2^{-1}) + ([W_T]-1)\Delta_d(t_1t_2)], \\ & \text{if } t_1 \text{ and } t_2 \text{ lie in the same torus.} \end{cases}$$

Finally, we observe that, if $T = T_\theta$, then $\kappa_T|\theta|^{-1/2}q^d = [T:T_d]$ (see (3.1) and 3.2)).

5. The Inversion of I_f.

In this paragraph, we use the kernel K_d computed in §4 to determine the Fourier ransform of the invariant distribution

$$I_f^{T_0}(t_0) = \int_G f(xt_0x^{-1})\, dx = \mu(T_0)|D(t_0)|^{-1/2}F_f^{T_0}(t_0),$$

here t_0 is a regular element in a compact Cartan subgroup T_0, dx is a Haar measure n G and $\mu(T_0)$ is the measure of the torus T_0 as determined by Weyl's Lemma.

If $f \in C_c^\infty(G)$ and Π is any irreducible unitary representation of G with haracter Θ_Π, we set (see §1) $\hat{f}(\Pi) = \int_G f(x)\Theta_\Pi(x)\, dx$. By Weyl's Lemma, we can rite

$$\hat{f}(\Pi) = [W_A]^{-1} \int_A |D(a)|^{1/2}F_f^A(a)\Theta_\Pi(a)\, da$$
$$+ \sum_T [W_T]^{-1} \int_T |D(t)|^{1/2}F_f^T(t)\Theta_\Pi(t)\, dt\ ,$$

here the last sum is taken over the conjugacy classes of compact tori in G.

We let D denote the collection of representations of the discrete series of G §3), and, if d is a positive integer, we write $\sum\limits_{\substack{\Pi \in D \\ \text{cond } \Pi \leq d}}$ to denote a sum taken ver the special representation, the four representations of the form $\Pi^{\pm}(\Phi,V)$ and hose representations $\Pi(\Phi,\psi,V)$, $\psi^2 \neq 1$, in the discrete series for which ond $\psi = c_\theta^{(h)}$, $h \leq d$.

__Theorem 5.1.__ Suppose that T_0 is a compact Cartan subgroup of G and that $t_0 \in (T_0)_{h_0} \setminus (T_0)_{h_0+1}$, $h_0 \geq 0$ (if $T_0 = T_\varepsilon$ or $T_\varepsilon^{\#}$, $h_0 = 0$ signifies that the eigenvalues $\lambda_0, \lambda_0^{-1}$ of t_0 satisfy $|1 + \lambda_0| = |1 + \lambda_0^{-1}| = 1$). Then, for $f \in C_c^\infty(G)$,

$$I_f^{T_0}(t_0) = \sum_{\Pi \in D} \overline{\Theta_\Pi(t_0)}\, \hat{f}(\Pi) + \frac{1}{2} \sum_{\Pi \in RPS_V} \overline{\Theta_\Pi(t_0)}\, \hat{f}(\Pi)$$

$$- \frac{q+1}{2q}\, \mu(A_1) \int_{\substack{\xi \in \hat{k}^\times \\ \xi | A_{h_0+1} = 1}} |\Gamma(\xi)|^{-2}\, \hat{f}(\xi)\, d\xi$$

$$+ \frac{q}{2}\, \mu(A_1)\, \kappa_{T_0}\, |D(t_0)|^{-1/2} \int_{\substack{\xi \in \hat{k}^\times \\ \xi | A_{h_0+1} = 1}} \hat{f}(\xi)\, d\xi ,$$

where κ_{T_0} is the constant appearing in Lemma 2.5, $d\xi$ is the Haar measure on $\hat{k}^\times$ which is dual to the measure da on A ($\simeq k^\times$) and $\Gamma(\xi)$ is the gamma function defined in [GG] and [ST].

__Proof.__ Take an integer $d > h_0$ and consider $\displaystyle\sum_{\substack{\Pi \in D \\ \text{cond}\, \Pi < d}} \overline{\Theta_\Pi(t_0)}\, \hat{f}(\Pi)$.

Substituting the above expression for $\hat{f}$ and using Theorem 4.6, we have

$$\sum_{\substack{\Pi \in D \\ \text{cond}\, \Pi < d}} \overline{\Theta_\Pi(t_0)}\, \hat{f}(\Pi) = [W_A]^{-1} \int_A |D(a)|^{1/2} F_f^A(a)\, K_d(a,t_0)\, da$$

$$+ \sum_T [W_T]^{-1} \int_T |D(t)|^{1/2} F_f^T(t) K_d(t,t_0)\, dt$$

$$= [W_A]^{-1} \int_A |D(a)|^{1/2} F_f^A(a) K_d(a,t_0)\, da$$

$$+ [W_{T_0}]^{-1} [T_0 : (T_0)_d] \int_{T_0} |D(t)|^{1/2} F_f^{T_0}(t) \sigma(t) \sigma(t_0) [\Delta_d(t^{-1}t_0) + ([W_{T_0}] - 1) \Delta_d(tt_0)]\, dt$$

$$- [W_{T_0}]^{-1} \int_{T_0} |D(t)|^{1/2} F_f^{T_0}(t) \sigma(t) \sigma(t_0)\, dt$$

$$+ [1 - W[T_0]^{-1}) \int_{T_0^{\#}} |D(t)|^{1/2} F_f^{T_0^{\#}}(t)\, \sigma(t) \sigma(t_0)\, dt ,$$

where $T_0^{\#}$ denotes the torus associated to T_0 as in (3.3).

Now, a few observations allow us to simplfy these last expressions. First of ll, $|D(t)|^{1/2}$, $F_f^{T_0}(t)$ and $\sigma(t)$ are locally constant on T_0'. Thus, by choosing d ufficiently large, it follows from the definition of Δ_d that we have

$$[W_{T_0}]^{-1}[T_0:(T_0)_d] \int_{T_0} |D(t)|^{1/2}F_f^{T_0}(t)\sigma(t)\sigma(t_0)[\Delta_d(t^{-1}t_0) + ([W_{T_0}]-1)\Delta_d(tt_0)]dt$$

$$= [W_{T_0}]^{-1}[T_0:(T_0)_d]\,|D(t_0)|^{-1/2}\mu((T_0)_d)\{F_f^{T_0}(t_0) + ([W_{T_0}]-1)F_f^{T_0}(t_0^{-1})\}.$$

ut $([W_{T_0}]-1)F_f^{T_0}(t_0) = ([W_{T_0}]-1)F_f^{T_0}(t_0^{-1})$, so the last expression can be written as

$$[T_0:(T_0)_d]\mu((T_0)_d)|D(t_0)|^{-1/2}\, F_f^{T_0}(t_0) = I_f^{T_0}(t_0).$$

Secondly, we note that for $t \in T_0$ or $T_0^{\#}$,

$$\pm\, \sigma(t)\sigma(t_0) = \Theta_{(\Phi,V)}(t)\,\overline{\Theta_{(\Phi,V)}(t_0)} = \Theta_{(\Phi_b,V)}(t)\,\overline{\Theta_{(\Phi_b,V)}(t_0)}\ ,$$

$\in k^x$. It then follows from (3.8) and the properties of $\kappa(\Phi,V)$ (see Appendix) hat

$$-[W_{T_0}]^{-1} \int_{T_0} |D(t)|^{1/2}F_f^{T_0}(t)\sigma(t)\sigma(t_0)\, dt$$

$$+ (\ 1- [W_{T_0}]^{-1}) \int_{T_0^{\#}} |D(t)|^{1/2}\, F_f^{T_0^{\#}}(t)\sigma(t)\sigma(t_0)\, dt$$

$$= \frac{1}{2} \sum_{\Pi \in RPS_V} \overline{\Theta_\Pi(t_0)}\ \hat{f}(\Pi)\ ,$$

here V is the quadratic extension asociated to T_0.

Putting these observations together, we obtain, for d sufficiently large,

$$I_f^{T_0}(t_0) = \sum_{\substack{\Pi \in D \\ \text{cond } \Pi < d}} \overline{\Theta_\Pi(t_0)}\ \hat{f}(\Pi) + \frac{1}{2} \sum_{\Pi \in RPS_V} \overline{\Theta_\Pi(t_0)}\ \hat{f}(\Pi)$$

$$-\ [W_A]^{-1} \int_A |D(a)|^{1/2}\ F_f^A(a)\ K_d(a,t_0)\ da.$$

rom Theorem 4.6, it follows that

$$-[W_A]^{-1} \int_A |D(a)|^{1/2} F_f^A(a) K_d(a,t_0)\, da$$

$$= [W_A]^{-1} \int_{A\setminus A_{h_0+1}} |D(a)|^{1/2} F_f^A(a) \frac{|\lambda|}{|\lambda - \lambda^{-1}||1 - \lambda|^2}\, da$$

$$+ \frac{[W_A]^{-1}}{|\lambda_0 - \lambda_0^{-1}|^2} \int_{A_{h_0+1}} |D(a)|^{1/2} F_f^A(a) \frac{1}{|\lambda - \lambda^{-1}|}\, da ,$$

where λ, λ^{-1} are the eigenvalues of $a \in A$ and λ_0, λ_0^{-1} are the eigenvalues of t_0.

As stated in [SS2], it is a simple matter to show that, for $a \in A \setminus A_{h_0+1}$,

$$\frac{|\lambda|}{|\lambda - \lambda^{-1}||1 - \lambda|^2} = -\frac{q+1}{2q}\, \mu(A_1) \int_{\substack{\xi \in \hat{k}^x \\ \xi|A_{h_0+1} = 1}} |\Gamma(\xi)|^{-2}\, \Theta_\xi(a)\, d\xi ,$$

where $d\xi$ is the measure on $\hat{k}^x$ dual to da. Furthermore, from (3.5) and the explicit values of the gamma function [ST], it is easy to see that, for $a \in A_{h_0+1}$,

$$\int_{\substack{\xi \in \hat{k}^x \\ \xi|A_{h_0+1} = 1}} |\Gamma(\xi)|^{-2}\, \Theta_\xi(a)\, d\xi = \frac{2}{|\lambda - \lambda^{-1}|} \int_{\substack{\xi \in \hat{k}^x \\ \xi|A_{h_0+1} = 1}} |\Gamma(\xi|^{-2}\, d\xi$$

$$= \frac{2}{|\lambda - \lambda^{-1}|}\, \mu(A_1)^{-1} \left(\frac{q}{q+1}\right) q^{2h_0+1} .$$

Substituting into the expression above, we get

$$-[W_A]^{-1} \int_A |D(a)|^{1/2} F_f^A(a) K_d(a,t_0)\, da = -\left(\frac{q+1}{2q}\right) \mu(A_1) \int_{\substack{\xi \in \hat{k}^x \\ \xi|A_{h_0+1} = 1}} |\Gamma(\xi)|^{-2}\, \hat{f}(\xi)\, d\xi$$

$$+ \left[q^{2h_0+1} + \frac{1}{|\lambda_0 - \lambda_0^{-1}|^2}\right] [W_A]^{-1} \int_{A_{h_0+1}} |D(a)|^{1/2} F_f^A(a) \frac{1}{|\lambda - \lambda^{-1}|}\, da .$$

To complete the proof of the theorem, we observe that

$$[W_A]^{-1} \int_{A_{h_0+1}} |D(a)|^{1/2} F_f^A(a) \frac{1}{|\lambda - \lambda^{-1}|}\, da = q^{-h_0} \mu(A_1) \frac{1}{2} \int_{\substack{\xi \in \hat{k}^x \\ \xi | A_{h_0+1} = 1}} \hat{f}(\xi)\, d\xi ,$$

nd that

$$[q^{2h_0+1} + \frac{1}{|\lambda_0 - \lambda_0^{-1}|^2}]\, q^{-h_0} = q\, \kappa_{T_0} |D(t_0)|^{-1/2} .$$

Since $\hat{f}(\Pi) = 0$ for all but a finite number of Π [SS1], we are done.

6. The Plancherel Formula for G.

In this paragraph, we use the inversion formula for I_f given in Theorem 5.1, ogether with the asymptotic formula (2.1) to derive the Plancherel formula for G. ur approach is similar in spirit to that of Harish-Chandra in the real case.

Theorem 6.1. For $f \in C_c^\infty(G)$,

$$\mu(K)\, f(1) = \sum_{\Pi \in D} \hat{f}(\Pi) d(\Pi) + \frac{1}{2}\left(\frac{q^2 - 1}{q}\right)\mu(A_1) \int_{\xi \in \hat{k}^x} |\Gamma(\xi)|^{-2} \hat{f}(\xi)\, d\xi ,$$

here $d(\Pi)$ is an integer which is the degree of the inducing representation if Π s a supercuspidal representation (see [SS2], [Sk1]) and, if $\Pi = \Pi_0$, the special epresentation, $d(\Pi_0) = q-1$.

Before giving the proof of Theorem 6.1, we make several observations about the haracters of the discrete series and the characters of the reducible principal eries. Suppose that T is a compact torus in G and let $V = V_T$. If $\Pi \in D$, we set $h_V(\Pi) = 1$ if Π is of the form $\Pi(\Phi,\psi,V)$, $\psi^2 \neq 1$, $\Pi^+(\Phi,V)$ or $\Pi^-(\Phi,V)$; $h_V(\Pi) = 0$, therwise. Then, it follows from the explicit formulas for the characters of the iscrete series, that there exists an integer d so that, for $t \in T_d$,

$$\Theta_\Pi(t) = \frac{-d(\Pi)}{q-1} + h_V(\Pi)\, \kappa(\Pi)\, \varepsilon(t) |D(t)|^{-1/2} \qquad (6.2)$$

if Π is a supercuspidal representation,

$$(6.3) \qquad \Theta_0(t) = \frac{-d(\Pi_0)}{q-1} = -1$$

if $\Pi = \Pi_0$, the special representation. Here, $\kappa(\Pi) = \kappa(\Phi,V)$ and $\varepsilon(t)$ is the function on T defined in §2. We also have, from (3.8),

$$(6.4) \qquad \Theta_{(\Phi,V)}(t) = \kappa(\Pi)\ \varepsilon(t)\ |D(t)|^{-1/2} .$$

Proof of Theorem 6.1. Fix a compact torus T and let $V = V_T$. Fix $f \in C_c^\infty(G)$. Then, since $\hat{f}(\Pi) = 0$ for almost all $\Pi \in D$ [SS2], by taking t sufficiently close to 1 in T, we may substitute the expressions (6.2), (6.3), (6.4) for $\Theta_\Pi(t)$ in the formula for I_f (Theorem 5.1). Furthermore, since $\xi \mapsto \hat{f}(\xi)$ has compact support [SS2], by taking t close to 1 (that is, h_0 large), we may replace the integral over $\xi|A_{h_0+1} = 1$ by the integral over $\xi \in \hat{k}^x$ in the formula for I_f. This yields

$$(6.5) \qquad I_f(t) = -\hat{f}(\Pi_0) + \sum_{\Pi \in D_0} \hat{f}(\Pi)\{\frac{-d(\Pi)}{q-1} + h_V(\Pi)\ \overline{\kappa(\Pi)}\ \varepsilon(t)|D(t)|^{-1/2}\}$$

$$+ \frac{1}{2} \sum_{\Pi \in RPS_V} \hat{f}(\Pi)\ \overline{\kappa(\Pi)}\ \varepsilon(t)\ |D(t)|^{-1/2}$$

$$- \frac{1}{2}\left(\frac{q+1}{q}\right) \mu(A_1) \int_{\hat{k}^x} |\Gamma(\xi)|^{-2}\ \hat{f}(\xi)\ d\xi$$

$$+ \frac{q}{2}\ \mu(A_1)\kappa_T|D(t)|^{-1/2} \int_{\hat{k}^x} \hat{f}(\xi)\ d\xi ,$$

where D_0 denotes the supercuspidal representations, and $I_f = I_f^T$.

We also have, from (2.1), for t close to 1,

$$(6.6) \qquad I_f(t) = -A_T\mu(T)f(1)$$

$$+ \mu(T)B_T|D(t)|^{-1/2} \sum_{\dim \mathcal{O} > 0} \frac{1}{2}\{1 + \varepsilon(t)\ \mathrm{sgn}_V(\mathcal{O})\}\Lambda_{\mathcal{O}}(f) .$$

Equating (6.5) and (6.6), we have, for t close to 1,

$$(6.7)\quad A_T\mu(T)f(1) - \hat{f}(\Pi_0) - \sum_{\Pi\in D_0}\hat{f}(\Pi)\frac{d(\Pi)}{q-1} + \frac{1}{2}\left(\frac{q+1}{q}\right)\mu(A_1)\int_{\hat{k}^x}|\Gamma(\xi)|^{-2}\hat{f}(\xi)\,d\xi$$

$$= |D(t)|^{-1/2}\{\mu(T)B_T \sum_{\dim\mathcal{O}>0}\frac{1}{2}[1+\varepsilon(t)\,\mathrm{sgn}_V(\mathcal{O})]\Lambda_{\mathcal{O}}(f)$$

$$-\sum_{\Pi\in D_0}[\hat{f}(\Pi)h_V(\Pi)\,\overline{\kappa(\Pi)}\varepsilon(t)] - \frac{1}{2}\sum_{\Pi\in RPS_V}[\hat{f}(\Pi)\,\overline{\kappa(\Pi)}\,\varepsilon(t)]$$

$$-\frac{q}{2}\mu(A_1)\kappa_T\int_{\hat{k}^x}\hat{f}(\xi)\,d\xi\} .$$

Now define $T^{\pm}$ to be the set of $t\in T'$ satisfying $\varepsilon(t) = \pm 1$ respectively. It is clear that every neighborhood of 1 in T intersects both T^+ and T^-. Thus, by taking t close to 1 in one of these sets and observing that $\lim_{t\to 1}|D(t)| = 0$, we conclude that both sides of (6.7) are zero. Using the fact that $\mu(T)A_T = \frac{\mu(K)}{q-1}$, we have the Plancherel formula for G.

7. The Fourier Transform of a Unipotent Orbit.

We now prove, using the results of §6, that the Fourier transforms of the distributions $\Lambda_{\mathcal{O}}$ exist and, moreover, we give an explicit formula for these transforms. We also determine the value of the constant B_T as given in §2.

As we stated in §2, the nontrivial unipotent orbits in G may be indexed by $k^x/(k^x)^2$, that is, each nontrivial unipotent orbit contains an element of the form $\begin{pmatrix}1 & \zeta\\ 0 & 1\end{pmatrix}$ with ζ unique modulo $(k^x)^2$. In this section, we write $\Lambda_\zeta(f)$ for $\Lambda_{\mathcal{O}}(f)$ if $\begin{pmatrix}1 & \zeta\\ 0 & 1\end{pmatrix}\in\mathcal{O}$. Also, if V is a quadratic extension of k, we denote by D_V the supercuspidal representations associated to V and we set $\kappa_V = \kappa_T$ if T is a compact torus in G and $V = V_T$.

Theorem 7.1. For V a quadratic extension of k, set

$$S_V = \mu(N\cap K)\left(\frac{q-1}{q}\right)\kappa_V^{-1}[2\sum_{\Pi\in D_V}\hat{f}(\Pi)\,\overline{\kappa(\Pi)} + \sum_{\Pi\in RPS_V}\hat{f}(\Pi)\,\overline{\kappa(\Pi)}] .$$

Then, for $f \in C_c^\infty(G)$ and $\zeta \in k^\times/(k^\times)^2$,

$$4\Lambda_\zeta(f) = \sum_V \mathrm{sgn}_V(\zeta)S_V + \mu(N \cap K)(q-1)\mu(A_1) \int_{\hat{k}^\times} \hat{f}(\xi)\, d\xi ,$$

where the sum is taken over the three quadratic extensions of k.

<u>Proof</u>. From the results of §6, we have, for T a compact torus in G and $V = V_T$

$$\mu(T)B_T \sum_{\dim \mathcal{O}>0} \frac{1}{2}\,[1 + \varepsilon(t)\ \mathrm{sgn}_V(\mathcal{O})]\ \Lambda_{\mathcal{O}}(f)$$

$$= \varepsilon(t)\{\sum_{\Pi \in D_V} \hat{f}(\Pi)\ \overline{\kappa(\Pi)} + \frac{1}{2} \sum_{\Pi \in RPS_V} \hat{f}(\Pi)\ \overline{\kappa(\Pi)}\}$$

$$+ \frac{q}{2}\ \mu(A_1)\kappa_T \int_{\hat{k}^\times} \hat{f}(\xi)\, d\xi .$$

It follows from the linear independence of 1 and $\varepsilon(t)$ that, for $f \in C_c^\infty(G)$,

$$\mu(T)B_T \sum_{\zeta \in k^\times/(k^\times)^2} \Lambda_\zeta(f) = q\ \mu(A_1)\ \kappa_T \int_{\hat{k}^\times} \hat{f}(\xi)\, d\xi \tag{7.2}$$

and

$$\mu(T)B_T \sum_{\zeta \in k^\times/(k^\times)^2} \mathrm{sgn}_V(\zeta)\ \Lambda_\zeta(f) = 2\sum_{\Pi \in D_V} \hat{f}(\Pi)\ \overline{\kappa(\Pi)} + \sum_{\Pi \in RPS_V} \hat{f}(\Pi)\ \overline{\kappa(\Pi)} . \tag{7.3}$$

Note that $\mu(T)^{-1}B_T^{-1}\kappa_T$ is independent of T (see §2). If we let V range through the three distinct quadratic extensions of k, then (7.3) leads to three equations of the form

$$\sum_{\zeta \in k^\times/(k^\times)^2} \mathrm{sgn}_V(\zeta)\ \Lambda_\zeta(f) = S_V .$$

Together with (7.2), this yields four equations which are easily solved to yield $\Lambda_\zeta(f)$ for any $\zeta \in k^\times/(k^\times)^2$. This completes the proof of the theorem except for the explicit form of B_T.

Now let $f_0 = \frac{1}{\mu(K)}$ times the characteristic function of K. Let A_0 denote the set of unramified characters in $\hat{k}^x$. It is easy to see that the function defined by $\xi \mapsto \hat{f}_0(\xi)$ is the characteristic function of A_0. Moreover, the measures da and dξ on A and $\hat{k}^x$ respectively have been normalized in such a way that $\mu(A_1)\mu(T) = (q-1)^{-1}$. Using (7.2) and the explicit formula for $\Lambda_{\mathcal{O}}$, we see immediately that $B_T = \mu(T)^{-1}\mu(N \cap K)^{-1}q(q-1)^{-1}\kappa_T$ as stated in §2.

Appendix -- Computation of the Kernel.

The following lemmas are useful in computing K_d. Most of the results in these lemmas were probably known to Gauss and others in one form or another.

Suppose that Φ is a non-trivial additive character on k and that $V = k(\sqrt{\theta})$ is a quadratic extension of k. If we set $B(Z,W) = \mathrm{tr}\, Z\overline{W} = 2xu - 2\theta yv$, $Z = x + \sqrt{\theta}\, y$, $W = u + \sqrt{\theta}\, v$, then the Fourier transform on V is given by $\hat{f}(Z) = \int_V f(W)\Phi(B(Z,W))\, d_\Phi W$, where $f \in L^1(V)$ and $d_\Phi W$ is an additive Haar measure on V normalized so that $\hat{\hat{f}}(Z) = f(-Z)$. We define

$$\kappa(\Phi,V) = \mathrm{P.V.} \int_V \Phi(N_{V/k}(Z))\, d_\Phi Z, \tag{A.1}$$

where P.V. denotes the principal value integral over V. The factor $\kappa(\Phi,V)$ arises naturally when the supercuspidal representations are constructed by the techniques of Weil (see [Sk1], [T]).

Lemma A.2. $|\kappa(\Phi,V)|^2 = 1$. Furthermore, if $b \in k^\times$, $b \notin N_{V/k}(V^\times)$, then

$$\kappa(\Phi_b,V) = -\kappa(\Phi,V).$$

Proof. Suppose that χ is an additive character on k with conductor R. Then $\Phi = \chi_a$ for some $a \in k^\times$ where $\chi_a(x) = \chi(ax)$. It is easy to see that

$$\kappa(\Phi,V) = |a|\,|\theta|^{1/2}\, \mathrm{P.V.} \iint_{kk} \chi_a(x^2 - \theta y^2)\, dxdy.$$

The results then follow immediately from [T], pp. 222-223.

Lemma A.3 (i) Suppose $\lambda \in C_\theta$, $\theta = \tau, \varepsilon\tau$, and h is a non-negative integer. Then

$$\sum_{\substack{\psi \in \hat{C}_\theta \\ \mathrm{cond}\,\psi = C_\theta^{(h+1)}}} \psi(\lambda) = 2q^{h+1}\Delta_{h+1}(\lambda) - 2q^h\Delta_h(\lambda)\ .$$

(ii) Suppose $\lambda \in C_\varepsilon$ and h is a positive integer. Then

$$\sum_{\substack{\psi \in \hat{C}_\varepsilon \\ \mathrm{cond}\,\psi = C_\varepsilon^{(h+1)}}} \psi(\lambda) = (q+1)q^h\Delta_{h+1}(\lambda) - (q+1)q^{h-1}\Delta_h(\lambda)\ .$$

(iii) Suppose $\lambda \in C_\varepsilon$. Then

$$\sum_{\substack{\psi \in \hat{C}_\varepsilon \\ \mathrm{cond}\,\psi = C_\varepsilon^{(1)} \\ \psi \neq \psi_0}} \psi(\lambda) = (q+1)\Delta_1(\lambda) - 1 - \psi_0(\lambda).$$

Proof. This follows directly from (3.1) and (3.2).

Another factor that arises in the character tables [SS1] is

$$S(\psi) = q^{1/2} \int_U \mathrm{sgn}_\theta(x)\psi\left(\frac{(1+\sqrt{\theta})\theta^{h-1}x}{(1-\sqrt{\theta})\theta^{h-1}x}\right) dx \ , \tag{A.4}$$

where $\theta = \tau,\ \varepsilon\tau$, $\psi \in \hat{C}_\theta$, cond $\psi = C_\theta^{(h)}$, $h \geq 1$. By using the Cayley transform ([Sa2], p. 413), we see that $\frac{(1+\sqrt{\theta})\theta^{h-1}x}{(1-\sqrt{\theta})\theta^{h-1}x} \in C_\theta^{(h-1)} \setminus C_\theta^{(h)}$,

and that $\psi\left(\frac{(1+\sqrt{\theta})\theta^{h-1}x}{(1-\sqrt{\theta})\theta^{h-1}x}\right)$ is constant on the cosets of U_1 (or P) in U. For simplicity, we write $\frac{(1+\sqrt{\theta})\theta^{h-1}x}{(1-\sqrt{\theta})\theta^{h-1}x} = (\theta^{h-1}x)\cdot\varphi$ as in [Sa2], p. 413.

Lemma A.5 $[S(\psi)]^2 = \mathrm{sgn}_\theta(-1)$. Furthermore, if h is a positive integer and $\lambda \in C_\theta^{(k)} \setminus C_\theta^{(k+1)}$, where $0 \leq k < h-1$ and $\theta = \tau$ or $\varepsilon\tau$, then

$$\sum_{\substack{\psi \in \hat{C}_\theta \\ \mathrm{cond}\,\psi = C_\theta^{(h)}}} \psi(\lambda)S(\psi) = 0 \ .$$

Proof. We have

$$[S(\psi)]^2 = q \int_U \int_U \mathrm{sgn}_\theta(xy)\ \psi(x\theta^{h-1}\cdot\varphi)\psi(y\theta^{h-1}\cdot\varphi)\ dxdy \ .$$

Making the substitution $x \to xy$ and simplifying, we get

$$q \int_U \mathrm{sgn}_\theta(x) \int_U \psi([\theta^{h-1}y(1+x)]\cdot\varphi)\ dydx \ .$$

If $|1+x| < 1$, $\psi([\theta^{h-1}y(1+x)]\cdot\varphi) = 1$ since $[\theta^{h-1}y(1+x)]\cdot\varphi \in C_\theta^{(h)}$. If $|1+x| = 1$, it follows from [Sa1] that $\int_U \psi([\theta^{h-1}y(1+x)]\cdot\varphi)\ dy = -1/q$.

Thus, we have $-\int_{\substack{x \in U \\ |1+x|=1}} \mathrm{sgn}_\theta(x)\ dx + (q-1)\int_{\substack{x \in U \\ |1+x|<1}} \mathrm{sgn}_\theta(x)\ dx = \mathrm{sgn}_\theta(-1).$

For the next part of the lemma, we have

$$\sum_{\substack{\psi \in \hat{C}_\theta \\ \text{cond}\,\psi = C_\theta^{(h)}}} \psi(\lambda)S(\lambda) = q^{1/2} \int_U \operatorname{sgn}_\tau(x) \sum_{\substack{\psi \in \hat{C}_\theta \\ \text{cond}\,\psi = C_\theta^{(h)}}} \psi(\lambda[\theta^{h-1}x \cdot \varphi])\, dx .$$

The hypotheses imply that $\lambda[\theta^{h-1}x \cdot \varphi]$ is in $C_\theta^{(k)} \setminus C_\theta^{(k+1)}$, so the result follows from Lemma A.3.

<u>Lemma A.6</u>. Suppose that $\lambda \in C_\tau^{(h)} \setminus C_\tau^{(h+1)}$, $h \geq 0$. Then

(i) $\operatorname{sgn}_\tau (\operatorname{tr}(\lambda-1)) = \operatorname{sgn}_\tau(-1)$,

(ii) $\operatorname{sgn}_{\varepsilon\tau}(\operatorname{tr}(\lambda-1)) = -\operatorname{sgn}_\tau(-1)$,

(iii) $$\sum_{\substack{\gamma \in C_\tau^{(h)}/C_\tau^{(h-1)} \\ \gamma \neq 1,\lambda,\lambda^{-1}}} \operatorname{sgn}_\tau(\operatorname{tr}(\lambda-\gamma)) = -1 - \operatorname{sgn}_\tau(-1),$$

(iv) $$\sum_{\substack{\gamma \in C_{\varepsilon\tau}^{(h)}/C_{\varepsilon\tau}^{(h+1)} \\ \gamma \neq 1}} \operatorname{sgn}_{\varepsilon\tau}(\operatorname{tr}\lambda - \operatorname{tr}\gamma) = -1 + \operatorname{sgn}_\tau(-1) .$$

The notation $\gamma \in C_\tau^{(h)}/C_\tau^{(h+1)}$, $\gamma \neq 1, \lambda, \lambda^{-1}$, means that γ runs through a complete set of coset representatives of $C_\tau^{(h+1)}$ in $C_\tau^{(h)}$ excluding the cosets defined by 1, λ, λ^{-1}.

<u>Proof</u>. (i) Let $\lambda = \dfrac{1+\sqrt{\tau}x}{1-\sqrt{\tau}x}$, $x \in k$, $|x| = q^{-h}$. Then $\operatorname{tr}(\lambda-1) = \dfrac{4\tau x^2}{1-\tau x^2}$ and $\operatorname{sgn}_\tau(\operatorname{tr}(\lambda-1)) = \operatorname{sgn}_\tau(\tau) = \operatorname{sgn}_\tau(-1)$.

(ii) $\operatorname{sgn}_{\varepsilon\tau}(\operatorname{tr}(\lambda-1)) = \operatorname{sgn}_{\varepsilon\tau}(\tau) = -\operatorname{sgn}_\tau(-1)$.

(iii) We write λ as above and $\gamma = \dfrac{1+\sqrt{\tau}y}{1-\sqrt{\tau}y}$, $y \in k$, $|y| = q^{-h}$. The sum then can be written $\displaystyle\sum_{\substack{y \in P^h/P^{h+1} \\ y \neq 0,x,-x}} \operatorname{sgn}_\tau(y^2 - x^2)$. If we set $x = \tau^h\eta$, $y = \tau^h\zeta$, $\eta,\zeta \in U$, we have

$$\sum_{\substack{\zeta \in R/P \\ \zeta \neq 0,\pm\eta}} \operatorname{sgn}_\tau(\zeta^2\eta^{-2} - 1) = \sum_{\substack{\alpha \in R/P \\ \alpha \neq 0,\pm 1}} \operatorname{sgn}_\tau(\alpha^2 - 1).$$

The result now follows from [D], p. 46.

(iv) We proceed as in (iii) except that we write

$$\gamma = \frac{1+\sqrt{\varepsilon\tau}y}{1-\sqrt{\varepsilon\tau}y}, \quad y \in k,\ |y| = q^{-h}.$$

he result again follows from [D], p. 46.

Lemma A.7. Suppose $\lambda_1 \in C_\tau^{(h)} \setminus C_\tau^{(h+1)}$, $\lambda_2 \in C_\tau^{(h)} \setminus C_\tau^{(h+1)}$, $h > 0$. Then

$$\sum_{\substack{\gamma \in C_\tau^{(h)}/C_\tau^{(h+1)} \\ \gamma \neq \lambda_1,\lambda_1^{-1},\lambda_2,\lambda_2^{-1}}} \operatorname{sgn}_\tau(\operatorname{tr}(\lambda_1-\gamma))\operatorname{sgn}_\tau(\operatorname{tr}(\lambda_2-\gamma))$$

$$+ \sum_{\gamma \in C_{\varepsilon\tau}^{(h)}/C_{\varepsilon\tau}^{(h+1)}} \operatorname{sgn}_{\varepsilon\tau}(\operatorname{tr}\lambda_1 - \operatorname{tr}\gamma)\operatorname{sgn}_{\varepsilon\tau}(\operatorname{tr}\lambda_2 - \operatorname{tr}\gamma)$$

$$= (-2+2q)\operatorname{sgn}_\tau\left(\frac{\lambda_1-\lambda_1^{-1}}{2\sqrt{\tau}}\right)\operatorname{sgn}_\tau\left(\frac{\lambda_2-\lambda_2^{-1}}{2\sqrt{\tau}}\right)[\Delta_{h+1}(\lambda_1\lambda_2^{-1}) + \operatorname{sgn}_\tau(-1)\Delta_{h+1}(\lambda_1\lambda_2)].$$

Proof. We first use the Cayley transform to transfer the problem to k. Write

$$\lambda_1 = \frac{1+\sqrt{\tau}x_1}{1-\sqrt{\tau}x_1}, \quad \lambda_2 = \frac{1+\sqrt{\tau}x_2}{1-\sqrt{\tau}x_2},$$

here $x_1, x_2 \in k$, $|x_1| = |x_2| = q^{-h}$. If we write $\gamma = \frac{1+\sqrt{\tau}y}{1-\sqrt{\tau}y}$ for $\gamma \in C_\tau^{(h)}/C_\tau^{(h+1)}$ nd $\gamma = \frac{1+\sqrt{\varepsilon\tau}y}{1-\sqrt{\varepsilon\tau}y}$ for $\gamma \in C_{\varepsilon\tau}^{(h)}/C_{\varepsilon\tau}^{(h+1)}$, the sum to be evaluated becomes

$$\sum_{\substack{y \in P^h/P^{h+1} \\ y \neq \pm x_1, \pm x_2}} \operatorname{sgn}_\tau(x_1^2-y^2)\operatorname{sgn}_\tau(x_2^2-y^2) + \sum_{y \in P^h/P^{h+1}} \operatorname{sgn}_{\varepsilon\tau}(x_1^2-\varepsilon y^2)\operatorname{sgn}_{\varepsilon\tau}(x_2^2-\varepsilon y^2).$$

riting $x_1 = \eta_1\tau^h$, $x_2 = \eta_2\tau^h$, $\eta_1, \eta_2 \in U$, and observing that $\operatorname{sgn}_\tau = \operatorname{sgn}_{\varepsilon\tau}$ on U, we et

$$\sum_{\substack{\alpha \in R/P \\ \alpha \neq \pm\eta_1, \pm\eta_2}} \operatorname{sgn}_\tau(\eta_1^2-\alpha^2)\operatorname{sgn}_\tau(\eta_2^2-\alpha^2) + \sum_{\alpha \in R/P} \operatorname{sgn}_{\varepsilon\tau}(\eta_1^2-\varepsilon\alpha^2)\operatorname{sgn}_{\varepsilon\tau}(\eta_2^2-\varepsilon\alpha^2)$$

$$= 1 + \sum_{\substack{\alpha \in F^\times \\ \alpha \neq \eta_1^2,\eta_2^2}} \operatorname{sgn}_\tau(\eta_1^2-\alpha)\operatorname{sgn}_\tau(\eta_2^2-\alpha)(1+\operatorname{sgn}_\tau(\alpha))$$

$$+ 1 + \sum_{\substack{\alpha \in F^{x} \\ \alpha \neq \eta_1^2,\, \eta_2^2}} \mathrm{sgn}_{\varepsilon\tau}(\eta_1^2 - \alpha)\mathrm{sgn}_{\varepsilon\tau}(\eta_2^2 - \alpha)(1 - \mathrm{sgn}_{\varepsilon\tau}(\alpha))$$

$$= 2 + 2 \sum_{\substack{\alpha \in F^{x} \\ \alpha \neq \eta_1^2,\eta_2^2}} \mathrm{sgn}_\tau[(\eta_1^2 - \alpha)(\eta_2^2 - \alpha)] ,$$

where F^x denotes the multiplicative group of R/P.

Now

$$\sum_{\substack{\alpha \in F^{x} \\ \alpha \neq \eta_1^2,\eta_2^2}} \mathrm{sgn}_\tau[(\eta_1^2 - \alpha)(\eta_2^2 - \alpha)] = \sum_{\substack{\alpha \in F^{x} \\ \alpha \neq \eta_1^2,\eta_2^2}} \mathrm{sgn}_\tau \left[(\alpha - \frac{\eta_1^2 + \eta_2^2}{2})^2 - (\frac{\eta_1^2 - \eta_2^2}{2})^2\right] .$$

If $\lambda_1\lambda_2$ or $\lambda_1\lambda_2^{-1}$ is in $C_\tau^{(h+1)}$, then $|\eta_1^2 - \eta_2^2| < 1$, that is, $\eta_1^2 \equiv \eta_2^2 \pmod{P}$. It is then immediate that the last sum is q-2. If $\lambda_1\lambda_2$ and $\lambda_1\lambda_2^{-1}$ are in $C_\tau^{(h)} \setminus C_\tau^{(h+1)}$, then $|\eta_1^2 - \eta_2^2| = 1$. Set $a = \frac{\eta_1^2 + \eta_2^2}{2}$, $b = \frac{\eta_1^2 - \eta_2^2}{2}$. Then, the last sum becomes

$$\mathrm{sgn}_\tau(-1)\{ \sum_{\substack{\alpha \in F^{x} \\ \alpha \neq a+b,a-b}} \mathrm{sgn}_\tau[1 - (\alpha - a)^2 b^{-2}]\}$$

$$= \mathrm{sgn}_\tau(-1) \begin{cases} 1 - \mathrm{sgn}_\tau(-1) + \sum_{\substack{\alpha \in F^{x} \\ \alpha \neq \pm 1}} \mathrm{sgn}_\tau(1 - \alpha^2), & |a| = 1, \\ \sum_{\substack{\alpha \in F^{x} \\ \alpha \neq \pm 1}} \mathrm{sgn}_\tau(1 - \alpha^2), & |a| < 1. \end{cases}$$

Since the condition $|a| < 1$ implies that $-1 \in (k^x)^2$, it follows from Lemma A.6 that we get -2 for the last expression in both cases.

Finally, we observe that

$$\mathrm{sgn}_\tau(\frac{\lambda_1 - \lambda_1^{-1}}{2\sqrt{\tau}})\, \mathrm{sgn}_\tau(\frac{\lambda_2 - \lambda_2^{-1}}{2\sqrt{\tau}}) = \mathrm{sgn}_\tau(\eta_1\eta_2).$$

Thus, if $\lambda_1\lambda_2^{-1} \in C_\tau^{(h+1)}$, we have $|\eta_1 - \eta_2| = |1 - \eta_1\eta_2^{-1}| < 1$, so that

$\mathrm{sgn}_\tau(\eta_1\eta_2) = 1$. If $\lambda_1\lambda_2 \in C_\tau^{(h+1)}$, then $|\eta_1 + \eta_2| = |1 + \eta_1\eta_2^{-1}| < 1$, so that $\mathrm{sgn}_\tau(\eta_1\eta_2) = \mathrm{sgn}_\tau(-1)$. This concludes the proof of the lemma.

Lemma A.8. Suppose $\lambda_1 \in C_\tau^{(h)} \setminus C_\tau^{(h+1)}$, $\lambda_2 \in C_{\varepsilon\tau}^{(h)} \setminus C_{\varepsilon\tau}^{(h+1)}$, $h \geq 0$. Then

$$\sum_{\substack{\gamma \in C_\tau^{(h)}/C_\tau^{(h+1)} \\ \gamma \neq \lambda_1, \lambda_1^{-1}}} \mathrm{sgn}_\tau(\mathrm{tr}(\lambda_1 - \gamma))\mathrm{sgn}_\tau(\mathrm{tr}\ \lambda_2 - \mathrm{tr}\ \gamma) + \sum_{\substack{\gamma \in C_{\varepsilon\tau}^{(h)}/C_{\varepsilon\tau}^{(h+1)} \\ \gamma \neq \lambda_2, \lambda_2^{-1}}} \mathrm{sgn}_\tau(\mathrm{tr}\ \lambda_1 - \mathrm{tr}\ \gamma)\mathrm{sgn}_\tau(\mathrm{tr}(\lambda_2 - \gamma)) = -2$$

Proof. We write $\lambda_1 = \dfrac{1 + \sqrt{\tau}x_1}{1 - \sqrt{\tau}x_1}$, $\lambda_2 = \dfrac{1 + \sqrt{\varepsilon\tau}x_2}{1 - \sqrt{\varepsilon\tau}x_2}$, $x_1, x_2 \in k$, $|x_1| = |x_2| = q^{-h}$ and proceed in a manner similar to the proof of Lemma A.7. The details will be left to the reader.

Lemma A.9. Suppose that $\lambda \in C_\varepsilon$, $|1 + \lambda| = 1$, and that ψ_0 is the character of order two on C_ε. Then

(i) $\psi_0(-1) = -\mathrm{sgn}_\tau(-1)$.

(ii) $\psi_0(\lambda) = \mathrm{sgn}_\tau(\lambda + \lambda^{-1} + 2)$.

(iii) If $|1 - \lambda| = 1$, then $\sigma(\lambda) = 1$.

Proof. It is well known that there is a cyclic subgroup H of order q+1 in C_ε such that C_ε is the direct product of H and $C_\varepsilon^{(1)}$. A complete set of coset representatives for $C_\varepsilon^{(1)}$ in C_ε is given by

$$\{1, -1, \frac{1 + \sqrt{\varepsilon}\varepsilon^r}{1 - \sqrt{\varepsilon}\varepsilon^r}\ ;\ r = 0,1,\ldots,q-2\}.$$

Since every element of $C_\varepsilon^{(1)}$ has a unique square root in $C_\varepsilon^{(1)}$, it follows that the kernel of ψ_0 is the subgroup of index two defined by $C_\varepsilon' = \{t^2 : t \in C_\varepsilon\}$.

To prove (i), we note that $-1 \in C_\varepsilon'$ if and only if $-1 = (\alpha + \sqrt{\varepsilon}\,\beta)^2$ for some $\alpha,\beta \in k$ with $\alpha^2 - \varepsilon\beta^2 = 1$. But this occurs if and only if $\alpha = 0$ and $-\varepsilon\beta^2 = 1$, which is possible if and only if $-1 \notin (k^x)^2$.

For (ii), we write $\lambda = \frac{1 + \sqrt{\varepsilon x}}{1 - \sqrt{\varepsilon x}}$, $x \in R$. If $x \in P$, then $\lambda \in C_\varepsilon^{(1)}$ and $\psi_0(\lambda) = \operatorname{sgn}_\tau(1 - \varepsilon x^2) = 1$. If $x \in U$, then $|1 - \lambda| = 1$ and $\psi_0(\lambda) = 1$ if and only if $\lambda = \frac{1 + \varepsilon x^2}{1 - \varepsilon x^2} + 2\sqrt{\varepsilon}\frac{x}{1 - \varepsilon x^2} = (\alpha + \sqrt{\varepsilon}\beta)^2$ for some $\alpha, \beta \in k$ with $\alpha^2 - \varepsilon\beta^2 = 1$. This is possible if and only if $\alpha^2 = 1/1-\varepsilon x^2$ so that $\operatorname{sgn}_\tau(\lambda + \lambda^{-1} + 2) = \operatorname{sgn}_\tau(\alpha^2) = 1$.

The proof of (iii) is immediate.

We now present some of the details involved in the actual computation of $K_d(t_1,t_2)$. There are a number of different cases that must be considered, and we have attempted to provide enough information to indicate both the total volume of computations and the essential guide lines along the way. In the following discussion t_1 and t_2 always denote regular elements in G. The eigenvalues of t_1 are written $\lambda_1, \lambda_1^{-1}$ and those of t_2 are written $\lambda_2, \lambda_2^{-1}$. For the compuations, we write $V = k(\sqrt{\tau})$, $V' = k(\sqrt{\varepsilon\tau})$, $V'' = k(\sqrt{\varepsilon})$.

Case I. $t_1 \in T_\tau$, $t_2 \in T_\tau$.

(I_1) $$\lambda_1 \in C_\tau^{(h_1)} \setminus C_\tau^{(h_1+1)}, \quad \lambda_2 \in C_\tau^{(h_2)} \setminus C_\tau^{(h_2+1)}, \quad 0 \le h_1 \le h_2 .$$

We first consider the case $0 < h_1 < h_2 < d$. Directly from [SS1], we have

$2K_d(t_1,t_2) =$

(a) $$2 + \{[\kappa(\Phi,V)\sigma(\lambda_1) - 1][\overline{\kappa(\Phi,V)}\sigma(\lambda_2) - 1] + [\kappa(\Phi,V)\sigma(\lambda_1) + 1][\overline{\kappa(\Phi,V)}\sigma(\lambda_2) + 1]\}$$

(b) $$+ \sum_{\Phi} \sum_{h=1}^{h_1} \sum_{\substack{\psi \in \hat{C}_\tau \\ \text{cond } \psi = C_\tau^{(h)}}} [-\tfrac{1}{2} q^h(\tfrac{q+1}{q}) + \kappa(\Phi,V)\sigma(\lambda_1)][-\tfrac{1}{2} q^h(\tfrac{q+1}{q}) + \overline{\kappa(\Phi,V)}\sigma(\lambda_2)]$$

c) $$+ \sum_{\Phi} \sum_{\substack{\psi \in \hat{C}_\tau \\ \text{cond } \psi = C_\tau^{(h_1+1)}}} \{ [\frac{q^{-1/2}}{2|\lambda_1 - \lambda_1^{-1}|} \sum_{\substack{\gamma \in C_\tau^{(h_1)}/C_\tau^{(h_1+1)} \\ \gamma \neq \lambda_1, \lambda_1^{-1}}}$$

$$\mathrm{sgn}_\tau(\mathrm{tr}(\lambda_1 - \gamma))\psi(\gamma) + \frac{1}{2}\kappa(\Phi,V)\sigma(\lambda_1)(\psi(\lambda_1) + \psi(\lambda_1^{-1}))]$$

$$\times [-\frac{1}{2} q^{h_1+1}(\frac{q+1}{q}) + \overline{\kappa(\Phi,V)}\sigma(\lambda_2)]\}$$

d) $$+ \sum_{\Phi} \sum_{h=h_1+2}^{h_2} \sum_{\substack{\psi \in \hat{C}_\tau \\ \text{cond } \psi = C_\tau^{(h)}}} \{ [\frac{1}{2}(\mathrm{sgn}_\tau(-1))^h \sigma(\lambda_1)\{\psi(\lambda_1)[\mathrm{sgn}_\tau(-1)S(\psi) + \kappa(\Phi,V)]$$

$$+ \psi(\lambda_1^{-1})[S(\psi) + \kappa(\Phi,V)]\}] \times [-\frac{1}{2} q^h(\frac{q+1}{q}) + \overline{\kappa(\Phi,V)}\sigma(\lambda_2)]\}$$

e) $$+ \sum_{\Phi} \sum_{\substack{\psi \in \hat{C}_\tau \\ \text{cond } \psi = C_\tau^{(h_2+1)}}} \{ [\frac{1}{2} (\mathrm{sgn}_\tau(-1))^{h_2+1} \sigma(\lambda_1)\{\psi(\lambda_1)[\mathrm{sgn}_\tau(-1)S(\psi) + \kappa(\Phi,V)]$$

$$+ \psi(\lambda_1^{-1})[S(\psi) + \kappa(\Phi,V)]\}]$$

$$\times [\frac{q^{-1/2}}{2|\lambda_2 - \lambda_2^{-1}|} \sum_{\substack{\gamma \in C_\tau^{(h_2)}/C_\tau^{(h_2+1)} \\ \gamma \neq \lambda_2, \lambda_2^{-1}}} \mathrm{sgn}_\tau(\mathrm{tr}(\lambda_2 - \gamma))\psi(\gamma^{-1})$$

$$+ \frac{1}{2} \overline{\kappa(\Phi,V)}\sigma(\lambda_2)(\psi(\lambda_2) + \psi(\lambda_2^{-1}))]\}$$

) $$+ \sum_{\Phi} \sum_{h=h_2+2}^{d} \sum_{\substack{\psi \in \hat{C}_\tau \\ \text{cond } \psi = C_\tau^{(h)}}} \{ [\frac{1}{2} (\mathrm{sgn}_\tau(-1))^h \sigma(\lambda_1)\{\psi(\lambda_1)[\mathrm{sgn}_\tau(-1)S(\psi)$$

$$+ \kappa(\Phi,V)] + \psi(\lambda_1^{-1})[S(\psi) + \kappa(\Phi,V)]\}]$$

$$\times [\frac{1}{2} (\mathrm{sgn}_\tau(-1))^h \sigma(\lambda_2)\{\psi(\lambda_2^{-1})[\mathrm{sgn}_\tau(-1)\overline{S(\psi)} + \overline{\kappa(\Phi,V)}] + \psi(\lambda_2)[\overline{S(\psi)} + \overline{\kappa(\Phi,V)}]\}]\}$$

$$\text{(g)}\quad + \sum_{\Phi}\sum_{h=1}^{h_1}\sum_{\substack{\psi\in\hat{C}_{\varepsilon\tau}\\ \mathrm{cond}\,\psi = C_{\varepsilon\tau}^{(h)}}} [-\tfrac{1}{2}\,q^{h}(\tfrac{q+1}{q})]^2$$

$$\text{(h)}\quad + \sum_{\Phi}\sum_{\substack{\psi\in\hat{C}_{\varepsilon\tau}\\ \mathrm{cond}\,\psi = C_{\varepsilon\tau}^{(h_1+1)}}} [\frac{q^{-1/2}}{2|\lambda_1-\lambda_1^{-1}|}\sum_{\gamma\in C_{\varepsilon\tau}^{(h_1)}/C_{\varepsilon\tau}^{(h_1+1)}} \mathrm{sgn}_{\varepsilon\tau}(\mathrm{tr}\,\lambda_1 - \mathrm{tr}\,\gamma)\psi(\gamma)]$$

$$\times [-\tfrac{1}{2}\,q^{h_1+1}(\tfrac{q+1}{q})]$$

$$\text{(i)}\quad + \sum_{h=1}^{h_1+1}\sum_{\substack{\psi\in\hat{C}_{\varepsilon}\\ \mathrm{cond}\,\psi = C_{\varepsilon}^{(h)}\\ \psi\neq\psi_0}} (-q^{h-1})^2 .$$

We first compute $\sum_{\Phi}$ and use the properties of $\kappa(\Phi,V)$ given in Lemma A.2. This yields

$$\text{(a)}\quad = 2 + 2[\sigma(\lambda_1)\sigma(\lambda_2) + 1] \ ;$$

$$\text{(b)}\quad = \sum_{h=1}^{h_1}\sum_{\substack{\psi\in\hat{C}_{\tau}\\ \mathrm{cond}\,\psi = C_{\tau}^{(h)}}} [\tfrac{1}{2}\,q^{2h}(\tfrac{q+1}{q})^2 + 2\sigma(\lambda_1)\sigma(\lambda_2)] \ ;$$

$$\text{(c)}\quad = \sum_{\substack{\psi\in\hat{C}_{\tau}\\ \mathrm{cond}\,\psi = C_{\tau}^{(h_1+1)}}} \{[-\tfrac{1}{2}\,q^{h_1+1}(\tfrac{q+1}{q})\,\frac{q^{-1/2}}{|\lambda_1-\lambda_1^{-1}|}\sum_{\substack{\gamma\in C_{\tau}^{(h_1)}/C_{\tau}^{(h_1+1)}\\ \gamma\neq\lambda_1,\lambda_1^{-1}}}$$

$$\mathrm{sgn}_{\tau}(\mathrm{tr}(\lambda_1-\gamma))\psi(\gamma)] + \sigma(\lambda_1)\sigma(\lambda_2)(\psi(\lambda_1)+\psi(\lambda_1^{-1}))\};$$

$$\text{(d)}\quad = \sum_{h=h_1+2}^{h_2}\sum_{\substack{\psi\in\hat{C}_{\tau}\\ \mathrm{cond}\,\psi = C_{\tau}^{(h)}}} (\mathrm{sgn}_{\tau}(-1))^h\sigma(\lambda_1)\{-\tfrac{1}{2}\,q^{h}(\tfrac{q+1}{q})S(\psi)[\psi(\lambda_1)\mathrm{sgn}_{\tau}(-1) + \psi(\lambda_1^{-1})]$$

$$+ \sigma(\lambda_2)[\psi(\lambda_1) + \psi(\lambda_1^{-1})]\} \ ;$$

e) $$= \sum_{\substack{\psi \in \hat{C}_\tau \\ \text{cond}\,\psi = C_\tau^{(h_2+1)}}} \frac{1}{2}(\mathrm{sgn}_\tau(-1))^{h_2+1}\sigma(\lambda_1)\{[\frac{q^{-1/2}}{|\lambda_2 - \lambda_2^{-1}|} \sum_{\substack{\gamma \in C_\tau^{(h_2)}/C_\tau^{(h_2+1)} \\ \gamma \neq \lambda_2, \lambda_2^{-1}}}$$

$$\mathrm{sgn}_\tau(\mathrm{tr}(\lambda_2 - \gamma))\psi(\gamma^{-1})][S(\psi)(\psi(\lambda_1)\mathrm{sgn}_\tau(-1) + \psi(\lambda_1^{-1}))]$$

$$+ \sigma(\lambda_2)(\psi(\lambda_1) + \psi(\lambda_1^{-1}))(\psi(\lambda_2) + \psi(\lambda_2^{-1}))\};$$

f) $$= \sum_{h=h_2+2}^{d} \sum_{\substack{\psi \in \hat{C}_\tau \\ \text{cond}\,\psi = C_\tau^{(h)}}} \sigma(\lambda_1)\sigma(\lambda_2)\{[\psi(\lambda_1\lambda_2^{-1}) + \psi(\lambda_1^{-1}\lambda_2)]$$

$$+ (\frac{1 + \mathrm{sgn}_\tau(-1)}{2})[\psi(\lambda_1\lambda_2) + \psi(\lambda_1^{-1}\lambda_2^{-1})]\};$$

g) $$= \frac{1}{2}(\frac{q+1}{q})^2 \sum_{h=1}^{h_1} \sum_{\substack{\psi \in \hat{C}_{\varepsilon\tau} \\ \text{cond}\,\psi = C_{\varepsilon\tau}^{(h)}}} q^{2h} \quad ;$$

h) $$= (-q^{h_1+1})(\frac{q+1}{q})(\frac{q^{-1/2}}{2|\lambda_1 - \lambda_1^{-1}|}) \sum_{\substack{\psi \in \hat{C}_{\varepsilon\tau} \\ \text{cond}\,\psi = C_{\varepsilon\tau}^{(h_1+1)}}} \sum_{\gamma \in C_{\varepsilon\tau}^{(h_1)}/C_{\varepsilon\tau}^{(h_1+1)}}$$

$$[\mathrm{sgn}_{\varepsilon\tau}(\mathrm{tr}\,\lambda_1 - \mathrm{tr}\,\gamma)\psi(\gamma)];$$

i) $$= \frac{2}{q^2} \sum_{h=1}^{h_1+1} \sum_{\substack{\psi \in \hat{C}_\varepsilon \\ \text{cond}\,\psi = C_\varepsilon^{(h)} \\ \psi \neq \psi_0}} q^{2h} .$$

Using Lemma A.3 and summing the resulting geometric series, we have

$$\text{(b)} \quad = \frac{(q+1)^2}{q^2+q+1}(q^{3h_1}-1) + 4\sigma(\lambda_1)\sigma(\lambda_2)(q^{h_1}-1) \;;$$

$$\text{(g)} \quad = \frac{(q+1)^2}{q^2+q+1}(q^{3h_1}-1) \;;$$

$$\text{(i)} \quad = -\frac{2(q^2+1)}{q^2+q+1} + \frac{2q^2(q+1)}{q^2+q+1}\, q^{3h_1} \,.$$

To compute (d) and (e), we first sum over ψ. Since $h \geq h_1 + 2$ and the elements at which the characters are evaluated lie in $C_\tau^{(h_1)} \setminus C_\tau^{(h_1+1)}$, it follows from Lemma A.3 and Lemma A.5 that (d) = (e) = 0. Similarly, it follows from Lemma A.3 that (f) = 0

For the computations of (c) and (h), we use Lemma A.3 and Lemma A.6. For example, in (c), we have

$$\sum_{\substack{\gamma \in C_\tau^{(h_1)}/C_\tau^{(h_1+1)} \\ \gamma \neq \lambda_1, \lambda_1^{-1}}} \operatorname{sgn}_\tau(\operatorname{tr}(\lambda_1 - \gamma)) \sum_{\substack{\psi \in \hat{C}_\tau \\ \operatorname{cond}\psi = C_\tau^{(h_1+1)}}} \psi(\gamma)$$

$$= \sum_{\gamma \neq \lambda_1, \lambda_1^{-1}} \operatorname{sgn}_\tau(\operatorname{tr}(\lambda_1 - \gamma))[2q^{h_1+1}\Delta_{h_1+1}(\gamma) - 2q^{h_1}\Delta_{h_1}(\gamma)]$$

$$= -2q^{h_1} \sum_{\gamma \neq \lambda_1, \lambda_1^{-1}, 1} \operatorname{sgn}_\tau(\operatorname{tr}(\lambda_1 - \gamma)) + 2q^{h_1}(q-1)\operatorname{sgn}_\tau(\operatorname{tr}(\lambda_1 - 1))$$

$$= 2q^{h_1}(1 + \operatorname{sgn}_\tau(-1)) + 2q^{h_1}(q-1)\operatorname{sgn}_\tau(-1) \quad \text{(by Lemma A.6)}$$

$$= 2q^{h_1}(1 + q\operatorname{sgn}_\tau(-1)).$$

ow, applying Lemma A.3 and observing that $\frac{1}{|\lambda_1 - \lambda_1^{-1}|} = q^{h_1+1/2}$, we get

$$\text{c)} \quad = -q^{3h_1}(q+1)[1 + q\,\mathrm{sgn}_\tau(-1)] - 4q^{h_1}\sigma(\lambda_1)\sigma(\lambda_2).$$

similar argument using Lemma A.3 and Lemma A.6 yields

$$\text{h)} = q^{3h_1}(q+1)[-1 + q\,\mathrm{sgn}_\tau(-1)] .$$

Finally, adding (a) through (i), we have

$$2K_d(t_1,t_2) = -2\sigma(t_1)\sigma(t_2) .$$

Next, we consider the case $0 < h_1 = h_2 < d$. From [SS1], we have

$$2K_d(t_1,t_2) = (a) + (b) + (c') + (f') + (g) + (h') + (i),$$

here (a), (b), (g) and (i) are as above,

$$\text{c')} = \sum_{\Phi} \sum_{\substack{\psi \in \hat{C}_\tau \\ \text{cond}\,\psi = C_\tau^{(h_1+1)}}} [\{[\frac{q^{-1/2}}{2|\lambda_1 - \lambda_1^{-1}|} \sum_{\substack{\gamma^1 \in C_\tau^{(h_1)}/C^{(h_1+1)} \\ \gamma^1 \neq \lambda_1,\lambda_1^{-1}}} \mathrm{sgn}_\tau(\mathrm{tr}(\lambda_1 - \gamma^1)\psi(\gamma^1)] + \frac{1}{2}\kappa(\Phi,V)\sigma(\lambda_1)(\psi(\lambda_1) + \psi(\lambda_1^{-1}))\}$$

$$\times \{[\frac{q^{-1/2}}{2|\lambda_2 - \lambda_2^{-1}|} \sum_{\substack{\gamma \in C_\tau^{(h_1)}/C_\tau^{(h_1+1)} \\ \gamma \neq \lambda_2,\lambda_2^{-1}}} \mathrm{sgn}_\tau(\mathrm{tr}(\lambda_2 - \gamma))\psi(\gamma^{-1})]$$

$$+ [\frac{1}{2}\overline{\kappa(\Phi,V)}\,\sigma(\lambda_2)(\psi(\lambda_2) + \psi(\lambda_2^{-1}))]\}],$$

f') has the same initial expression as (f) above, but the value is different because of the position of t_1 and t_2, and

$$(h') = \sum_{\Phi} \sum_{\substack{\psi \in \hat{C}_{\varepsilon\tau} \\ \text{cond } \psi = C_{\varepsilon\tau}^{(h_1+1)}}} \{[\frac{q^{-1/2}}{2|\lambda_1 - \lambda_1^{-1}|} \sum_{\gamma^1 \in C_{\varepsilon\tau}^{(h_1)}/C_{\varepsilon\tau}^{(h_1+1)}} \mathrm{sgn}_{\varepsilon\tau}(\mathrm{tr}\ \lambda_1 - \mathrm{tr}\ \gamma^1)\psi(\gamma^1)]$$

$$\times [\frac{q^{-1/2}}{2|\lambda_2 - \lambda_2^{-1}|} \sum_{\gamma \in C_{\varepsilon\tau}^{(h_1)}/C_{\varepsilon\tau}^{(h_1+1)}} \mathrm{sgn}_{\varepsilon\tau}(\mathrm{tr}\ \lambda_2 - \mathrm{tr}\ \gamma)\ \psi(\gamma^{-1})]\}.$$

From Lemma A.3 we have immediately

$$(f') = \sum_{h=h_2+2}^{d} \sum_{\substack{\psi \in \hat{C}_\tau \\ \text{cond } \psi = C_\tau^{(h)}}} \sigma(\lambda_1)\sigma(\lambda_2)\{[\psi(\lambda_1\lambda_2^{-1}) + \psi(\lambda_1^{-1}\lambda_2)]$$

$$+ (\frac{1 + \mathrm{sgn}_\tau(-1)}{2})\ [\psi(\lambda_1\lambda_2) + \psi(\lambda_1^{-1}\lambda_2^{-1})]\}$$

$$= 4\sigma(\lambda_1)\sigma(\lambda_2)\{q^d[\Delta_d(\lambda_1\lambda_2^{-1}) + (\frac{1 + \mathrm{sgn}_\tau(-1)}{2})\Delta_d(\lambda_1\lambda_2)]$$

$$- q^{h_1+1}[\Delta_{h_1+1}(\lambda_1\lambda_2^{-1}) + (\frac{1 + \mathrm{sgn}_\tau(-1)}{2})\Delta_{h_1+1}(\lambda_1\lambda_2)]\} .$$

If we sum over Φ and use Lemma A.2 and Lemma A.3, we get $(C') = (C_1') + (C_2')$,

where

$$(C_1') = \frac{q^{h_1}}{|\lambda_1 - \lambda_1^{-1}||\lambda_2 - \lambda_2^{-1}|} \{ \sum_{\substack{\gamma \in C_\tau^{(h_1)}/C_\tau^{(h_1+1)} \\ \gamma \neq \lambda_1, \lambda_1^{-1}, \lambda_2, \lambda_2^{-1}}} \mathrm{sgn}_\tau(\mathrm{tr}(\lambda_1 - \gamma)\mathrm{sgn}_\tau(\mathrm{tr}(\lambda_2 - \gamma))$$

$$- \frac{1}{q} \sum_{\substack{\gamma, \gamma^1 \in C_\tau^{(h_1)}/C_\tau^{(h_1+1)} \\ \gamma \neq \lambda_1, \lambda_1^{-1}, \gamma^1 \neq \lambda_2, \lambda_2^{-1}}} \mathrm{sgn}_\tau(\mathrm{tr}(\lambda_1 - \gamma))\mathrm{sgn}_\tau(\mathrm{tr}(\lambda_2 - \gamma^1))\},$$

and

$$(C_2') = \sigma(\lambda_1)\sigma(\lambda_2) \sum_{\substack{\psi \in \hat{C}_\tau \\ \text{cond } \psi = C_\tau^{(h_1+1)}}} [\psi(\lambda_1^{-1}\lambda_2) + \psi(\lambda_1\lambda_2)]$$

$$= 2q^{h_1}\sigma(\lambda_1)\sigma(\lambda_2)\{q[\Delta_{h_1+1}(\lambda_1^{-1}\lambda_2) + \Delta_{h_1+1}(\lambda_1\lambda_2)] - 2\} .$$

ote that summing over Φ in (h') just introduces a factor of 2. Then, using Lemma .3, we get

$$(h') = \frac{q^{h_1}}{|\lambda_1 - \lambda_1^{-1}||\lambda_2 - \lambda_2^{-1}|} \sum_{\gamma \in C_{\varepsilon\tau}^{(h_1)}/C_{\varepsilon\tau}^{(h_1+1)}} \mathrm{sgn}_{\varepsilon\tau}(\mathrm{tr}\,\lambda_1 - \mathrm{tr}\,\gamma)\mathrm{sgn}_{\varepsilon\tau}(\mathrm{tr}\,\lambda_2 - \mathrm{tr}\,\gamma)$$

$$- \frac{1}{2} \sum_{\gamma,\gamma^1 \in C_{\varepsilon\tau}^{(h_1)}/C_{\varepsilon\tau}^{(h_1+1)}} \mathrm{sgn}_{\varepsilon\tau}(\mathrm{tr}\,\lambda_1 - \mathrm{tr}\,\gamma)\mathrm{sgn}_{\varepsilon\tau}(\mathrm{tr}\,\lambda_2 - \mathrm{tr}\,\gamma^1).$$

ow, from Lemma A.6, it follows that the second sum in both (C_1') and (h') is equal o one. Thus, using Lemma A.7 and the fact that $|\lambda_1 - \lambda_1^{-1}||\lambda_2 - \lambda_2^{-1}| = q^{-2h_1-1}$, e have

$$(C_1') + (h') = -2q^{3h_1}(q+1) + 2q^{h_1+1}\sigma(\lambda_1)\sigma(\lambda_2)[\Delta_{h_1+1}(\lambda_1\lambda_2^{-1}) + \mathrm{sgn}_\tau(-1)\Delta_{h_1+1}(\lambda_1\lambda_2)].$$

Finally, adding (a), (b), $(C_1') + (h')$, (C_2'), (f'), (g) and (i), we see that

$$2K_d(t_1,t_2) = -2\sigma(t_1)\sigma(t_2) + 4\sigma(t_1)\sigma(t_2)q^d[\Delta_d(t_1t_2^{-1}) + (\frac{1 + \mathrm{sgn}_\tau(-1)}{2})\Delta_d(t_1t_2)].$$

In either of the cases computed so far, it is easy to see that the same result s obtained when $h_1 = 0$, or, in the first case, when $0 \leq h_1 < d \leq h_2$. The only emaining case is $d \leq h_1 \leq h_2$. Here, we have

$$2K_d(t_1,t_2) = 2\sigma(t_1)\sigma(t_2)[2q^d - 1] + 2(\frac{q+1}{q})\, q^{3d}\ .$$

his follows from a simple computation and will not be used in the sequel.

The remaining cases in Theorem 4.6 can be computed in a similar fashion. We oint out that it is unlikely that this type of computation can be carried out for ore general groups.

Bibliography

[Ca] W. Casselman, The Steinberg character as a true character, PSPM XXVI, AMS, Providence, 1973, pp. 413-418.

[D] L. E. Dickson, Linear Groups, Dover, New York, 1958.

[Fr] S. Franklin, The Reducible Principal Series of SL(2) over a p-adic Field, Thesis, University of Chicago, 1971.

[GG] I. M. Gel'fand and M. I. Graev, Representations of a group of the second order with elements from a locally compact field, Uspehi Mat. Nauk = Russian Math. Surveys 18(1963), 29-100.

[Sa1] P. J. Sally, Jr., Invariant subspaces and Fourier-Bessel transforms on the p-adic plane, Math. Ann. 174(1967), 247-264.

[Sa2] P. J. Sally, Jr., Unitary and uniformly bounded representations of the two by two unimodular group over local fields, Amer. J. Math. 90(1968), 406-443.

[Sc] R. Scott, The Fourier Transform of Orbital Integrals on GL(2) over a p-adic Field, Thesis, University of Chicago, 1983.

[Sk1] J. A. Shalika, Representations of the Two by Two Unimodular Group over Local Fields, Thesis, The Johns Hopkins University, 1966.

[Sk2] J. A. Shalika, A theorem on semisimple p-adic groups, Annals of Math. 95(1972), 226-242.

[Sk3] J. A. Shalika, On the space of cusp forms of a p-adic Chevalley group, Annals of Math. 92(1970), 262-278.

[SS1] P. J. Sally, Jr. and J. A. Shalika, Characters of the discrete series of representations of SL(2) over a local field, Proc. Nat. Acad. Sci. U. S. A. 61(1968), 1231-1237.

[SS2] P. J. Sally, Jr. and J. A. Shalika, The Plancherel formula for SL(2) over a local field, Proc. Nat. Acad. Sci. U. S. A. 63(1969), 661-667.

[ST] P. J. Sally, Jr. and M. H. Taibleson, Special functions on locally compact fields, Acta Math. 116(1966), 279-309.

[T] S. Tanaka, On irreducible unitary representations of some special linear groups of the second order, I, Osaka J. Math. 3(1966), 217-227.

Additional References for Orbital Integrals on p-adic Groups

[C1] L. Clozel, Sur une conjecture de Howe - I, preprint.

[F] D. Flath, A comparison of the automorphic representations of GL(3) and its twisted forms, Pacific J.Math. 97(1981), 373-402.

[F1] Y. Flicker, The Trace Formula and Base Change for GL(3), SLN 927, Springer, Berlin, 1982.

[H1] R. Howe, Two conjectures about reductive p-adic groups, PSPM XXVI, AMS, 1973, pp. 377-380.

[H2] R. Howe, The Fourier transform and germs of characters (case of GL_n over a p-adic field), Math. Ann. 208(1974), 305-322.

[HC1] Harish-Chandra, Harmonic Analysis on Reductive p-adic Groups, SLN 162, Springer, Berlin, 1970.

[HC2] Harish-Chandra, Harmonic analysis on reductive p-adic groups, PSPM XXVI, AMS, Providence, 1973, pp. 167-192.

[HC3] Harish-Chandra, Admissible distributions on reductive p-adic groups, Lie Theories and Their Applications,Queen's Papers in Pure and Applied Mathematics, Queen's University, Kingston, Ontario, 1978, pp. 281-347.

[K1] R. Kottwitz, Orbital integrals and base change, PSPM XXXIII, AMS, 1979, Part 2, pp. 185-192.

[K2] R. Kottwitz, Orbital integrals on GL_3, Amer. J. Math. 102(1980), 327-384.

[K3] R. Kottwitz, Unstable orbital integrals on SL(3), Duke Math. J. 48(1981), 649-664.

[L1] R. P. Langlands, Base Change for GL(2), Princeton, 1980.

[L2] R. P. Langlands, Les debuts d'une formule des traces stables, preprint.

[L3] R. P. Langlands, Orbital integrals on forms of SL(3), Amer. J. Math. 105(1983), 465-506.

[LL] J. P. Labesse and R. P. Langlands, L-indistinguishability for SL(2), Can. J. Math. 31(1979), 726-785.

[R] R. Ranga Rao, Orbital integrals in reductive groups, Annals of Math. 96(1972), 505-510.

[Re1] J. Repka, Shalika's germs for p-adic GL(n): the leading term, preprint.

[Re2] J. Repka, Shalika's germs for p-adic GL(n), II: the subregular term, preprint.

[Re3] J. Repka, Germs associated to regular unipotent classes in p-adic SL(n), preprint.

[Ro1] J. Rogawski, An application of the building to orbital integrals, Compositio Math. 42(1981), 417-423.

[Ro2] J. Rogawski, Representations of GL(n) and division algebras over a p-adic field, Duke Math. J. 50(1983), 161-196.

[Ro3] J. Rogawski, Some remarks on Shalika germs, preprint.

[Si] A. Silberger, Introduction to Harmonic Analysis on Reductive p-adic Groups, Princeton, 1979.

[V] M-F. Vigneras, Caractérisation des intégrales orbitales sur un groupe réductif p-adique, J. Fac. Sci., University of Tokyo 28(1981), 945-961.

Department of Mathematics
University of Chicago
Chicago, IL 60637

Department of Mathematics
Johns Hopkins University
Baltimore, MD 21218